Wynn Kapit / Lawrence M. Elson

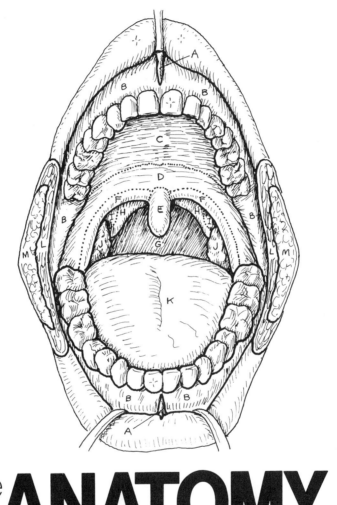

The ANATOMY
COLORING BOOK

SECOND EDITION

HarperCollins*Publishers*

I promise my wife Lauren, and my sons, Eliot and Neil, to make up for so much lost time.

Wynn Kapit

For my wife Joan, with love, who once again provided the ambience and means by which this book has been re-created. For Jennifer, Chris, and Gina, with love, who by just being there made it all worthwhile.

To Wynn Kapit, a uniquely gifted designer and illustrator; to Archie Morris, for keeping me fit; to Carolyn Scott, who made it possible for me to finish this book.

To the million-plus colorers of the first edition, who have wonderfully corroborated and given real meaning to our work. We thank you.

Larry Elson

ABOUT THE AUTHORS

Wynn Kapit, designer/illustrator of the book, is also the designer/illustrator of *The Physiology Coloring Book*, and the author/illustrator of *The Geography Coloring Book*.

Lawrence M. Elson, Ph.D., provided sketches, planned the content and organization, and wrote the text for the book. Dr. Elson has taught anatomy in medical school and college, and continues to teach health care providers and attorneys at seminars around the country. He is a consultant and lecturer throughout the United States and Canada on the anatomical bases and causation of injury. He is the author of *It's Your Body*, and co-author of *The Human Brain Coloring Book*.

The Anatomy Coloring Book, Second Edition
Copyright © 1993 by Wynn Kapit and Lawrence M. Elson

ISBN 0-06-455016-8

95 96 97   9

TABLE OF CONTENTS

IV ARTICULAR SYSTEM

V MUSCULAR SYSTEM

VI CARDIOVASCULAR SYSTEM

VII LYMPHATIC SYSTEM

VIII IMMUNE (LYMPHOID) SYSTEM

IX RESPIRATORY SYSTEM

X DIGESTIVE SYSTEM

XI URINARY SYSTEM

XII REPRODUCTIVE SYSTEM

XIII ENDOCRINE SYSTEM

XIV NERVOUS SYSTEM

XV INTEGUMENTARY SYSTEM

PREFACE

The Anatomy Coloring Book (ACB) has been an immense success since its inception in 1976 and publication in 1977. By personal contacts and letters, we have found that directed coloring of body structures has increased the learning curve, opened reservoirs of visual memory, and facilitated instant recall in students at secondary schools, colleges, and graduate and professional schools throughout the world. *The Anatomy Coloring Book* has been translated into French, Spanish, Portuguese, Japanese, and Italian.

Thousands of colorers have advised and encouraged us, including coaches, trainers, teachers, paramedics, body workers, court reporters, attorneys, insurance claims adjusters, judges, and students and practitioners of dentistry and dental hygiene, nursing, medicine/surgery, chiropractice, podiatry, massage therapy, physical therapy, occupational therapy, and exercise therapy. Informal inquirers, and those with impairments seeking better understanding of their bodies, have found new insights about themselves through periodic coloring of ACB plates. In addition to enhancing professional growth, this book can and does accommodate all who seek self-realization through understanding the structure and function of their physical being.

In this new edition of the ACB, we are responding to those who took the time to point out errors, and who told us what they liked and what they needed. Included are many new illustrations with a crisp, more precise look. Numerous small drawings have been added relating organs to microscopic views of tissues, in addition to pictures of specific body postures induced by the muscles being colored. The text has been rewritten, and timely material has been added. An extensive index has been added to facilitate finding information. A glossary of definitions and explanations, both anatomical and medical, has been added. This has special significance in a book such as this where text space is severely limited. For those concerned with the innervation of skeletal muscle and related spinal cord/root segments, a listing is provided in Appendix B.

A scan of the contents of this new edition will reveal that 19 new plates have been added to the original 142. Perhaps most significantly, the lymphatic system in the first edition has been completely revised and expanded here, separating the lymphatic system (Plate 83) from the lymphoid or immune system (Plates 84–89). Changes in our understanding of the immune system have developed at a mind-boggling rate over the last 10 years, and these changes are reflected in these seven plates. A plate bringing together the effects of HIV infection on the function of critical cells of the immune system (Plate 90) brings practical meaning to having colored the preceding plates on the lymphatic and lymphoid systems.

ACKNOWLEDGEMENTS

The authors are in debt to many people who helped make this book happen. Our reviewers gave us a good opportunity to reflect on and to correct the text, and we are grateful to them: Drs. Don Matthies, Jack Wagner, and Gerald N. Waagen for content, and Ms. Maureen Savage and Mr. Edward Loskamp for secondary (high) school student and nonscientist perspectives, respectively. Special thanks to Dr. Andrew Lichtman, assistant professor of pathology at Harvard Medical School, and co-author of the text *Cellular and Molecular Immunology*, for reviewing and criticizing the immune system and AIDS plates. I am grateful to Dr. John Gullett, a specialist in infectious diseases, for his review as well. Professors Richard Simmons and John Boyd encouraged one of the authors (LME) to undertake the immune system in this book, and their support was much appreciated. A special nod of appreciation goes to Jill Breedon, proofreader, and the staff of the TypeStudio in Santa Barbara. Their superb and accurate work made the complicated job of typesetting and proofing far more enjoyable to handle than one might otherwise expect.

The human body, like all organisms on this planet, reflects in its organization, structure, and function, universal principles and truths. We need to learn more about these relationships, as they have tremendous personal and social implications. We offer this book as a means of gaining further insight into these relationships, and as an opportunity to see just how remarkable we all really are.

WYNN KAPIT
Santa Barbara, California

LARRY ELSON
Napa Valley, California

EXTREMELY IMPORTANT TIPS
(On how to get the most out of this book)

1. Please review this section before coloring the book. The short time required to read through these tips will enable you to get the most benefit from the book. After learning the meaning of a few symbols used throughout, and reading some basic instructions on how to proceed, common sense will normally dictate the manner in which you color each plate. It is advisable to occasionally review these notes until such time that you feel completely at home with the coloring format.

2. Look over the table of contents. Note the arrangement of plates, organized into introductory, general body organization, systemic and regional organization, and specific system topics are organized according to body system (systemic anatomy). If you are unfamiliar with the study of anatomy, be sure to color the plates on terminology (1 and 2) and introduction to systems and regions (13–16) first.

3. Turn now to any plate of the book and note the following:

   a. At the top of each plate are coloring notes (CN) which provides specific guidance in coloring that particular plate. Be sure to read these before coloring. The CN will usually recommend certain colors for specific subscripts, direct the order of coloring, and explain any ambiguities that might arise. Whether you color the plate first and then read the text, or read the text first and then color, is your decision . . . whichever works best for you.

   b. A glance at the front or back cover will illustrate the basic principle of this coloring format: each "title" (a term in colorable outline letters) followed by a small, lettered subscript should receive a different color. After coloring the title, you should then search through the illustration(s) and color any structure identified with that subscript, using the same color as the title. *Unless you run out of colors, you should not repeat that color for any other subscripted title or structure.* Occasionally, title subscripts will appear with a numbered superscript, e.g., A^1, A^2, and so on. These titles and related structures get the same color because of the strong relationship between the structures.

   c. Take care to color the titles first. In many instances the subscript identifying a structure is embedded within the structure. If you use a dark color for the structure, you may obscure the identifying subscript. This can be avoided by simply coloring the title first, and then the related structure.

   d. Related plates (cross references) are noted under the plate number in the upper right corner ("see . . .").

4. The more colors you have at your disposal, the more effective, as well as enjoyable, your coloring will be. A minimum of twelve colors, including gray, is essential. Lighter colors are preferable because they are less likely to cover up surface detail or identifying subscripts. Gray is an important color for column headings and miscellaneous uses. Whether you use felt-tipped pens or colored pencils makes no difference, provided their points are sharp enough for detail work. *Crayons won't do.* Buying colors individually at art supply or stationery stores will enable you to choose from a lighter range, as well as provide the opportunity to replace individual colors when they are exhausted.

5. This book does not have to be colored in any specific order, but the individual chapters (indicated by Roman numerals, I, II, and so on) should be colored in numerical order. This is because there are times in which you will be asked to repeat colors used on previous plates, creating a color-coordinated set of interrelated plates on a region or structure. However, you may find that certain plates deal with material the technical nature of which exceeds your needs; in these cases, you may wish to skip the plate(s).

6. Generally, you will be able to choose colors as you wish. Occasionally, you will be asked to use specific colors recommended for structures that are universally colored one color in atlases: red for arteries, blue for veins, purple for capillaries, yellow for nerves, and green for lymph vessels and lymph nodes. It is usually preferable to employ lighter colors for the largest areas, and darker or brighter colors for the smallest areas.

7. Dark outlines provide the boundaries between areas receiving different colors (identified by different subscripts). In some illustrations, there are repeated identical structures; for example, numbers of cells, vessels, lobules, and so on. Not every one of those structures may be labeled. You should, however, color all of them unless otherwise indicated by the "don't color" symbol. In those plates with a lot of detail, the identification of different subscripted areas/structures and "no color" areas will have to be pursued diligently before coloring, to prevent mistakes.

8. Symbols used throughout are:

-¦- = don't color   ✳ = color gray

N.S. = not shown   • = color black

◻◻◻◻ = outline of a structure seen below the surface of another

9. In the text and titles, the following abbreviations may precede or follow the names of the structures identified, e.g., POST. AURICULAR M., BRACHIAL A., SCALENUS MED. M.:

A.	= Artery
Ant.	= Anterior
Br	= Branch
Inf.	= Inferior
Int.	= Internal
Lat.	= Lateral
Lig.	= Ligament
M., Mus.	= Muscle
Med. (preceding term)	= Medial
Med. (after term)	= Medius
N.	= Nerve
Post.	= Posterior
Sup.	= Superior, superficial
Sys.	= System
Tr.	= Tract
V.	= Vein

INTRODUCTION
ANATOMIC PLANES & SECTIONS

CN: (1) Color the four body planes and related sections of the body in very light colors. (2) Take care not to color in areas marked by "do not color" (-'¡-) symbols.

Study of the human body (anatomy) requires visualization of internal regions or parts. Dissection (dis, apart; sect-, cut) is the name given to preparing the body for internal inspection. One method of dissection permits consistent visual orientation by cutting the body into parts or sections along fixed lines of reference called planes. Two of these planes are oriented along the long axis of the body or body part, one perpendicular to the other (longitudinal sections). The third plane is oriented perpendicular to or across the long axis of the body or body part (cross section). Such planes have application in medical imaging studies, such as computerized tomography (CT) and magnetic resonance imaging (MRI). Here the body interior is imaged as computerized "slices" of the body in sagittal, coronal and transverse planes.

MEDIAN A

The midline, longitudinal plane dividing the head and torso into right/left halves. The presence of the vertebral column and spinal cord is characteristic of the median plane of the torso. The median plane is the middle sagittal plane.

SAGITTAL B

The longitudinal plane dividing the body into left and right parts. It is parallel to the median plane and may be applied to the head, torso and limbs.

CORONAL, FRONTAL C

The longitudinal plane dividing the body into equal or unequal front and back parts. In CT and MRI, the term "coronal" is used by radiologists.

CROSS, TRANSVERSE D

The transverse plane dividing the body or body parts into upper and lower segments. This plane is a cross section perpendicular to the longitudinal planes. Transverse planes of the body, called axial or transaxial sections by radiologists, are commonly seen in CT and MRI studies of the body.

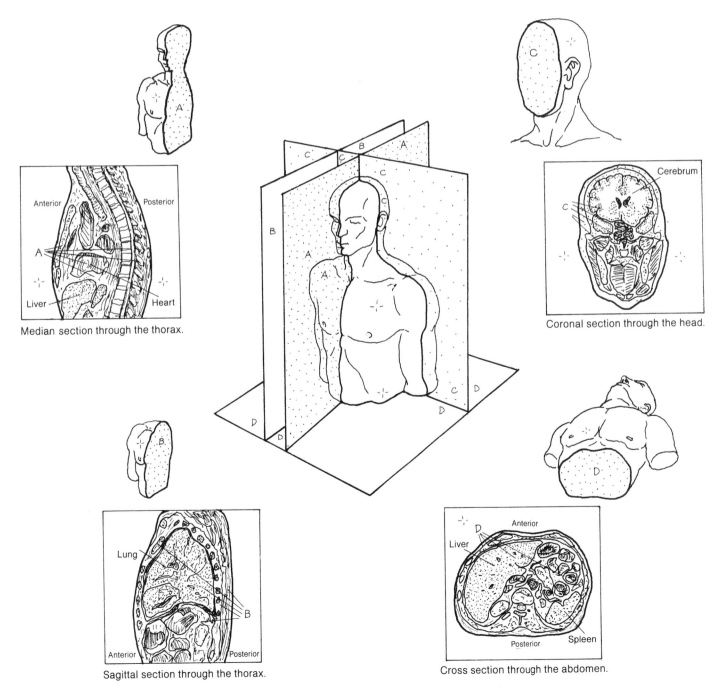

Median section through the thorax.

Coronal section through the head.

Sagittal section through the thorax.

Cross section through the abdomen.

INTRODUCTION
TERMS OF POSITION & DIRECTION

CN: (1) Use bright or dark colors for emphasis.
(2) Color the arrows but not the illustrations.

Terms of position and direction describe the relationship of one organ to another, usually along one of the three body planes illustrated in the previous plate. To avoid confusion, these terms are related to the standard *anatomical position:* body standing erect, limbs extended, palms of the hands forward.

CRANIAL, SUPERIOR, ROSTRAL A

These terms refer to a structure being closer to the head or higher than another structure of the body. See the quadruped in the right corner for a related application of the term "cranial."

ANTERIOR, VENTRAL B

These terms refer to a structure being more in front than another structure in the body. The term "anterior" is preferred. See the quadruped for another application of the term "ventral."

POSTERIOR, DORSAL C

These terms refer to a structure being more in back than another structure in the body. The term "posterior" is preferred. See the quadruped for another application of the term "dorsal."

MEDIAL D

This term refers to a structure that is closer to the median plane than another structure in the body. "Medial" is *not* synonymous with "median."

LATERAL E

This term refers to a structure that is further away from the median plane than another structure in the body.

PROXIMAL F

Employed only with reference to the limbs, this term refers to a structure being closer to the median plane or root of the limb than another structure in the limb.

DISTAL G

Employed only with reference to the limbs, this term refers to a structure being further away from the median plane or the root of the limb than another structure in the limb.

CAUDAL, INFERIOR H

These terms refer to a structure being closer to the feet or the lower part of the body than another structure in the body. See the quadruped for a related application of the term "caudal."

SUPERFICIAL I DEEP J

The term "superficial" is synonymous with external, and the term "deep" with internal. Related to the reference point on the chest wall, structure closer to the surface of the body is superficial; structure further away from the surface is deep.

IPSILATERAL K CONTRALATERAL L

The term "ipsilateral" means "on the same side" (in this case, as the reference point); "contralateral" means "on the opposite side" (of the reference point).

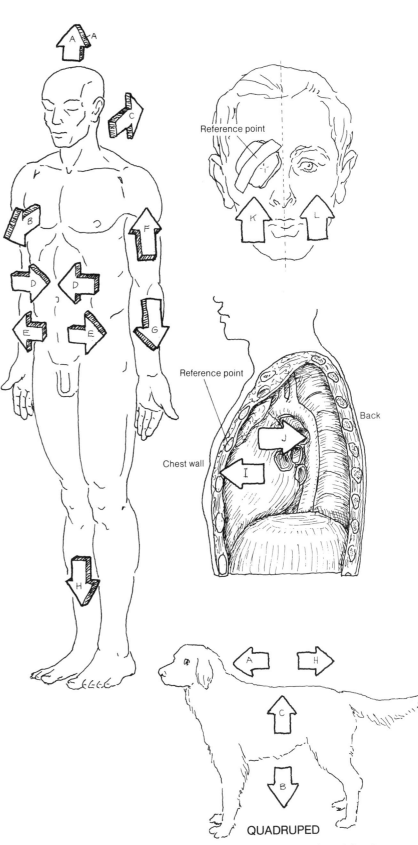

Reference point

Reference point

Back

Chest wall

QUADRUPED

The quadruped presents four points of direction: head end (cranial), tail end (caudal), belly side (ventral), back side (dorsal). In the biped (e.g., human), the ventral side is also anterior, the dorsal side is also posterior, the cranial end is also superior, and the caudal end is inferior.

1. ORGANIZATION OF THE BODY
THE GENERALIZED CELL

CN: Color gray the variety of cell shapes at upper left. Use lightest colors for A, B, C, D, F and G. (1) Small circles representing ribosomes (H) are found throughout the cytoplasm (F) and on the rough endoplasmic reticulum (G¹); color those larger areas, including the ribosomes, first, and then color over the ribosomes again with a darker color. Each organelle shown is just one of many found in the living cell.

CELL SHAPES: *

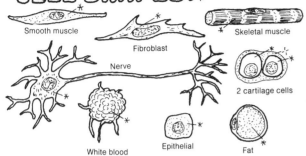

Smooth muscle

Fibroblast

Nerve

Skeletal muscle

2 cartilage cells

White blood

Epithelial

Fat

ORGANELLES: *
CELL MEMBRANE A
MICROVILLI B
NUCLEAR MEMBRANE C
NUCLEOPLASM D
NUCLEOLUS E
CYTOPLASM F
ENDOPLASMIC RETICULUM
SMOOTH₇ G ROUGH G'
RIBOSOME H
GOLGI COMPLEX I
MITOCHONDRION J
VACUOLE / PINOCYTOTIC
VESICLE K
LYSOSOME L
CENTRIOLE M
MICROTUBULE N
MICROFILAMENT N'
CELL INCLUSION O

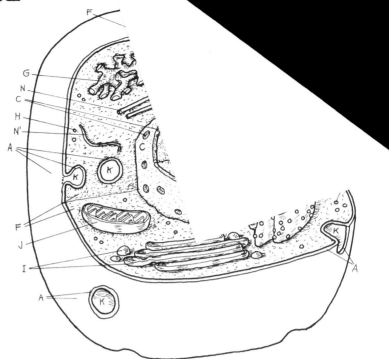

The cell is the basic structural and functional unit of all living things. Living things are characterized by the ability to reproduce and grow, metabolize (transformation or production/consumption of energy), and adapt to limited changes in their internal and external environment. Body structure lacking these characteristics, such as connective tissue fibers, is not considered to be "alive." Body structure more complex than a cell consists of a collection of cells and their products.

The activities of cells constitute the life process, and include ingestion, assimilation, and digestion of nutrients, and excretion of the residue; respiration; synthesis and degradation of materials; movement; and excitability or response to stimuli. The impairment or cessation of these activities in normal cells, whether caused by trauma, infection, tumors, degeneration, or congenital defects, is the basis of a disordered or disease process.

The chemical composition of a cell is generally about 15% protein, 3% lipids, 1% carbohydrates, 1% nucleic acids and minerals, and 80% water (by volume). These compounds are integrated together into *organelles*, the working components of the cell. The basic function of cells is to produce protein, the structure of which is determined by DNA. The manifestation of this activity is the characteristic function of the cell (e.g., formation, repair, and breakdown of structure, secretion, absorption, contraction, conduction of electrochemical impulses, and so on).

Cell membrane: the limiting membrane of the cell; retains internal structure; permits exportation and importation of materials. Composed primarily of lipid and protein, and a smaller amount of carbohydrate.

Microvilli: finger-like extensions of the cell membrane covering the free surface of certain epithelial cells; they increase the surface area of the cell, enhancing secretion/absorption.

Nuclear membrane: porous membrane of similar construction to the cell membrane; the limiting membrane of the nucleus, separating it from the cytoplasm; regulates passage of molecules.

Nucleoplasm: the ground substance of the nucleus, containing the chromatin or thin threads of genetic material (DNA and related protein). During cell division, the chromatin transforms into chromosomes

Nucleolus: a mass of largely RNA (and some DNA and protein) in the nucleus producing units of RNA which combine in the cytoplasm to form ribosomes.

Cytoplasm: the ground substance of the cell less the nucleus. Contains organelles and inclusions listed below.

Smooth/rough endoplasmic reticulum (ER): membrane-lined tubules to which ribosomes may be attached (rough ER; flattened tubules) or not (smooth ER; rounded tubules). Rough ER is concerned with transport of protein synthesized at the ribosomes. Smooth ER synthesizes complex molecules called steroids in some cells; stores calcium ions in muscle; breaks down toxins in liver.

Ribosome: the site of protein synthesis where amino acids are strung in sequence as directed by messenger RNA from the nucleus.

Golgi complex: flattened membrane-lined sacs which bud off small vesicles from the edges; collect secretory products and package them for export or cell use.

Mitochondrion: membranous, oblong structure in which the inner membrane is convoluted like a maze. Energy for cell operations is generated here through a complex series of reactions between oxygen and products of digestion.

Vacuoles/pinocytotic vesicles: membrane-lined containers which can merge with one another or other membrane-lined structure, such as the cell membrane. They function as transport vehicles.

Lysosome: membrane-lined container of enzymes with great capacity to break down structure, especially ingested foreign substances.

Centriole: bundle of microtubules in the shape of a short barrel; usually seen paired, perpendicular to one another. They give rise to spindles used by migrating chromatids during cell division.

Microtubule: microtubules are formed of protein and provide structural support for the cell.

Microfilament: microfilaments are support structures formed of protein different from that of microtubules. In skeletal muscle, the proteins actin and myosin are examples of thin and thick microfilaments.

Cell inclusion: aggregation of material within the cell that is not a functional part (organelle) of the cell, e.g., glycogen, fat, and so on.

I. ORGANIZATION OF THE BODY
CELL DIVISION / MITOSIS

CN: Use the colors you used on Plate 3 for cell membrane, nuclear membrane, nucleolus, and centriole for those titles on this plate, even though the previous letter labels may be different. Use contrasting colors for E-E² and F-F², and gray for D-D¹ to distinguish the latter from those with the contrasting colors. (1) Begin with the cell in interphase, reading the related text and completing each cell before going on to the next. (2) Color gray the name of each stage and its appropriate arrow of progression. Note that in interphase, the chromatin material within the nuclear membrane is in a thread-like state; color over the entire area with the appropriate color. Note that the starting chromatin (D* in interphase) is colored differently in the daughter cells (E², F²); it is the same chromatin.

CELL MEMBRANE A
NUCLEAR MEMBRANE B
NUCLEOLUS C
CHROMATIN D* / CHROMOSOME D¹*
CHROMATID E / CHROMOSOME E¹
CHROMATIN E²
CHROMATID F / CHROMOSOME F¹
CHROMATIN F²
CENTROMERE G
CENTRIOLE H
ASTER I
SPINDLE J

The ability to reproduce its kind is a characteristic of living things. Cells reproduce in a process of duplication and division called mitosis. Epithelial and connective cells reproduce frequently; mature muscle cells not so frequently; mature nerve cells rarely if at all. Overactive mitoses may result in the formation of an encapsulated tumor; uncontrolled mitoses, associated with invasiveness and metastases, is called cancer.

As the main cellular changes during mitosis occur in the nucleus and surrounding area, only these parts of the cell are illustrated here. We are showing here how the nuclear chromatin (diffuse network of DNA and related protein), once duplicated, transforms into 46 chromosomes which divide into paired subunits (92 chromatids), and how those chromatids separate and move into opposite ends of the dividing cell, forming the 46 chromosomes of each of the newly formed daughter cells. For clarity, we show only 4 pairs of chromatids and chromosomes. The phases of the observed nuclear changes during mitosis are:

Interphase: the longest period of the reproductive cycle; the phase between successive divisions. Duplication of DNA (in chromatin) occurs during this phase. The dispersed chromatin (D*) here is a network of fine fibrils, not visible as discrete entities in the nucleoplasm. The cell membrane, nucleus, and nucleolus are intact. The centrioles are paired and adjacent to one another at one pole of the cell.

Prophase: the dispersed chromatin (D*) thickens, shortens, and coils to form condensed chromatin or chromosomes (D¹*). Each chromosome consists of 2 chromatids (E and F) connected by a centromere (G). Each chromatid has the equivalent amount of DNA of a chromosome. In the latter part of this phase, the nuclear membrane breaks up and dissolves, as does the nucleolus. The centrioles, having duplicated during interphase, separate, each pair going to opposite poles of the cell. They project microtubules called asters.

Metaphase: strands of spindle fibers project across the cell center from paired centrioles. The chromatids attach to the spindle fibers at the centromere, and line up in the center, half (46) on one side, half (46) on the other.

Anaphase: the centromeres divide, each daughter centromere attached to one chromatid. Each centromere is drawn to the ipsilateral pole of the cell, along the track of the spindle fiber, and taking its chromatid with it. The separated chromatids now constitute chromosomes. Anaphase ends when the daughter chromosomes arrive at their respective poles (46 on each side).

Telophase: here the cell pinches off in the center, forming 2 daughter cells, each identical to the mother cell. The cytoplasm and organelles had duplicated earlier and are segregated each into their respective newly-forming cells. As the nucleus is reconstituted, and the nuclear membrane and nucleolus reappear in each new cell, the chromosomes fade into dispersed chromatin and the centromere disappears. Complete cleavage of the parent cell into daughter cells terminates the mitotic process. Each daughter cell enters interphase to start the process anew. The process of cell division serves to increase cell numbers, not change cellular content.

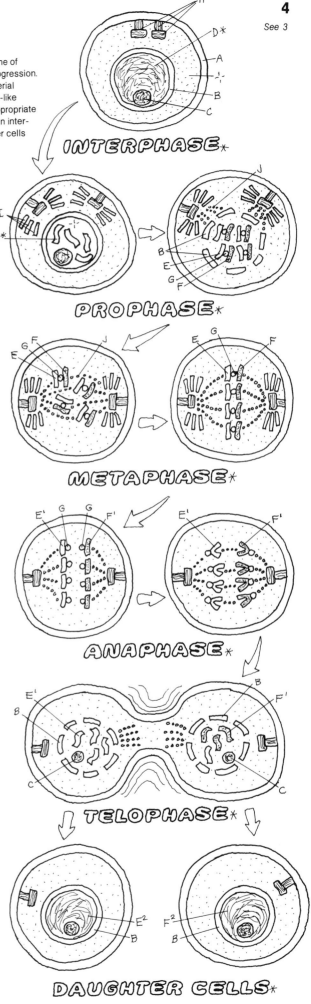

INTERPHASE*

PROPHASE*

METAPHASE*

ANAPHASE*

TELOPHASE*

DAUGHTER CELLS*

I. ORGANIZATION OF THE BODY
TISSUES: EPITHELIUM

CN: Use very light colors throughout. (1) Color the arrows pointing to the location of the epithelial tissues in the body organs.

There are four basic tissues of the body: epithelial, connective, muscle, and nervous. Epithelial tissues (epithelium) form the body's surface (skin), the surfaces of the body's cavities and their contained viscera, glands, and all tubular organs, e.g., ducts and vessels. Neuroepithelia convey sensations. Epithelia are arranged into single (simple) or several (stratified) layers; their cells are bound together by specialized fibers and substances (e.g., the basement membrane). Epithelial tissues are generally sensitive but avascular, and receive their nutrition by diffusion.

SIMPLE EPITHELIUM∗

Surface tissue functioning in filtration, diffusion, secretion, and absorption.

SQUAMOUS_A

Simple squamous epithelia line the heart cavities and the internal surfaces of all blood and lymph vessels (endothelia), the air cells of the lung, filtration capsules and thin tubules in the kidney, and the major body cavities (mesothelia). Rapid diffusion of gases in solution are characteristic activities in these cells.

CUBOIDAL_B

Simple cuboidal epithelia are generally secretory cells, and make up glands throughout the body, tubules of the kidney, terminal bronchioles of the lungs, and ducts of the reproductive tracts.

COLUMNAR_C

Simple columnar epithelia line the gastrointestinal tract and are concerned with secretion and absorption. Their free (apical) surface may be covered with finger-like projections of cell membrane called microvilli, increasing the cell's surface area for secretion/absorption.

PSEUDOSTRATIFIED COLUMNAR_D

This tissue consists of simple columnar cells bunched together with irregularly placed nuclei giving the appearance of multiple cell layers. However, each of the cells is attached to the basement membrane. This tissue lines ducts of the reproductive tracts and air conduction pathways of the respiratory tract. They often exhibit cilia on their free surfaces and contain unicellular goblet-shaped (secretory) cells. The cilia collectively move surface material by virtue of undulating power strokes.

STRATIFIED EPITHELIUM ∗

Stratified epithelia are generally resistant to damage by wear and tear because of ready replacement of cells. Passive diffusion through these layers is slow but not impossible.

STRATIFIED SQUAMOUS_E

These layers of cells line the skin, oral cavity, pharynx, vocal folds, esophagus, vagina, and anus. The basal cells are columnar and germinal. The outermost layers of skin epithelia are fibrous-like, flat, desiccated, non-nucleated cells containing keratin (a scleroprotein).

TRANSITIONAL_F

Multiple layers of cells lining the urinary tract. In the empty (contracted) bladder, the fibromuscular layer is contracted due to resting tension of muscle cells, and the surface layer of rounded cells is closely concentrated, creating a bumpy surface. With distension of the bladder, all the cells stretch out to form a smooth, thin surface. The bladder can store volumes of urine up to 1000 milliliters or so.

GLANDULAR EPITHELIUM∗

Glandular cells produce and secrete/excrete materials of varying composition, e.g., sweat, milk, sebum, cerumen, hormones, enzymes, and so on. Specialized contractile epithelial cells (myoepithelia) encourage discharge of the glandular material.

EXOCRINE_G

Exocrine glands (e.g., sweat, sebaceous, pancreatic, mammary, and so on) arise as outpocketings of epithelial lining tissue, retain a duct to the free surface of the cavity or skin, and excrete/secrete some substance. Secretory portions may have one of several shapes (tubular, coiled, alveolar) connected to one or more ducts.

ENDOCRINE_H

Endocrine glands arise as epithelial outgrowths but lose their connections to the surface during development. They are intimately associated with a dense capillary network and secrete their products into them. See Plate 128 for examples of these glands.

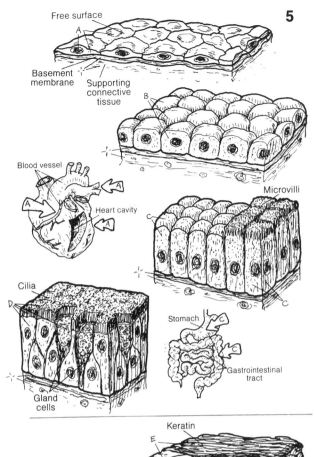

Free surface

Basement membrane · Supporting connective tissue

Blood vessel

Heart cavity

Microvilli

Cilia

Gland cells

Stomach

Gastrointestinal tract

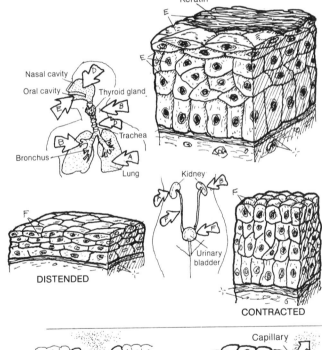

Keratin

Nasal cavity

Oral cavity · Thyroid gland

Bronchus

Trachea

Lung

Kidney

Urinary bladder

DISTENDED

CONTRACTED

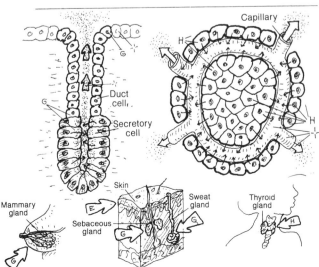

Capillary

Duct cell

Secretory cell

Skin

Mammary gland

Sebaceous gland

Sweat gland

Thyroid gland

I. ORGANIZATION OF THE BODY
TISSUES: FIBROUS CONNECTIVE TISSUES

CN: Use yellow for C and C¹, and red I. (1) Begin with the illustration at middle left, and the related titles (A through K). The titles and borders of the microscopic sections of dense regular/irregular c.t. (F¹, F²) receive the color of collagen (F) as that is the dominant structure in both tissues. (2) Do not color the matrix.

The connective tissues (c.t.) connect, bind, and support body structure. They consist of variable numbers of cells, fibers, and ground substance (fluid, viscous sol/gel, or mineralized). At the microscopic level (here illustrated at about 600 x magnification), connective tissues range from blood (cells/fluid), through the fibrous tissues (cells/fibers/variable matrix) to the more stiff supporting tissues (cells/fibers/dense matrix) of cartilage and mineralized bone. Connective tissue can be seen at visible levels of body organization as well, in fascial layers of the body wall, tendons, ligaments, bone, and so on. This plate introduces the fibrous connective tissues (c.t. proper).

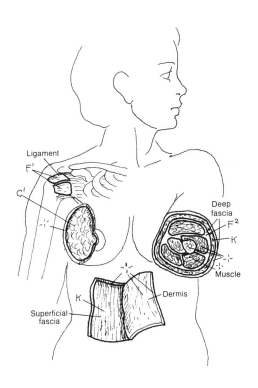

CELLS:*
FIBROBLAST A
MACROPHAGE B
FAT CELL C
PLASMA CELL D
MAST CELL E

FIBERS:*
COLLAGEN F
ELASTIC G
RETICULAR H

MATRIX, GROUND SUBSTANCE I·¡·

CAPILLARY J

LOOSE, AREOLAR C.T. K

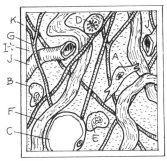

Loose, areolar connective tissue is characterized by many cells, a loose, irregular arrangement of fibers, and a moderately viscous fluid matrix. Fibroblasts secrete the fibers and ground substance of this tissue. Mobile macrophages engulf cell debris, foreign matter, and microorganisms. Fat cells, storing lipids, may be seen in small numbers or large (adipose tissue). Plasma cells secrete antibodies in response to infection. Mast cells contain heparin and other secretory products, some of which initiate allergic reactions when released. Numerous other cells may transit the loose fibrous tissues, including white blood cells (leukocytes). Collagen (linkages of protein exhibiting great tensile strength) and elastic fibers (made of the protein elastin) are the fibrous support elements in this tissue. Reticular tissue is a smaller form of collagen, forming supporting networks around cell groups of the blood-forming tissues, the lymphoid tissues, and adipose tissue. The matrix (consisting largely of water with glycoproteins and glycosaminoglycans in solution) is the intercellular ground substance in which all of the above function; it is fluid-like in the fibrous tissue. Numerous capillaries roam throughout this tissue. Loose connective tissue found deep to the skin is called superficial fascia, subcutaneous tissue, or hypodermis. It is found deep to the epithelial tissues of mucous and serous membranes of hollow organs.

ADIPOSE C.T. C¹

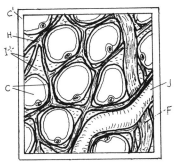

Aggregations of fat cells, supported by reticular and collagenous fibers, and closely associated with both blood and lymph capillaries, constitute adipose tissue. The storage/release of fat in/from adipose tissue is regulated by hormones (including nutritional factors) and nervous stimuli. It is a source of fuel, an insulator, mechanical padding, and stores fat-soluble vitamins. Adipose tissue is located primarily in the superficial fasciae (largely breast, buttock, anterior abdominal wall, arm, and thigh), yellow marrow, and the surface of serous membranes.

DENSE REGULAR C.T. F¹

DENSE IRREGULAR C.T.

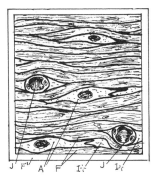

Dense, parallel-arranged, masses of collagenous/elastic fibers form ligaments and tendons that are powerfully resistant to axially loaded tension forces, yet permitting some stretch. Tendons/ligaments contain few cells; largely fibroblasts. Elastic, dense regular ligaments are found in the posterior neck and between vertebrae; the tendocalcaneus is the largest elastic structure (tendon or ligament) in the body, storing energy used in gait.

Dense, irregularly arranged masses of interwoven collagenous (and some elastic) fibers in a viscous matrix form capsules of joints, envelop muscle tissue (deep fasciae), encapsulate certain visceral organs (liver, spleen, and others) and largely make up the dermis of the skin. It is impact resistant (bearing stress omnidirectionally), contains few cells, and is minimally vascularized.

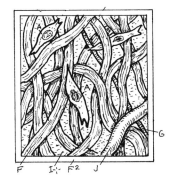

TISSUES: SUPPORTING CONNECTIVE TISSUES

CN: Use the same colors as used on the previous plate for collagen (D) and elastic (E) fibers. Use a light tan or yellow for F and red for L. Use light colors for A, B, G, I, and M. Complete the upper material before coloring the bone section.

CARTILAGE:*
CHONDROCYTEA
LACUNAB
MATRIXC·¹
COLLAGEN FIBERD
ELASTIC FIBERE

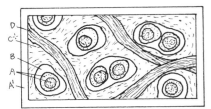

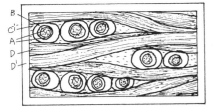

The supporting connective tissues consist of cartilage and/or bone. Microscopic sections of cartilage tissue reveal cells (*chondrocytes*) in small cavities (*lacunae*) surrounded by a specialized, hard but flexible *matrix* consisting of water electrochemically bound to proteoglycans and very fine *collagen fibers*. Cartilage is avascular; it receives its nutrition by diffusion. It generally does not repair well after injury but does replace itself with wear, as on joint surfaces.

Bone is unique for its mineralized matrix (average bone is 65% mineral, 35% organic tissue by weight). Bone forms the skeleton of the body; it is a reservoir of calcium; it acts as an anchor for muscles, tendons, and ligaments; it harbors many internal viscera, including the central nervous system; it assists in the mechanism of respiration, and is a center of blood-forming (hemopoietic) activity and fat storage.

HYALINE CARTILAGE A¹
Hyaline cartilage is a flexible, avascular, insensitive, compressible cartilage, characterized by tiny pores. Its major significance is covering bone ends at synovial joints (articular cartilage). Joint movement enhances nutrition of the 1–3 mm thick articular cartilage, by pushing synovial fluid through the pores. Hyaline cartilage also supports the nose, contributes to the nasal septum, and is the main structural support of the larynx and lower respiratory tract. It forms the cartilage model for some bones in embryonic/fetal development; it is often a part of the intermediate framework (callus) in the healing process of fractured bone. Non-articular cartilage is generally ensheathed by perichondrium, a vascular fibrous tissue.

ELASTIC CARTILAGE E¹
Elastic cartilage is essentially hyaline cartilage with elastic fibers and a slightly different type of collagen. It supports the external ear and contributes to the support of the larynx (epiglottis). It is remarkably flexible; test it on your external ear.

FIBROCARTILAGE D¹
Fibrocartilage is dense fibrous tissue interspersed with chondrocytes in a reduced amount of intercellular matrix. It is found in intervertebral discs, the sacroiliac joint, pubic symphysis, and in several sites of ligamentous attachment to bone. Fibrocartilage enhances resistance to tensile and impact forces.

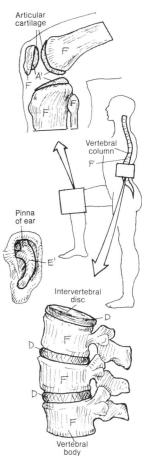

Articular cartilage

Vertebral column

Pinna of ear

Intervertebral disc

Vertebral body

BONE F
PERIOSTEUMF¹
COMPACT BONEG
HAVERSIAN SYS.:*
HAV. CANALH
LAMELLAEG¹
OSTEOCYTEI
LACUNAB
CANALICULIJ
VOLKMANN CANALK
BLOOD VESSELL
SPONGY BONEM

Microscopic sections of *bone* consist of osteocytes in *lacunae*, supported by collagen fibers in a mineralized (calcium hydroxyapatite) matrix. Other bone cells (not shown) include bone forming cells (osteoprogenitor cells, osteoblasts) and bone-absorbing cells (osteoclasts). *Compact bone* is the outer, impact-resistant, weight-bearing shell of bone. It is surrounded on its outer surface by a fibrous, vascular, cellular *periosteum*. The matrix of compact bone occurs in two patterns: concentric layers (*lamellae*) with a central canal (*haversian system/canal*) arranged in columns; and layers between and around haversian systems (circumferential system). The canals are interconnected by *volkmann canals*; both conduct *blood vessels*. The bone internal to compact bone is trabecular, characterized by irregular and interwoven bony beams (*spongy bone*). These beams are constantly reorienting in response to the stress imposed on them. Unlike cartilage, bone is well-vascularized; bone cells reach for vascular nutrition by multiple long cellular processes threading through small canals (*canaliculi*), giving the cells an insect-like appearance.

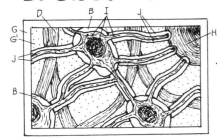

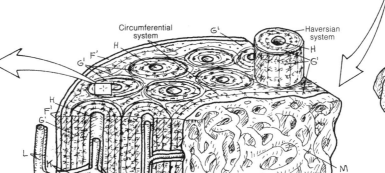

Circumferential system

Haversian system

TISSUES: ENDOCHONDRAL OSSIFICATION

CN: Use the same colors as used on the previous plate for hyaline cartilage (A), periosteal bone (B) which was compact bone on Plate 7, and endochondral bone (E) which was spongy bone. Use red for D. Complete each stage before going on to the next. Do not color the periosteum which appears adjacent to periosteal bone in step 3 and continues to the end. Color the small shapes (E) that appear in the epiphyses and, to a lesser extent, the diaphyses (views 5-8). They represent spongy (cancellous) bone of endochondral origin.

Bone development occurs by intramembranous and/or endo-chondral ossification. Here we show longitudinal sections of developing long bone, demonstrating both forms of ossification, but emphasizing endochondral bone growth.

The endochondral process begins at about 5 weeks of post-fertilization age with formation of cartilage models (bone pro-totypes) from embryonic connective tissue. Subsequently (over the next 16–25 years), the cartilage is largely replaced by bone. The rate and duration of this process largely determines a person's standing height. Intramembranous bone development begins in embryonic connective tissue (membrane) and does not involve replacement of cartilage. The flat cranial bones, the clavicle, and the bone collar surrounding the shaft of cartilage models develop in this fashion.

Endochondral ossification begins with a *hyaline cartilage* model (1). As the cartilage structure grows, its central part dehydrates. The cartilage cells there begin to degenerate: enlarge, die and calcify (2). Concurrently, *blood vessels* bring bone-forming cells to the waist of the cartilage model and a collar of bone is formed around the cartilage shaft (2) within the membranous perichondrium (intramembranous ossification). This vascular, cellular, fibrous membrane around the bone collar is now called periosteum. The new bone collar (*periosteal bone*) becomes a supporting tubular shaft for the cartilage model, with a core of degenerating, calcifying cartilage (3).

Blood vessels from the fibrous periosteum penetrate the bone collar, enter the cartilage model (periosteal bud), and proliferate, conducting periosteal osteoblasts into the cartilage model (4). Starting at about 8 weeks post-fertilization, these bone-forming cells line up along peninsulas of *calcified cartilage* at the extremes of the shaft (*diaphysis*) and secrete new bone (5). The calcified cartilage degenerates and is absorbed into the blood: endochondral bone has now replaced the cartilage. The two sites of this activity are called primary centers of ossification (5). The direction of growth at these sites is toward the ends of the developing bone. The calcified cartilage and some endo-chondral bone of the diaphysis is subsequently absorbed, forming the medullary or marrow cavity (5). This cavity of the developing tubular bone shaft becomes filled with gelatinous red marrow in the fetus. Productive primary (diaphyseal) centers of ossification are well established at birth.

Beginning in the first few years after birth, secondary centers of ossification begin at the ends or *epiphyses* as blood vessels penetrate the cartilage there (6). The healthy cartilage between the epiphyseal and diaphyseal centers of ossification becomes the *epiphyseal plate* (7). It is the growth of this cartilage that is responsible for bone lengthening; it is the gradual replacement of this cartilage by bone cells in the metaphysis (7) that thins this plate and ultimately permits fusion of the epiphyseal and diaphyseal ossification centers (8), ending longitudinal bone growth (at 12–20 years of age). Dense areas of bone at the fusion site may remain into maturity (*epiphyseal line*). Epiphyseal bone is less structured (irregular beams) than that of the diaphysis (organized columns or osteons), and in maturity is called spongy or cancellous bone (recall Plate 7).

Intramembranous ossification of the diaphyseal shaft (bone collar to compact bone) is responsible for the widening of developing long bone. The ossification process is regulated by growth hormone (from the pituitary gland) and the sex hormones.

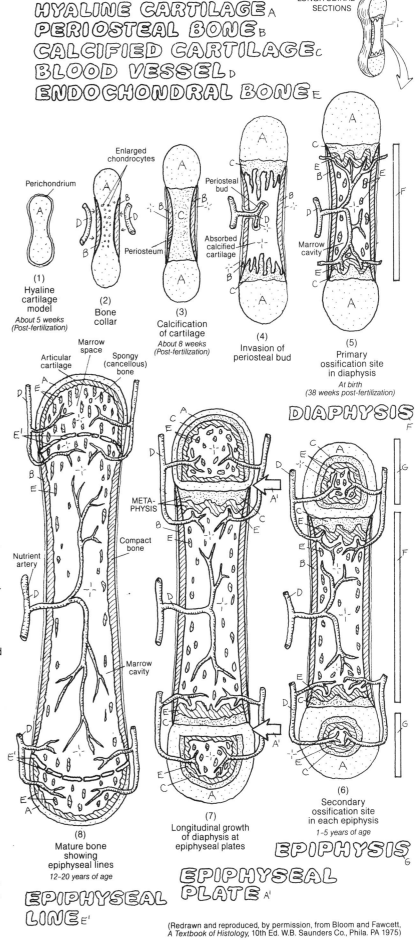

HYALINE CARTILAGE A
PERIOSTEAL BONE B
CALCIFIED CARTILAGE C
BLOOD VESSEL D
ENDOCHONDRAL BONE E

ALL VIEWS ARE LONGITUDINAL SECTIONS

(1) Hyaline cartilage model
About 5 weeks (Post-fertilization)

(2) Bone collar

(3) Calcification of cartilage
About 8 weeks (Post-fertilization)

(4) Invasion of periosteal bud

(5) Primary ossification site in diaphysis
At birth (38 weeks post-fertilization)

DIAPHYSIS

(6) Secondary ossification site in each epiphysis
1-5 years of age

(7) Longitudinal growth of diaphysis at epiphyseal plates

(8) Mature bone showing epiphyseal lines
12-20 years of age

EPIPHYSIS

EPIPHYSEAL PLATE A¹

EPIPHYSEAL LINE E¹

Enlarged chondrocytes

Perichondrium

Periosteum

Periosteal bud

Absorbed calcified cartilage

Marrow cavity

Articular cartilage

Marrow space

Spongy (cancellous) bone

META-PHYSIS

Compact bone

Nutrient artery

Marrow cavity

(Redrawn and reproduced, by permission, from Bloom and Fawcett, *A Textbook of Histology*, 10th Ed. W.B. Saunders Co., Phila. PA 1975)

TISSUES: MUSCLE

Muscle tissue, one of the four basic tissue types of the body, consists of muscle cells ("fibers") and their fibrous connective tissue coverings. There are three kinds of muscle tissues: skeletal, cardiac, and smooth. Muscle tissue shortens (contracts) in response to nerve, nerve-like, or hormonal stimulation. Depending on their attachments, skeletal muscles move bones at joints, constrict cavities, and move the skin; cardiac muscle compresses a heart cavity or orchestrates the sequence of cardiac muscle contraction; and smooth muscle moves the contents of cavities by rhythmic contractions, constricts vessels they surround, and moves hairs/closes pores of the skin. The surrounding *connective tissue* transfers the force of contraction from cell to cell, and supports the muscle fibers and the many blood *capillaries* and nerves that supply them.

CN: Use red for C and your lightest colors for B, E, G, and I. (1) The sarcolemma (F), which covers each skeletal and cardiac muscle cell, is colored only at the cut ends. The plasmalemma (F¹), which covers each smooth muscle cell, is colored only at the cut ends. (2) The nuclei of cardiac and smooth muscle cells, located deep within the cells, are to be colored only at the cut ends (A). (3) One of the intercalated discs (I) of the cardiac cells has been separated to reveal its structure (schematically). (4) The cellular views are microscopic.

NUCLEUS A
CONNECTIVE TISSUE B
CAPILLARY C
MITOCHONDRION D

SKELETAL/STRIATED MUSCLE, E CELL E¹
SARCOLEMMA F

Skeletal muscle cells are long, striated, and *multi-nucleated*, formed of myofibrils, *mitochondria*, and other organelles within the cytoplasm (sarcoplasm). Each cell is enveloped in cell membrane called *sarcolemma*. Collections of muscle cells make up the belly of a muscle. The highly vascularized skeletal muscles contribute greatly to the size and shape of the body. Skeletal muscles attach to bones or other muscles at their tendinous ends. Between bony attachments, muscles cross one or more joints, moving them. Muscles always pull . . . they never push. Skeletal muscle contractions consist of rapid, brief shortenings, often generating considerable force. Each contracting cell shortens maximally. Three kinds of skeletal muscle fibers are recognized: red (small, dark, long acting, slow contracting, postural muscle fibers with oxygen-rich myoglobin and many mitochondria), white (relatively large, pale, anaerobic, short acting, fast contracting muscle fibers with few mitochondria), and intermediate fibers. With exercise, fast fibers can convert to slow; slow fibers can convert to fast. Contraction of skeletal muscle requires nerves (innervation). Without a nerve supply (denervation), skeletal muscle cells cease to shorten; without reinnervation, the cells will die. A denervated portion of muscle loses its tone and becomes flaccid. In time, the entire muscle will become smaller (atrophy). Muscle contraction is generally under voluntary control, but the brain involuntarily maintains a degree of contraction among the body's skeletal muscles (muscle tone). After injury, skeletal muscle cells can regenerate from myoblasts with moderate functional significance; such regeneration may also occur in association with muscle cell hypertrophy in response to training/exercise.

CARDIAC/STRIATED MUSCLE, G CELL G¹
INTERCALATED DISC H

Cardiac muscle cells make up the heart muscle. They are branched, striated cells with one or two centrally located nuclei and a sarcolemma surrounding the sarcoplasm. They are connected to one another by junctional complexes called *intercalated discs*. Their structure is similar to skeletal muscle, but less organized. Cardiac muscle is highly vascularized; its contractions are rhythmic, strong, and well regulated by a special set of impulse-conducting muscle cells, not nerves. Rates of contraction of cardiac muscle are mediated by the autonomic (visceral) nervous system, the nerves of which increase/decrease heart rate. Cardiac muscle is probably not capable of regeneration.

VISCERAL/SMOOTH MUSCLE, I CELL I¹
PLASMALEMMA F¹

Smooth muscle cells are long, tapered cells with centrally placed nuclei. Each cell is surrounded by a *plasmalemma* (cell membrane). These cells are smooth (non-striated). Myofibrils are not seen; the myofilaments intersect with one another in a pattern less organized than that seen in skeletal muscle. Smooth muscle cells occupy the walls of organs with cavities (viscera) and serve to propel the contents along the length of those cavities by slow, sustained, often powerful rhythmic contractions (consider menstrual or intestinal cramps). Smooth muscle cells, oriented perpendicular to the flow of tubular contents, act as gates (sphincters) in specific sites, regulating the flow, as in delaying the flow of urine. Well-vascularized, smooth muscle fibers contract in response to both autonomic nerves and hormones. They are also capable of spontaneous contraction. Regeneration of smooth muscle, to some extent, is possible after injury.

Tendon

Muscle belly

Tendon

Intermediate fiber

Red fiber

Myofibrils

White fiber

Endomysium

Left ventricle

Myofibrils

Endomysium

Gastrointestinal tract

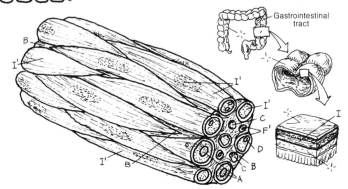

TISSUES: SKELETAL MUSCLE STRUCTURE

CN: Use the same colors used on Plate 9 for sarcolemma (A) and mitochondrion (D). Use the same color used on the skeletal muscle cell for the myofibril (E) here. Use light colors for G and J, a dark color for H, and very dark colors for F and K. The cell nucleus is not shown here. (1) Begin with the drawing of the arm. (2) Color the parts of the muscle cell in the central illustration; note the presence of mitochondria (D) between the myofibrils. (3) Color the parts of the exposed (lowest) myofibril and the color-related letters, bands, lines, zone. Note that the cut end of this myofibril receives the color E, for identification purposes, and is part of the A band of the sarcomere adjacent to the one to be colored. (4) Color the relaxed and contracted sarcomere, the filaments, and the mechanism for contraction, noting the color relationship with the myofibril and its parts.

A part of a *skeletal muscle cell* is shown with the *sarcolemma* opened to reveal some cellular contents. The most visible of the contents are the *myofibrils*, the contractile units of the cell. They are enveloped by a flat tubular *sarcoplasmic reticulum* (SR) that, in part, regulates the distribution of calcium ions (Ca^{++}) into the myofibrils. Inward tubular extensions of the sarcolemma, called the *transverse tubule system* (TTS), run transversely across the SR, at the level of the Z lines of the myofibrils. The TTS, containing stores of sodium ions (Na^+) and calcium ions (Ca^{++}), conducts electrochemical excitation to the myofibrils from the sarcolemma. *Mitochondria* provide energy for the cell work.

The myofibrils consist of myofilaments: *thick filaments* (largely myosin) with heads that project outward as *cross bridges*, and *thin filaments* (largely actin) composed of two interwoven strands. These two filament types are arranged into contractile units each of which is called a *sarcomere*. Each myofibril consists of several, radially arranged sarcomeres. At the end of each sarcomere, the thin filaments are permanently attached to the *Z line*, which separates one sarcomere from the next. The relative arrangement of the thick and thin filaments in the sarcomere creates light (I, H) and dark (A) bands/zone and the *M line*, all of which contribute to the appearance of cross striations in skeletal (and cardiac) muscles.

Shortening of a myofibril occurs when the thin filaments slide toward the center (H zone), bringing the Z lines closer together in each sarcomere. The filaments do not shorten; the myosin filaments do not move. The close relationship of the TTS to the Z lines suggests that this site is the "trigger area" for induction of the sliding mechanism. This sliding motion is induced by *cross bridges* (heads of the immovable thick filaments) that are connected to the thin filaments. Activated by high energy bonds from ATP, the paddle-like cross bridges swing in concert toward the H zone, drawing the thin filaments with them. The sarcomere shortens as the opposing thin filaments meet or even overlap at the M line.

Occurring simultaneously in all or most of the myofibrils of a muscle cell, shortening of sarcomeres translates to a variable shortening of the resting length of the muscle cell. Repeated in hundreds of thousands of conditioned muscle cells of a professional athlete, the resultant contractile force can pull a baseball bat through an arc sufficient to send a hardball a hundred meters or more through the air.

SKELETAL MUSCLE CELL:*
SARCOLEMMA A
SARCOPLASMIC RETICULUM B
TRANSVERSE TUBULE SYS. C
MITOCHONDRION D

MYOFIBRIL E
SARCOMERE F
I BAND G
THIN FILAMENT (ACTIN) G'
Z LINE F'
A BAND H
THICK FILAMENT (MYOSIN) H'
CROSS BRIDGE I
H ZONE J
M LINE K

TISSUES: NERVOUS

CN: Use a light color for A. Note the small arrows which indicate direction of impulse conduction. The neurons of the peripheral nervous system shown at lower left are illustrated in the orientation of the left upper limb, although highly magnified.

NEURON*
CELL BODY A
PROCESS(ES) (c-)
DENDRITE B
AXON C

Nucleolus

Nucleus

Cytoplasm

Nissl substance (ER)

Nervous tissue consists of *neurons* (nerve cells) and *neuroglia*. Neurons generate and conduct electrochemical impulses by way of neuronal (cellular) *processes*. Neuroglia are the supporting, non-impulse generating/conducting cells of the nervous system. The main, nucleus-bearing part of the neuron is the *cell body*. Its cytoplasm contains the usual cell organelles. Uniquely, the endoplasmic reticulum occurs in clusters called Nissl substance. Neurons do not undergo mitosis after birth, compromising their ability to regenerate after injury. Neuronal growth consists of migration and arborization of processes. Neurons are the impulse-conducting cells of the brain and spinal cord (central nervous system or CNS) and the spinal and cranial nerves (peripheral nervous system or PNS).

TYPES OF NEURONS *

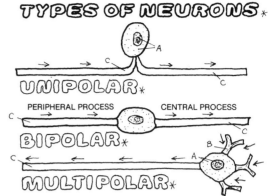

UNIPOLAR*

PERIPHERAL PROCESS — CENTRAL PROCESS

BIPOLAR*

MULTIPOLAR*

Neurons fall into three structural categories based on numbers of processes ("poles"). Processes that are highly branched (arborized) and uncovered are called *dendrites*. Slender, long, minimally branched processes are called *axons*. Within each category, there is a great variety of shape and size of neurons. *Unipolar* neurons have or appear to have (pseudounipolar) one process which splits near its cell body into a central and peripheral process. Both processes conduct impulses in the same direction, and each is termed an axon (see the sensory neuron at lower left). *Bipolar* neurons have two (central and peripheral) processes, called axons, conducting impulses in the same direction (see Plate 131). *Multipolar* neurons have three or more processes, one of which is an axon (see PNS motor neuron at lower left, and CNS neuron at lower right).

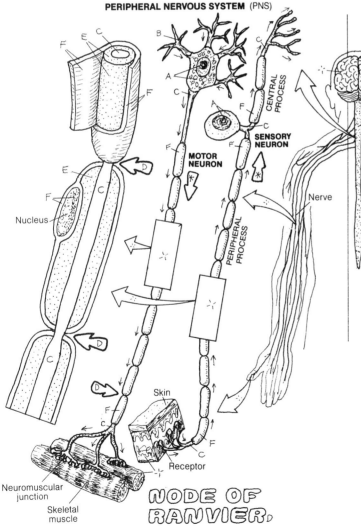

PERIPHERAL NERVOUS SYSTEM (PNS)

CENTRAL PROCESS

SENSORY NEURON

MOTOR NEURON

PERIPHERAL PROCESS

Nucleus

Neuromuscular junction

Skeletal muscle

Skin

Receptor

NODE OF RANVIER D

AXON COVERINGS *
MYELIN E
SCHWANN CELL F

CENTRAL NERVOUS SYSTEM (CNS)

Brain

Spinal cord

Nerve

Blood vessel

Blood vessel

NEUROGLIA *
PROTOPLASMIC ASTROCYTE G
FIBROUS ASTROCYTE H
OLIGODENDROCYTE I
MICROGLIA J

Most axons are enveloped in one or more (up to 200) layers of an insulating phospholipid (*myelin*) that enhances impulse conduction rates. In the CNS (lower right), myelin is produced by *oligodendrocytes*; in the PNS (lower left), by *Schwann cells*. All axons of the PNS are ensheathed by the cell membranes of Schwann cells (neurilemma) but not necessarily myelin. The gaps between Schwann cells are *nodes of Ranvier*, making possible rapid node-to-node impulse conduction. Schwann cells make possible axonal regeneration in the PNS. Significant axonal regeneration in the CNS has not been observed.

Neuroglia exist in both the CNS and PNS (Schwann cells). *Protoplasmic astrocytes* occur primarily in gray matter (dendrites, cell bodies) of the CNS, *fibrous astrocytes* in the white matter (myelinated axons). Their processes attach to both neurons and blood vessels and may offer metabolic, nutritional and physical support. They may play a role in the blood brain barrier. Oligodendrocytes are smaller than astrocytes, have fewer processes, and are seen near neurons. *Microglia* are the small scavenger cells of the brain and spinal cord.

INTEGRATION OF TISSUES

This plate has one goal: to aid you in visually integrating the four basic tissues into somatic (body wall) and visceral (cavity-containing organs) structure. Concentrate on how the four tissues are arranged in each example of body structure. Consider the general function of each tissue in the overall function of the part/organ. There are an infinite number of functionally related variations in the way these four tissues form a discrete construction of the soma and viscera of the body.

CN: Use yellow for D and light, contrasting colors for A and B, and a medium brown for C. The various vessels that are shown in these tissues—arteries and veins above, and arterioles, venules, capillaries, and lymph vessels below—are not to be colored, as they are made up of more than one basic tissue. Note that within deep fascia, arteries are generally paired with veins.

SOMATIC STRUCTURE
EPITHELIAL TISSUE *
SKIN (OUTER LAYER)ₐ
CONNECTIVE TISSUE *
SKIN (DEEP LAYER)в
SUPERFICIAL FASCIA в¹
DEEP FASCIA в²
LIGAMENT в³
BONE в⁴
PERIOSTEUM в⁵
MUSCLE TISSUE *
SKELETAL MUSCLE c
NERVOUS TISSUE *
NERVE ᴅ

Somatic structure, making up the skin-covered musculoskeletal frame of the body, is concerned with stability, movement, and protection. Its construction reflects these functions. The outermost covering of the body wall everywhere is a protective keratinized *stratified squamous epithelial tissue*, constituting the *outer layer of skin* (epidermis). Other epithelial tissues in somatic structure are the inner layers of blood vessels, and the glands (not shown). Connective tissue layers of the body wall include the *deep layer of skin* (dermis), consisting of dense, irregular fibrous *connective tissue*; and the sub-adjacent, variously mobile, subcutaneous *super-*

ficial fascia (loose connective and adipose tissues), containing cutaneous nerves, small vessels, and occasional large veins. *Deep fascia*, a more vascular, sensitive, dense, irregular fibrous tissue, ensheaths skeletal muscle (myofascial tissue) as well as the supporting nerves and vessels. *Ligaments* (dense regular connective tissue) bind *bone* to bone by way of *periosteum* (vascular, cellular, dense, irregular, fibrous tissue). Skeletal *muscles* and their *nerves* are packaged in groups, separated by slippery septa of deep fascia securing neurovascular structure. The fibrous investments of skeletal muscle converge to form tendons of the muscle.

VISCERAL STRUCTURE
EPITHELIAL TISSUE *
MUCOSAL LINING ₐ¹
GLAND ₐ²
SEROSA (OUTER LAYER) ₐ³
CONNECTIVE TISSUE *
LAMINA PROPIA в⁶
SUBMUCOSA в⁷
SEROSA (INNER LAYER) в⁸
MUSCLE TISSUE *
SMOOTH MUSCLE c¹
NERVE TISSUE *
NERVE CELLS ᴅ¹

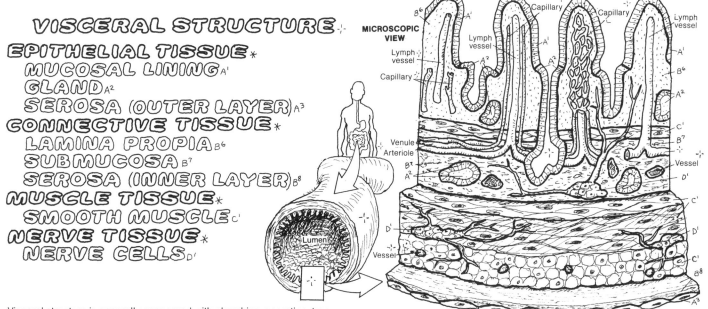

Visceral structure is generally concerned with absorbing, secreting, trapping, and/or moving food, air, secretions, and/or waste in its cavities. *Epithelial tissue* is the innermost layer (*mucosal lining*) of the thin and pliable visceral wall. It faces the lumen (cavity of the viscus); it is often a single layer of cells (esophagus, urinary tract, and reproductive tract excepted) and deals with the contents of the visceral cavity. *Glands*, unicellular or larger in the mucosa or submucosa, are epithelial, as are the inner layers of blood and lymph vessels. The mucosa includes a sub-epithelial layer of loose fibrous tissue (*lamina propria*), supporting mobile cells, glands, vessels and *nerves*. The deepest layer of the mucosa (when

present) is a thin *smooth muscle* layer moving finger-like projections (villi) of the mucosal surface. Deep to the mucosa is a dense fibrous tissue (*submucosa*), replete with large vessels and small nerves/nerve cells (intramural ganglia) supplying the mucosa. Deeper yet, two or three layers of smooth muscle (tunica muscularis), innervated by local nerve cells, move the visceral wall in peristaltic contractions. The outermost layer of the gastrointestinal tract is the slippery *serosa*: an *outer* secretory simple squamous epithelial layer and an *inner* supporting layer of light fibrous tissue.

CN: Use light colors on this and the next plate.
Color the entire skeleton (A); only the knee joint and shoulder
joint show joint capsules (A¹). Color the entire musculature (B)
brown. Color all the vessels and heart (C): arteries and heart
red, veins blue. Color all lymphatic vessels (D) green. Color all
the nerves, as well as the brain and spinal cord (E), yellow. Do
not color the background of the rectangular insets representing
the endocrine system (F). Choose a skin color for the integument system (G).

Collections of similar cells constitute tissues. The four
basic tissues are integrated into body wall and visceral
structures/organs. A *system* is a collection of organs and
structures sharing a common function. Organs and structures of a single system occupy diverse regions within
the body and are not necessarily grouped together.

SKELETAL SYS. A
ARTICULAR SYS. A'

The *skeletal system* consists of the skeleton of bones
and their periosteum, and the ligaments which secure the
bones at joints. By extension, this system could include
the varied fasciae which ensheath the body wall/skeletal
muscles and contribute to the body's structural stability.
The *articular system* comprises the joints, both movable
and fixed, and the related structures, including joint
capsules, synovial membranes, and discs/menisci.

MUSCULAR SYS. B

The *muscular system* includes the skeletal muscles
which move the skeleton, the face, and other structures,
and give form to the body; the cardiac muscle of the
heart walls; and the smooth muscle of the walls of viscera and vessels, and in the skin.

CARDIOVASCULAR SYS. C

The *cardiovascular system* consists of the 4-chambered
heart, arteries conducting blood to the tissues, capillaries
through which nutrients, gases, and molecular material
pass to and from the tissue, and veins returning blood
from the tissues to the heart. Broadly interpreted, the
cardiovascular system includes the lymphatic system.

LYMPHATIC SYS. D

The *lymphatic system* is a system of vessels assisting
the veins in recovering the body's tissue fluids and
returning them to the heart. The body is about 60% water,
and the veins alone are generally incapable of meeting
the demands of tissue drainage. Lymph nodes filter
lymph and are located throughout the body.

NERVOUS SYS. E

The *nervous system* consists of impulse-generating/con-
ducting tissue organized into a central nervous system
(brain and spinal cord), and a peripheral nervous system
(nerves) that includes the visceral (autonomic) nervous
system involved in involuntary "fight or flight" and vegetative responses.

ENDOCRINE SYS. F

The *endocrine system* consists of glands which secrete
chemical agents (hormones) into the tissue fluids and
blood, affecting the function of multiple areas of the body.
Many of these glands are under some control by the
brain (hypothalamus). Hormones help maintain balanced
metabolic functions in many of the body's systems.

INTEGUMENT. SYS. G

The *integumentary system* is the skin, replete with
glands, sensory receptors, vessels, immune cells and
antibodies, and layers of cells and keratin resisting
environmental factors harmful to the body.

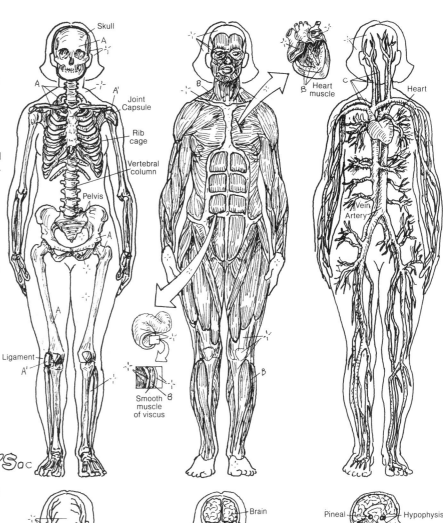

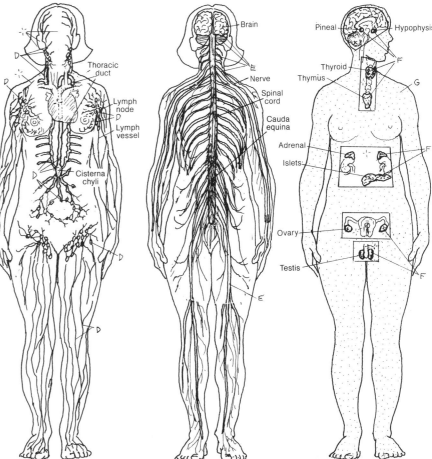

CN: Use different light colors from those used on the preceding plate.

RESPIRATORY SYS. H

The *respiratory system* consists of the upper (nose through larynx) and lower respiratory tract (trachea through the air spaces of the lungs). Most of the tract is airway; only the air spaces (alveoli) and very small bronchioles exchange gases between alveoli and the lung capillaries.

DIGESTIVE SYS. I

The *digestive system* is concerned with the breakdown, digestion and assimilation of food, and excretion of the residua. Its tract begins with the mouth, continues down to the abdomen wherein it takes a convoluted course to open again at the anus. Associated glands include the liver and pancreas, including the biliary system (gall bladder and related ducts).

URINARY SYS. J

The *urinary system* is concerned with the conservation of water and maintenance of a neutral acid-base balance in the body fluids. The kidneys are the main functionaries of this system; residual fluid (urine) is excreted through ureters to the urinary bladder for retention, and discharged to the outside through the urethra.

IMMUNE / LYMPHOID SYS. K

The *lymphoid system* consists of organs concerned with body defense: thymus, bone marrow, spleen, lymph nodes, tonsils, and smaller aggregates of lymphoid tissue. This system, including a diffuse arrangement of immune-related cells throughout the body, is concerned with resistance to invasive microorganisms, and the removal of damaged and/or otherwise abnormal cells.

REPRODUCTIVE SYS./ FEMALE L

The female reproductive system is concerned with the secretion of sex hormones, production and transportation of germ cells (ova), receipt and transport of male germ cells to the fertilization site, maintenance of the developing embryo/fetus, and initial sustenance of the newborn.

REPRODUCTIVE SYS./ MALE M

The male reproductive system is concerned with the secretion of male sex hormones, formation and maintenance of germ cells (sperm), and transport of germ cells to the female genital tract.

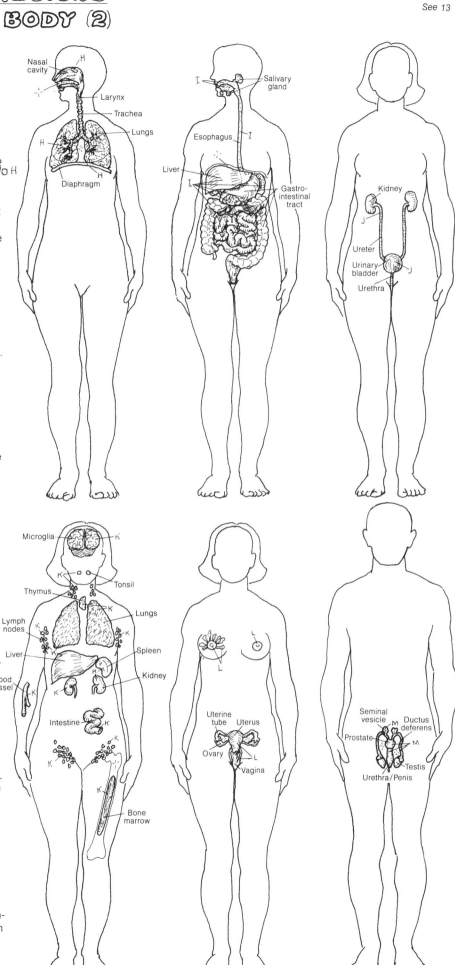

CN: You will most likely have to repeat colors among the arrows shown here. Consider a coloring pattern based on major regions (shades of red for head and neck, blues for upper limb, brown for lower limb, and so on).

HEAD/NECK:*
CRANIAL A
ORBITAL B
FACIAL C
MANDIBULAR D
CERVICAL E
SUPRACLAVICULAR F

UPPER LIMB:*
DELTOID G
AXILLARY H
BRACHIAL I
CUBITAL J
ANTECUBITAL K
ANTEBRACHIAL L
CARPAL M

THORAX:*
PECTORAL N
SCAPULAR O

ABDOMINOPELVIC:*
ABDOMINAL P
PELVIC Q
INGUINAL R
PUDENDAL S

BACK:*
THORACIC T
LUMBAR U
SACROILIAC V

LOWER LIMB:*
GLUTEAL W
FEMORAL X
PATELLAR Y
POPLITEAL Z
CRURAL 1.
MALLEOLAR 2.
TARSAL 3.

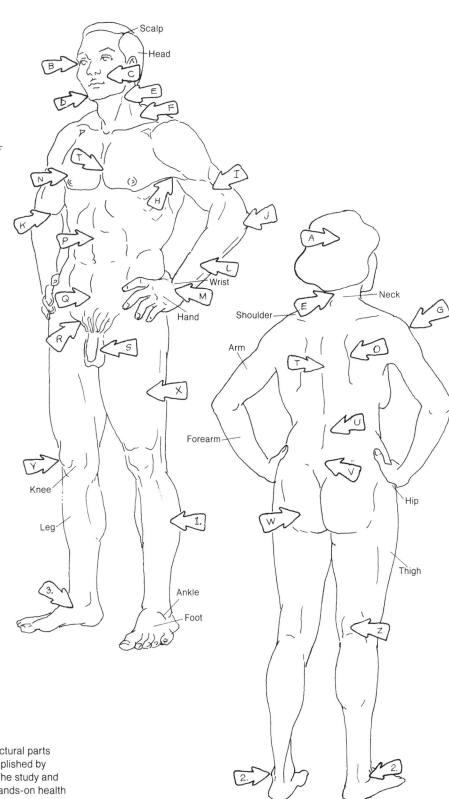

Regional anatomy is the organization of human structural parts by regions. The study of regional anatomy is accomplished by dissection of body parts in an embalmed cadaver. The study and awareness of regional anatomy is fundamental to hands-on health care providers who examine and evaluate the human body. Within each region, there are usually sub-divisions or sub-regions, e.g., within the cranial region, there are frontal, parietal, temporal, and occipital sub-regions. Characteristically, each region is composed of structures representing several systems.

CN: Except for H, use light colors throughout. (1) Note that the linings for closed body cavities are all colored gray. (2) In the open visceral cavities shown below, the linings receive a darker color (H), and the closed cavities have their linings omitted.

CLOSED BODY CAVITIES:
CRANIAL_A / DURA MATER_A'*
VERTEBRAL_B / DURA MATER_B'*
THORACIC_C / PLEURA_C'*
ABDOMINOPELVIC_D / PERITONEUM_D'*

Closed body cavities are not open to the outside of the body. The *cranial cavity* is located in the upper part of the skull and houses the brain and related structures. It is continuous with the *vertebral cavity* (canal) within the vertebral column. The vertebral cavity contains the spinal cord and related vessels and nerve roots. The tough, fibrous layer lining the cranial and vertebral cavities is called the *dura mater.* Portions of the dura envelop the brain and spinal cord; other parts are folded to form dividers (dural septa, not shown) separating parts of the brain.

The *thoracic cavity* is located in the thorax, surrounded by the rib cage and related muscles; its posterior wall is the vertebral column, its floor is the muscular thoracic diaphragm. The thoracic cavity is divided by a central set of structures (mediastinum) into left and right cavities for the lungs. These cavities are lined with a thin layer of simple squamous, secretory epithelium supported by fibrous tissue. Such a lining secretes a watery (serous) fluid and is called a serous membrane or serosa. The serosa lining the thoracic cavities for the lungs is called *pleura,* a subject to be developed in Plate 95.

The *abdominopelvic cavity* is located anterior to the posterior abdominal wall. It is surrounded anteriorly and laterally by muscle layers, the lower ribs and related muscles, and the bones of the pelvis. Its upper and lower boundaries are muscular (respiratory and pelvic) diaphragms. The abdominopelvic cavity contains the abdominal and pelvic viscera. The abdominal cavity is largely lined with a serous membrane called *peritoneum,* a subject to be developed in Plate 102.

There are a number of other cavities within the body that are closed, including the joint cavities, heart, vessels, ventricles of the brain, cavities/ducts of the eye and internal ear, and the potential cavities created by the foldings of serosal membranes (Plates 65, 95, and 102).

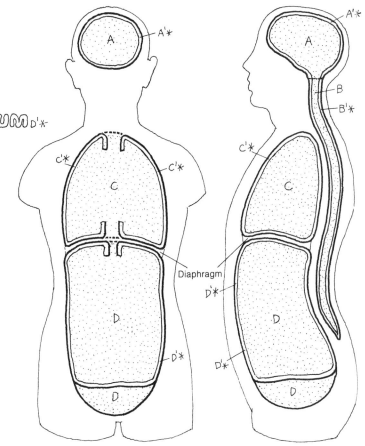

Diaphragm

OPEN VISCERAL CAVITIES:
RESPIRATORY TRACT_E
URINARY TRACT_F
DIGESTIVE TRACT_G
MUCOSA_H

Open cavities, located within closed cavities, are open to the outside of the body. They are generally tubular cavities of viscera: the digestive, respiratory, reproductive, and urinary tracts. The digestive tract opens at the mouth superiorly and at the anus inferiorly. The respiratory tract opens at the mouth and nose. The reproductive tracts (not shown) open at the perineum; the urinary tract opens at the perineum as well. Not included in this category are glands of the skin and viscera whose ducts open on to the surface of the skin.

A variably-thick membrane covered with mucus (mucous membrane or *mucosa*) lines open cavities. It is continuous with skin at orifices. The mucosa is the working membrane of open cavities (secretion, absorption, protection), dealing with molecules placed before it (food, air, fluid, and so on).

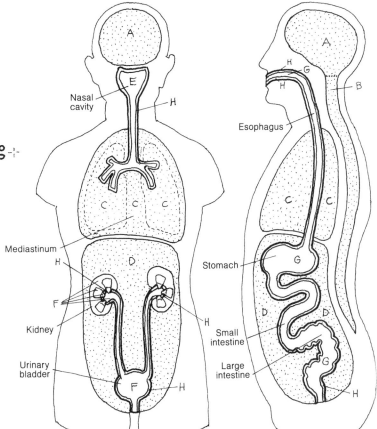

Nasal cavity

Esophagus

Mediastinum

Stomach

Kidney

Small intestine

Urinary bladder

Large intestine

III. SKELETAL SYSTEM
ANATOMY OF A LONG BONE

CN: Use light blue for C, a "bone" color for D, very light colors for E and F, yellow for I, and red for J. The title red marrow (H) is not to be colored as the red marrow in this bone is not shown, having been replaced by yellow marrow during maturity. Only part of the yellow marrow in the medullary canal is shown.

Bone is a living, vascular structure, composed of organic tissue (cells, fibers, extra-cellular matrix, vessels, nerves—about 35% of a bone's weight), and mineral (calcium hydroxyapatite—about 65% of a bone's weight). Bone functions as a support structure, a site of attachment for skeletal muscle, ligaments, tendons, and joint capsules, a source of calcium, and a significant site of blood cell development (hematopoiesis) for the entire body. Here we show a long bone, specifically the femur, the bone of the thigh.

EPIPHYSIS A
The *epiphysis* is the end of a long bone or any part of a bone separated from the main body of an immature bone by cartilage. It is formed from a secondary site of ossification. It is largely cancellous bone, and its articulating surface is lined with 3–5 mm of hyaline (articular) cartilage. The epiphysis is supplied by vessels from the joint capsule.

DIAPHYSIS B
The *diaphysis* is the shaft or central part of a long bone. It has a marrow-filled cavity (*medullary cavity*) surrounded by compact bone which is lined externally by *periosteum* and internally by *endosteum* (not shown). The diaphysis is formed from one or more primary sites of ossification, and is supplied by one or more *nutrient arteries*.

ARTICULAR CARTILAGE C
The only remaining evidence of an adult bone's cartilaginous past, *articular cartilage* is smooth, slippery, porous, malleable, insensitive, and bloodless. It is massaged by movement, permitting absorption of synovial fluid, oxygen and nutrition. Articular (hyaline) cartilage is also nourished by vessels from the subchondral bone. Bones of a synovial joint make physical contact at their cartilaginous ends. The degenerative process of arthritis involves the breakdown and fibrillation of articular cartilage.

PERIOSTEUM D
Periosteum is a fibrous, cellular, vascular and highly sensitive life support sheath for bone, providing nutrient blood for bone cells and a source of osteoprogenitor cells throughout life. It does not cover articular cartilage.

CANCELLOUS (SPONGY) BONE E
Cancellous (spongy) *bone* consists of interwoven beams (trabeculae) of bone in the epiphyses of long bones, the bodies of the vertebrae, and other bones without cavities. The spaces among the trabeculae are filled with red or yellow marrow and blood vessels. Cancellous bone forms a dynamic latticed truss capable of mechanical alteration (reorientation, construction, destruction) in response to the stresses of weight, postural change, and muscle tension.

COMPACT BONE F
Compact bone is dense bone characterized in long bones by microscopic hollow cylinders of bone (haversian systems) interwoven with non-cylindrical lamellae of bone. It forms the stout walls of the diaphysis of long bones and the thinner outer surface of other bones where there is no articular cartilage, e.g., the flat bones of the skull. Blood vessels reach the bone cells by a system of integrated canals.

MEDULLARY CAVITY G
The *medullary cavity* is the cavity of the diaphysis. It contains marrow: red in the young, turning to yellow in many long bones in maturity. It is lined by endosteal tissue (thin connective tissue with many osteoprogenitor cells).

RED MARROW H
Red marrow is a red, gelatinous substance composed of a red and white blood cells in a variety of developmental forms (hematopoietic tissue) and specialized capillaries (sinusoids) enmeshed in reticular tissue. In adults, red marrow is generally limited to the sternum, vertebrae, ribs, hip bones, clavicles, and cranial bones.

YELLOW MARROW I
Yellow marrow is fatty connective tissue and no longer productive of blood cells. It replaces red marrow in the epiphyses and medullary cavities of long bones, and cancellous bone of other bones.

NUTRIENT ARTERY J /BRANCHES J'
The *nutrient artery* is the principal artery and major supplier of oxygen and nutrients to the shaft or body of a bone; its branches snake through the labyrinthine canals of the haversian systems and other tubular cavities of bones.

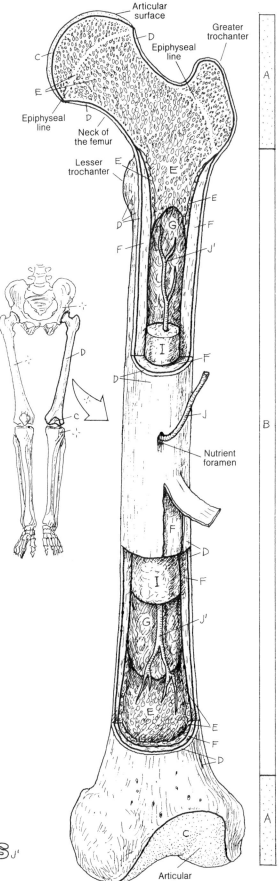

ANTERIOR VIEW
(Left femur)

Coronal section through proximal epiphysis and dissection of medullary cavity

Articular surface

Greater trochanter

Epiphyseal line

Epiphyseal line

Neck of the femur

Lesser trochanter

Nutrient foramen

Articular surface

CN: Use light but contrasting colors for A and B.
(1) First color the axial skeleton A in all three views.
Do not color the intercostal spaces.
(2) Color the appendicular skeleton B.
Note that the bones labeled A are drawn in a lighter line than B.
(3) Color the arrows identifying bone shape/classification.

CLASSIFICATION OF BONES: *
LONG c
SHORT D
FLAT E
IRREGULAR F
SESAMOID G

Bones have a variety of shapes and defy classification by shape; yet such a classification generally exists. *Long bones* are clearly longer in one axis than in another; they are characterized by a medullary cavity, a hollow diaphysis of compact bone, and at least two epiphyses; e.g., femur, phalanx. *Short bones* are roughly cube-shaped; they are predominantly cancellous bone with a thin cortex of compact bone; no cavity; e.g., carpal and tarsal bones. *Flat bones* (cranial bones, ribs) are generally more flat than round, and *irregular bones* (scapula, vertebrae) have two or more different shapes; e.g., the scapula, with a flat surface, and irregular-shaped spine. Bones not specifically long or short fit this latter category. *Sesamoid bones* are developed in tendons (e.g., patellar tendon); they are mostly bone, often mixed with fibrous tissue and cartilage. They have a cartilaginous articular surface facing an articular surface of an adjacent bone; they may be part of a synovial joint ensheathed within the fibrous joint capsule. They are generally pea-sized, and are almost always found in certain tendons/joint capsules in hands and feet, and occasionally in other articular sites of the upper and lower limbs. The largest is the patella, in the tendon of quadriceps femoris. Sesamoid bones resist friction and compression, enhance joint movement, and may assist local circulation.

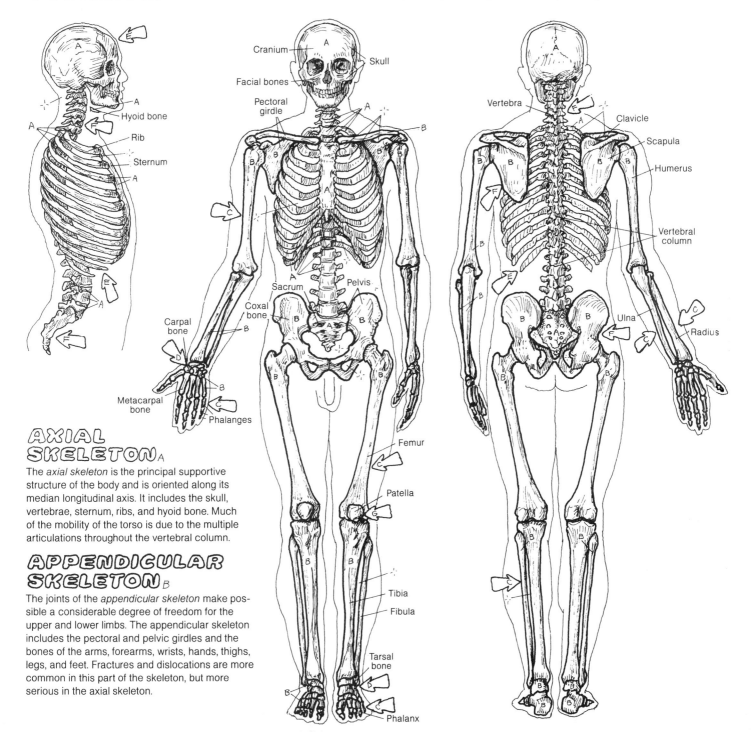

AXIAL SKELETON A

The *axial skeleton* is the principal supportive structure of the body and is oriented along its median longitudinal axis. It includes the skull, vertebrae, sternum, ribs, and hyoid bone. Much of the mobility of the torso is due to the multiple articulations throughout the vertebral column.

APPENDICULAR SKELETON B

The joints of the *appendicular skeleton* make possible a considerable degree of freedom for the upper and lower limbs. The appendicular skeleton includes the pectoral and pelvic girdles and the bones of the arms, forearms, wrists, hands, thighs, legs, and feet. Fractures and dislocations are more common in this part of the skeleton, but more serious in the axial skeleton.

III. SKELETAL SYSTEM
BONES OF THE SKULL (1)
CRANIAL (8):*
OCCIPITAL (1)ₐ PARIETAL (2)ʙ FRONTAL (1)c
TEMPORAL (2)ᴅ ETHMOID (1)ᴇ SPHENOID (1)ꜰ
FACIAL (14):*
NASAL (2)ɢ VOMER (1)ʜ LACRIMAL (2)ɪ
ZYGOMATIC (2)ᴊ PALATINE (2)ᴋ MAXILLA (2)ʟ
MANDIBLE (1)ᴍ INFERIOR NASAL CONCHA (2)ɴ

CN: Work with this plate and the next one at the same time. Save the brightest colors for the smallest bones; use light colors on bones with surface detail. Work one bone at a time, coloring it *where* it appears in any of the 7 views shown on this and the next plate. (1) In the anterior view, do not color the darkened areas in the orbits and nasal cavity.

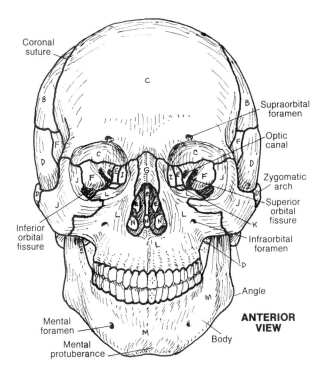

Coronal suture

Supraorbital foramen

Optic canal

Zygomatic arch

Superior orbital fissure

Infraorbital foramen

Inferior orbital fissure

Angle

Mental foramen

Mental protuberance

Body

ANTERIOR VIEW

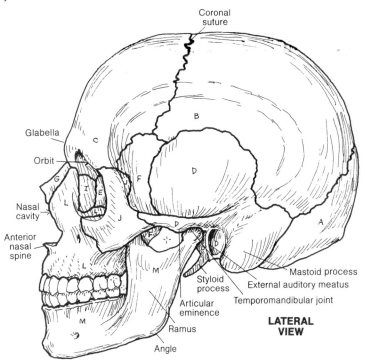

Coronal suture

Glabella

Orbit

Nasal cavity

Anterior nasal spine

Styloid process

Articular eminence

Ramus

Angle

Mastoid process

External auditory meatus

Temporomandibular joint

LATERAL VIEW

The skull is composed of *cranial bones* (forming a vault for the brain) and *facial bones* (giving origin to the muscles of facial expression and providing buttresses protecting the brain). Except for the temporomandibular joint (a synovial joint), all bones are connected by generally immovable fibrous sutures.

The orbit is composed of 7 bones, has 3 significant fissures/canals, and is home to the eye and related muscles, nerves, and vessels. The most delicate of the skull bones is at the medial orbital wall (I). The external nose is largely cartilaginous and is, therefore, not part of the bony skull.

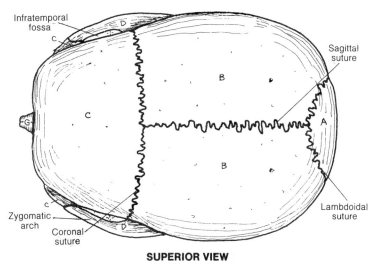

Infratemporal fossa

Sagittal suture

Lambdoidal suture

Zygomatic arch

Coronal suture

SUPERIOR VIEW

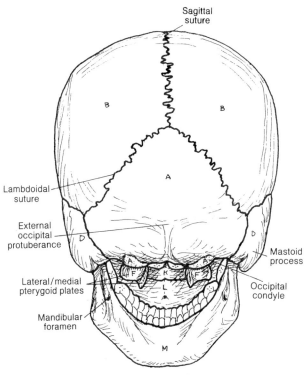

Sagittal suture

Lambdoidal suture

External occipital protuberance

Lateral/medial pterygoid plates

Mandibular foramen

Mastoid process

Occipital condyle

POSTERIOR VIEW

BONES OF THE SKULL (2)
CRANIAL:* OCCIPITAL_A PARIETAL_B FRONTAL_C
TEMPORAL_D ETHMOID_E SPHENOID_F
FACIAL:* NASAL_G VOMER_H ZYGOMATIC_J PALATINE_K
MAXILLA_L INFERIOR NASAL CONCHA_N

CN: Use the same colors as were used on Plate 19. (1) Color the three views simultaneously. (2) In the lower views, pay close attention to the many foramina that are left uncolored. (3) Notice but don't color the small drawing below that identifies the large fossae of the skull interior to its left. Try to visualize those fossae in the larger view.

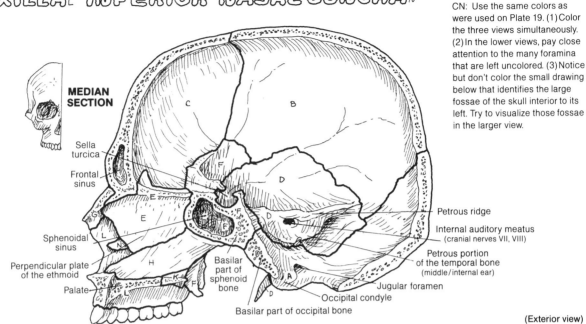

MEDIAN SECTION

Sella turcica
Frontal sinus
Sphenoidal sinus
Perpendicular plate of the ethmoid
Palate

Petrous ridge
Internal auditory meatus (cranial nerves VII, VIII)
Petrous portion of the temporal bone (middle / internal ear)
Jugular foramen
Occipital condyle
Basilar part of sphenoid bone
Basilar part of occipital bone

You are looking into the interior of the right side of the skull. The *vomer* and perpendicular plate of the *ethmoid* contribute significantly to the nasal septum, and hide from view the conchae on the lateral wall of the nasal cavity.

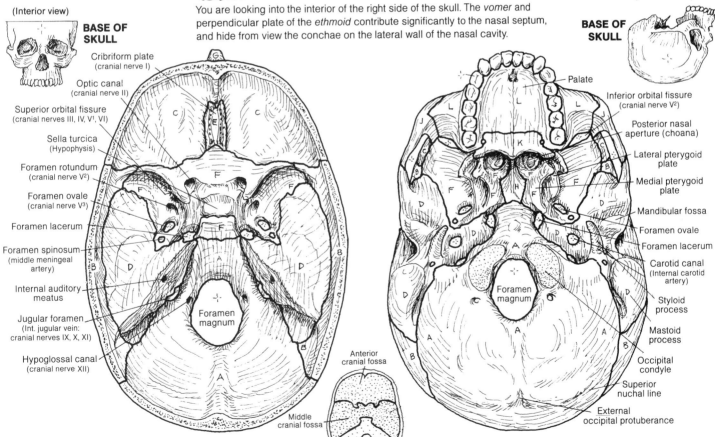

(Interior view)
BASE OF SKULL

Cribriform plate (cranial nerve I)
Optic canal (cranial nerve II)
Superior orbital fissure (cranial nerves III, IV, V¹, VI)
Sella turcica (Hypophysis)
Foramen rotundum (cranial nerve V²)
Foramen ovale (cranial nerve V³)
Foramen lacerum
Foramen spinosum (middle meningeal artery)
Internal auditory meatus
Jugular foramen (Int. jugular vein: cranial nerves IX, X, XI)
Hypoglossal canal (cranial nerve XII)
Foramen magnum

Anterior cranial fossa
Middle cranial fossa
Posterior cranial fossa

(Exterior view)
BASE OF SKULL

Palate
Inferior orbital fissure (cranial nerve V²)
Posterior nasal aperture (choana)
Lateral pterygoid plate
Medial pterygoid plate
Mandibular fossa
Foramen ovale
Foramen lacerum
Carotid canal (Internal carotid artery)
Styloid process
Mastoid process
Occipital condyle
Superior nuchal line
External occipital protuberance
Foramen magnum

You are looking into the cranial cavity from above. The anterior cranial fossa contains the frontal lobes of the cerebrum (brain); the olfactory tracts lie over the cribriform plates and receive the olfactory nerves. The middle cranial fossa embraces the temporal lobes of the cerebrum; note the numerous foramina/canals for cranial nerves and vessels which enter/exit the cavity. The posterior cranial fossa retains the cerebellum and the brain stem, along with related cranial nerves and vessels which enter/exit the cavity.

The occipital condyles articulate with the facets of the atlas or first cervical veretebra. The muscular pharyngeal wall attaches around the posterior nasal apertures. The lateral pterygoid plate offers attachment for certain muscles of mastication. The foramen magnum transmits the lower brain stem/spinal cord and the vertebral arteries. Much of the occipital bone posterior to the foramen magnum is a site of attachment for large muscle bundles making up the posterior cervical (paraspinal) musculature.

CN: Use gray for D, yellow for H, and light colors for the rest, especially C, T, L, S, and Co. (1)Begin with regions of the column and the three examples of vertebral disorders at lower left. (2)Color the motion segment and its role in flexion and extension. (3)Color the vertebral foramina and canal. (4)Color the example of a protruding intervertebral disc pressing on a spinal nerve.

REGIONS:*
CERVICAL C
THORACIC T
LUMBAR L
SACRAL S
COCCYGEAL Co

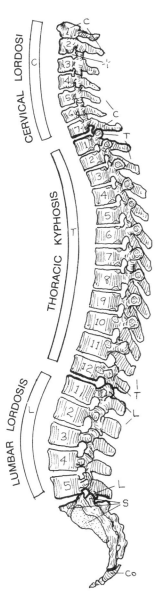

The vertebral column has 24 individual vertebrae arranged in *cervical, thoracic,* and *lumbar* regions; the *sacral* and *coccygeal* vertebrae are fused (sacrum/coccyx). Numbers of vertebrae in each region are remarkably constant; rarely S1 may be free or L5 may be fused to the sacrum (transitional vertebrae). The seven mobile cervical vertebrae support the neck and the 3-4 kg (6-8 lb) head. The cervical spine is normally curved *(cervical lordosis)* secondary to the development of postural reflexes about three months after birth. The 12 thoracic vertebrae support the thorax, head, and neck. They articulate with 12 ribs bilaterally. The thoracic spine is congenitally curved *(kyphosis)* as shown.

The five lumbar vertebrae support the upper body,

torso, and low back. The column of these vertebrae is curved *(lumbar lordosis)* due to the onset of walking at 1-2 years of age. The sacrum is the keystone of a weightbearing arch involving the hip bones. The sacral/coccygeal curve is congenital. The variably numbered 1-5 coccygeal vertebrae are usually fused, although the first vertebra may be movable.

Vertebral curvatures may be affected (usually exaggerated) by posture, activity, obesity, pregnancy, trauma, and/or disease; these conditions are named the same as the normal curves. There may normally be a slight lateral curvature to the spine often due to dominant handedness; a significant, possibly disabling, lateral curve *(scoliosis)* may occur for many reasons.

MOTION SEGMENT:*
VERTEBRA L
JOINTS:*
INTERVERTEBRAL DISC A
POSTERIOR (FACET) B
LIGAMENT D*
VERTEBRA L²

VERTEBRAL FORAMEN E
VERTEBRAL CANAL E'
INTERVERTEBRAL FORAMEN F

Each pair of individual, unfused vertebrae constitutes a *motion segment,* the basic movable unit of the back. Combined movements of motion segments underlie movement of the neck, middle and low back. Each pair of vertebrae in a motion segment, except C1-C2, is attached by three joints: a partly movable, *intervertebral disc* anteriorly, and a pair of gliding synovial *facet* (zygapophyseal) *joints* posteriorly. *Ligaments* secure the bones together and encapsulate the facet joints (joint capsules). The *vertebral* or neural *canal,* a series of *vertebral foramina,* transmits the spinal cord and related coverings, vessels, and nerve roots. Located bilaterally between each pair of vertebral pedicles are passageways, each called an *intervertebral foramen,* transmitting spinal nerves, their coverings/vessels, and some vessels to the spinal cord.

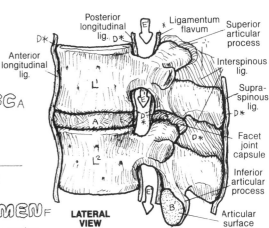

Posterior longitudinal lig.
Anterior longitudinal lig.
Ligamentum flavum
Superior articular process
Interspinous lig.
Supraspinous lig.
Facet joint capsule
Inferior articular process
Articular surface
LATERAL VIEW

Flexion
Extension

Superior view
Posterior view
Pedicle
Pedicle
Lateral view

INTERVERTEBRAL DISC A
ANNULUS FIBROSUS A'
NUCLEUS PULPOSUS G
SPINAL NERVE H

The intervertebral disc consists of the *annulus fibrosus* (concentric, interwoven collagenous fibers integrated with cartilage cells) attached to the vertebral bodies above and below, and the more central *nucleus pulposus* (a mass of degenerated collagen, proteoglycans, and water). The discs make possible movement between vertebral bodies. With aging, the discs dehydrate and thin, resulting in a loss of height. The cervical and lumbar discs, particularly, are subject to early degeneration from one or more of a number of causes. Weakening and/or tearing of the annulus can result in a broad-based bulge or a localized (focal) protrusion of the nucleus and adjacent annulus; such an event can compress a spinal nerve root as shown.

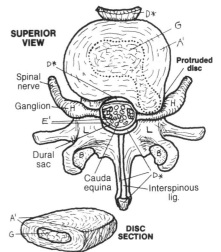

SUPERIOR VIEW
Protruded disc
Spinal nerve
Ganglion
Dural sac
Cauda equina
Interspinous lig.
DISC SECTION

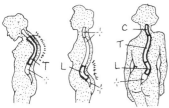

VERTEBRAL DISORDERS

"KYPHOSIS"   "LORDOSIS"   "SCOLIOSIS"

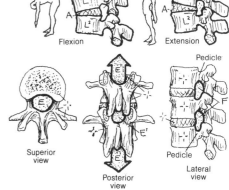

CN: Use red for M and use the same colors as were used on Plate 21 for C and T. Use dark colors for N, O, and P. (1) Begin with the parts of a cervical vertebra. Color the atlas and axis and note they have been given separate colors to distinguish them from other cervical vertebrae. (2) Color the parts of a thoracic vertebra and then the thoracic portion of the vertebral column. Note the three different facet/demifacet colors.

CERVICAL VERTEBRA c
BODY c'
PEDICLE B
TRANSVERSE PROCESS G
ARTICULAR PROCESS H
FACET H'
LAMINA I
SPINOUS PROCESS J

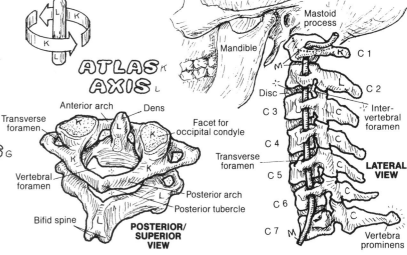

ATLAS K
AXIS L

Anterior arch — Dens
Transverse foramen
Facet for occipital condyle
Transverse foramen
Vertebral foramen
Posterior arch
Posterior tubercle
Bifid spine

POSTERIOR/ SUPERIOR VIEW

Cranium
Mastoid process
Mandible
Disc
C 1
C 2
C 3
Inter-vertebral foramen
C 4
C 5
LATERAL VIEW
C 6
C 7
Vertebra prominens

VERTEBRAL ARTERY M

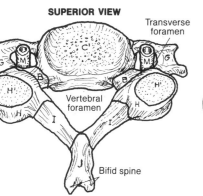

SUPERIOR VIEW

Transverse foramen
Vertebral foramen
Bifid spine

TYPICAL CERVICAL (C4) VERTEBRA

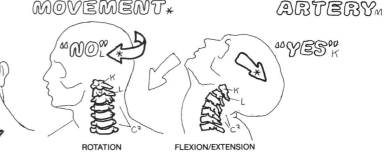

MOVEMENT *

"NO"
"YES"

ROTATION
FLEXION/EXTENSION

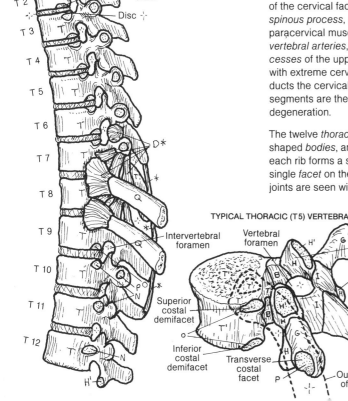

LATERAL VIEW

T 1
T 2
Disc
T 3
T 4
T 5
T 6
D*
T 7
*
T 8
T 9
Intervertebral foramen
T 10
T 11
Superior costal demifacet
T 12
Inferior costal demifacet
Transverse costal facet
Outline of rib

TYPICAL THORACIC (T5) VERTEBRA

Vertebral foramen
Superior articular process
Inferior articular process
LATERAL/ SUPERIOR VIEW

The small seven *cervical vertebrae* support and move the head and neck, supported by ligaments and strap-like paracervical (paraspinal) muscles. The ring-shaped *atlas* (C1) has no body; thus there are no weight-bearing discs between the occiput and C1, and between C1 and C2 (the *axis*). Head weight is transferred to C3 by the large *articular processes* and *facets* of C1 and C2. The atlantooccipital joints, in conjunction with the C3-C7 facet joints, permit a remarkable degree of flexion/extension ("yes" movements). The dens of C2 projects into the anterior part of the C1 ring, forming a pivot joint, enabling the head and C1 to rotate almost 90° ("no" movements). Such rotational capacity is permitted by the relatively horizontal orientation of the cervical facets. The C3-C6 vertebrae are similar; C7 is remarkable for its prominent *spinous process*, easily palpated. The anteriorly directed cervical curve and the extensive paracervical musculature preclude palpation of the other cervical spinous processes. The *vertebral arteries*, enroute to the brain stem, pass through foramina of the *transverse processes* of the upper six cervical vertebrae. These vessels are subject to stretching injuries with extreme cervical rotation of the hyperextended neck. The cervical vertebral canal conducts the cervical spinal cord and its coverings (not shown). The C4-5 and C5-6 motion segments are the most mobile of the cervical region and are particularly prone to disc/facet degeneration.

The twelve *thoracic vertebrae*—characterized by long, slender spinous processes, heart-shaped *bodies*, and nearly vertically oriented *facets*—articulate with *ribs* bilaterally. In general, each rib forms a synovial joint with two *demifacets* on the bodies of adjacent vertebrae and a single *facet* on the transverse process of the lower vertebra. Variations of these costovertebral joints are seen with T1, T11, and T12.

THORACIC VERTEBRA T
BODY T'
FACET N
DEMIFACET O
TRANSVERSE FACET P
RIB Q
LIGAMENT D*

LUMBAR, SACRAL, & COCCYGEAL VERTEBRAE

CN: Use the same colors as were used on the previous two plates for C, T, L, E, F, A, S, and Co. (1) Begin with the three large views of lumbar vertebrae. (2) Color the different planes of articular facets. (3) Color the four views of the sacrum and coccyx. Note that the central portion of the median section receives the vertebral canal color (E¹).

The five *lumbar vertebrae* are the most massive of all the individual vertebrae, their thick processes securing the attachments of numerous ligaments and muscles/tendons. Significant flexion and extension of the lumbar and lumbosacral motion segments, particularly at L4–L5 and L5–S1, are possible. At about L1, the spinal cord terminates and the cauda equina (bundle of lumbar, sacral, and coccygeal nerve roots; see Plate 21) begins. The lumbar *intervertebral foramina* are large. Transiting nerve roots/sheaths take up only about 50% of the volume of these foramina. *Disc* and facet degeneration is common in the L4-5 and L5-S1 segments; reduction of space for the nerve roots increases the risk of nerve root irritation/compression (radiculitis/radiculopathy). Occasionally, the L5 vertebra is partially or completely fused to the sacrum (sacralized L5). The S1 vertebra may be partially or wholly non-fused (lumbarized S1), resulting in essentially six lumbar vertebrae.

LUMBAR VERTEBRA L
VERTEBRAL FORAMEN E
VERTEBRAL CANAL E¹
INTERVERTEBRAL FORAMEN F
INTERVERTEBRAL DISC A

PLANES OF ARTICULAR FACETS:
CERVICAL C
THORACIC T
LUMBAR L

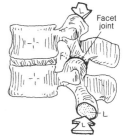

The planes (orientation) of the articular facets determine the direction and influence the degree of motion segment movement. The plane of the *cervical facets* is angled coronally off the horizontal plane about 30°. Considerable freedom of movement of the cervical spine is permitted in all planes (sagittal, coronal, horizontal). The *thoracic facets* lie more vertically in the coronal plane, and are virtually non-weightbearing. The range of motion here is significantly limited in all planes, less so in rotation. The plane of the *lumbar facets* is largely sagittal, resisting rotation of the lumbar spine, transitioning to a more coronal orientation at L5-S1. The L4-L5 facet joints permit the greatest degree of lumbar motion in all planes.

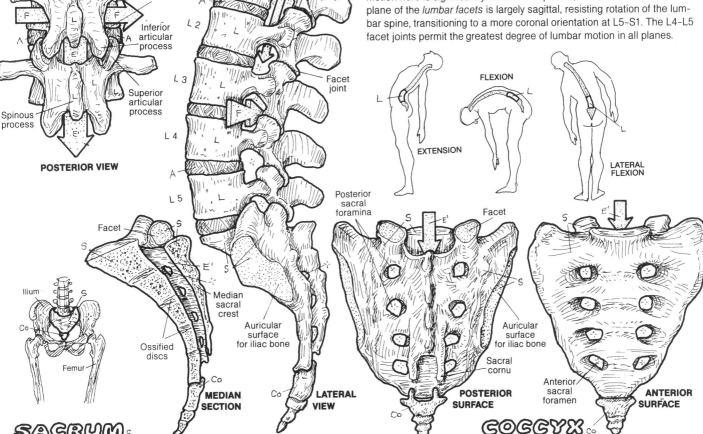

SACRUM S

COCCYX Co

The *sacrum* consists of five fused vertebrae; the intervertebral discs are largely replaced by bone. The sacral (vertebral) canal contains the terminal sac of the dura mater (dural sac, thecal sac) to S2 and the sacral nerve roots, which transit the sacral foramina. The sacrum joins with the ilium of the hip bone at the auricular surface, forming the sacroiliac joint.

The sacrum and the ilia of the hip bones form an arch for the transmission and distribution of weightbearing forces to the heads of the femora. It is a strong arch, and the sacrum is its keystone. The *coccyx* consists of 2-4 tiny individual or partly fused, rudimentary vertebrae. The first coccygeal vertebra is the most completely developed.

CN: Use the same colors as were used on Plate 22 for true ribs, thoracic vertebrae, demifacets, and transverse process facets. Use bright colors for A-C. (1) Color the anterior view of the bony thorax. Color each rib completely before going on to the next. (2) Color the posterior view in the same manner. (3) Color the lateral view of the bony thorax. (4) When coloring the drawings of a rib and the sites of articulation, note that the rib facets (drawn with dotted lines) are to be colored even though they are on the underside of the rib.

STERNUM:*
MANUBRIUM A
BODY B
XIPHOID PROCESS C

12 RIBS:*
7 TRUE D
5 FALSE E
(2 FLOATING) E'

COSTAL CARTILAGE (10) F
THORACIC VERTEBRA (12) G

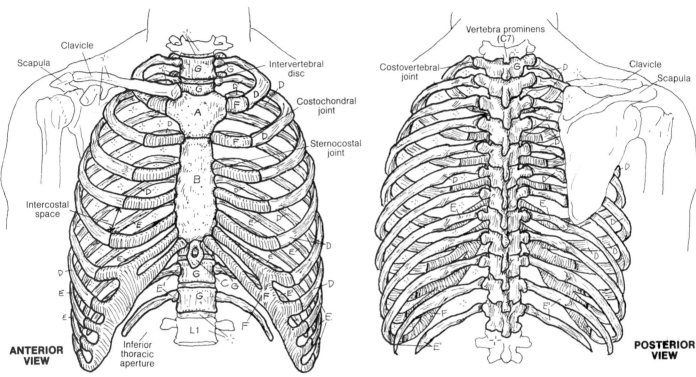

ANTERIOR VIEW

POSTERIOR VIEW

The *bony thorax* is the skeleton of the chest, representing a fairly mobile set of structures important to respiration and harboring the heart, lungs, and other significant organs. The superior thoracic aperture (thoracic inlet; often incorrectly termed thoracic outlet in a clinical context) transmits the esophagus, trachea, nerves, and important ducts and vessels. The inferior thoracic aperture is virtually sealed by the thoracic diaphragm. The space between ribs is the intercostal space, and contains three layers of muscle and fasciae, and intercostal vessels and nerves. Collective rib movement is responsible for about 25% of the respiratory effort.

The fibrocartilaginous joint between the *manubrium* and the *body* of the *sternum* (sternal angle, sternomanubrial joint) makes subtle hinge-like movements during respiration. The xiphoid makes a fibrocartilaginous (xiphisternal) joint with the body of the sternum. The sternum is largely cancellous bone containing red marrow. The *costal cartilages*, representing unossified cartilage models of the anterior ribs, articulate with the sternum by gliding-type synovial joints (sternocostal joints; except for the first joint, which is not synovial). All ribs form synovial joints with the thoracic vertebrae (costovertebral joints). Within each of these joints, the rib (2 through 9) forms a synovial joint with a *demifacet* of the upper vertebral body and with a demifacet of the lower body (costocorporeal joints). In addition, the tubercle of the rib articulates with a cartilaginous *facet* at the tip of the transverse process of the lower vertebra (costotransverse joint). Ribs 1, 10, 11, 12 each join with one vertebra instead of two; ribs 11 and 12 have no costotransverse joints. *True ribs* (1-7) articulate directly with the sternum. *Ribs 8-12* are called *false ribs*; ribs 8-10 articulate indirectly with the sternum (via cartilages connecting to the 7th costal cartilage); ribs 11 and 12 (also called *floating ribs*) end in the muscular abdominal wall.

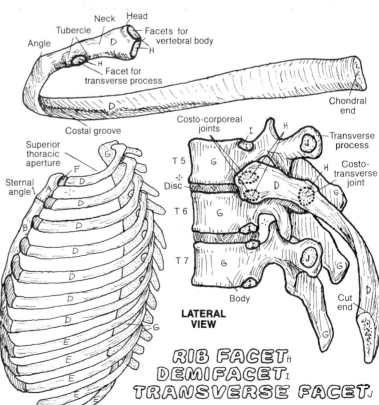

LATERAL VIEW

RIB FACET H
DEMIFACET I
TRANSVERSE FACET J

PECTORAL GIRDLE: CLAVICLE, SCAPULA,
ARM BONE: HUMERUS_C

The mobility of the upper limb is largely dependent upon the *pectoral girdle* whose only bony attachment to the axial skeleton is via the sternoclavicular joint (saddle type synovial joint with disc). Distally, the *clavicle* articulates with the acromion of the *scapula* (acromioclavicular joint, a gliding type synovial joint). The clavicle forces the scapula backward and outward, creating the shoulder; in its role as a strut, it is subject to fracture. The scapula is moored to the axial skeleton by muscles, giving it considerable mobility on the upper back (scapulo-thoracic motion). Largely packaged in muscle, the scapula fractures infrequently. The supraspinatus muscle/tendon passing under the acromion and coracoacromial ligament is subject to irritation (impingement syndrome). The glenoid fossa of the scapula is shallow, and the glenohumeral joint (shoulder; ball and socket, synovial) is relatively insecure. The glenohumeral ligaments/joint capsule are lax, and are reinforced by a musculotendinous cuff. Given these "rotator cuff" muscles, the *humerus* has excellent mobility at the shoulder joint. The humerus is vulnerable to fracture at the surgical neck, mid-shaft, and medial epicondyle.

CN: Use very light colors. (1) Color each view completely before going on to the next. (2) Color the ligaments of the shoulder region (inset) in gray. Note the ligaments at the top of the plate which should also be colored.

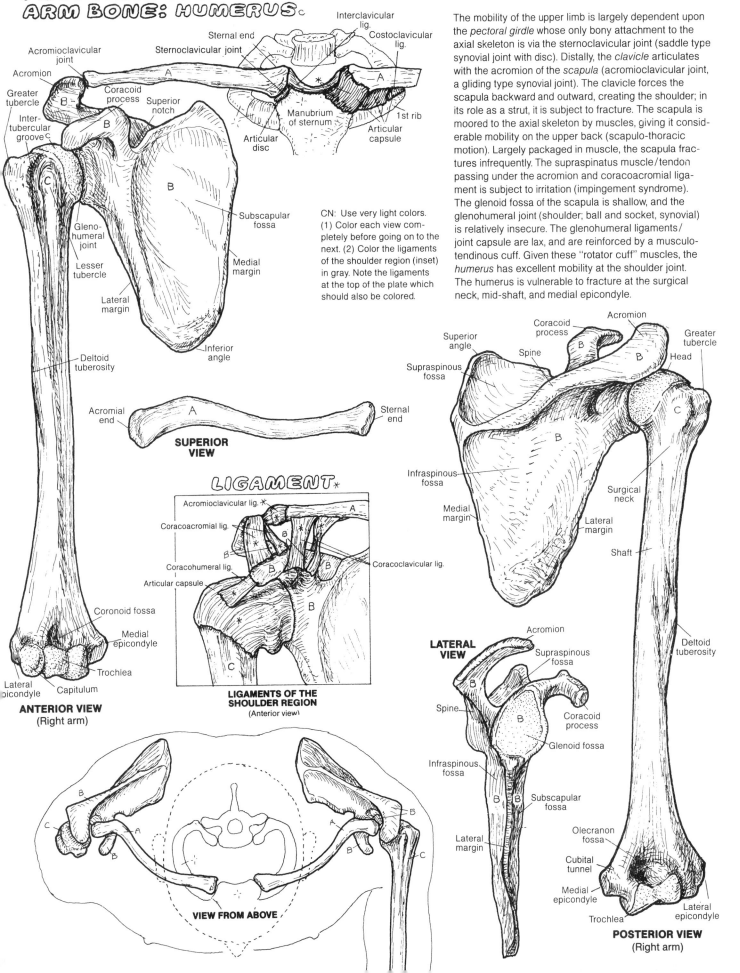

SUPERIOR VIEW

LIGAMENT

LIGAMENTS OF THE SHOULDER REGION
(Anterior view)

ANTERIOR VIEW
(Right arm)

LATERAL VIEW

POSTERIOR VIEW
(Right arm)

VIEW FROM ABOVE

FOREARM BONES
ULNA_A
RADIUS_B

CN: Though the humerus is not colored, the titles and arrows (C) that reflect its participation in the elbow joint should be colored with the same color the bone received on Plate 25. (1) Color the two large views, including the interosseous membrane (gray). (2) Color the four views of the elbow joint. (3) Color the ligaments of the region.

The presence of two bones in the forearm make possible the diverse movements seen at the elbow and reflected in hand motion. The *ulna*, the major, stabilizing forearm bone at the elbow, narrows distally to form an inconsequential joint with the radius (distal radioulnar joint; synovial, pivot-type). The *radius*, smaller above, widens and thickens distally to form the major joint at the wrist (radiocarpal joint; synovial, biaxial, ellipsoid-type). At the elbow, the ulna forms a hinge type synovial *humeroulnar joint* with the trochlea of the humerus, and the radius forms a pivot-type synovial *radiohumeral joint* with the capitulum of the humerus. These joints share the same joint capsule with the proximal *radioulnar joint* (synovial, pivot type) between the radial notch of the ulna and the radial head. The three joints constitute the elbow (cubital) joint.

Rotation of the radius at the elbow (involving two of the three joints at the elbow) rotates the forearm, wrist, and hand without moving the ulna. Movement of the hand to a palm-forward (up) position is supination: movement of the hand to a palm-back (down) position is pronation.

After coloring and studying the supination/pronation and elbow movement diagrams, try this: place the fingers of your left hand on your right olecranon (bump at posterior elbow), elbow flexed so that the palm of your right hand is up (supine). Now rotate (pronate) your right hand so your palm turns away from you, facing down. Move your right hand back and forth in this manner, feeling that the olecranon does not move during these motions. Further, stare at the styloid process of the radius at the base of the right thumb and note that it rotates back and forth with the thumb. You have just demonstrated that the radius moves around the ulna during pronation/supination, and that joint movement occurs at the radiohumeral and proximal radioulnar joints.

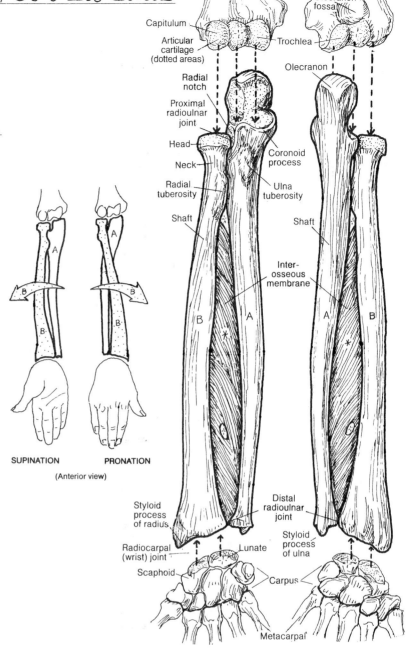

SUPINATION   **PRONATION**

(Anterior view)

ANTERIOR VIEW
(Right arm)

POSTERIOR VIEW
(Right arm)

3 JOINTS AT THE RIGHT ELBOW:*
HUMERO-_C ULNAR_A
RADIO-_B HUMERAL_C
RADIO-_B ULNAR_A

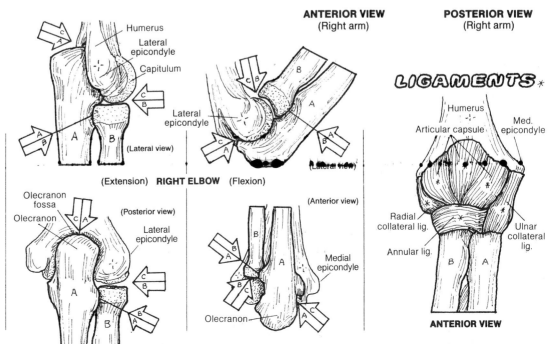

(Extension) **RIGHT ELBOW** (Flexion)

LIGAMENTS *

ANTERIOR VIEW

CN: Use two light colors other than those used on Plates 25 and 26 for I and J. (1)Color each bone, or bone group, in all three major views simultaneously. Note the hand drawings which demonstrate movements at the joints. (2)Color the bones and ligament of the carpal tunnel. You may wish to color those bones in their location in the hand to the left.

CARPALS (8):*
SCAPHOID_A LUNATE_B TRIQUETRUM_C PISIFORM_D
TRAPEZIUM_E TRAPEZOID_F CAPITATE_G HAMATE_H
METACARPALS (5)_I PHALANGES (14)_J

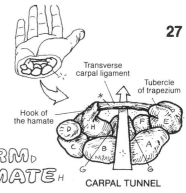

Transverse carpal ligament
Tubercle of trapezium
Hook of the hamate
CARPAL TUNNEL

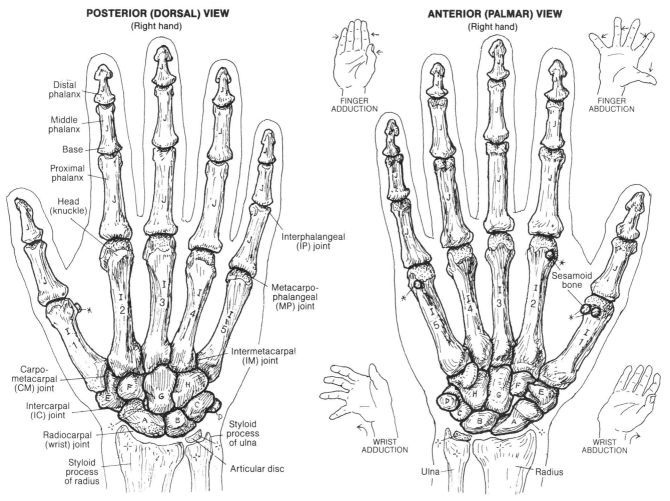

POSTERIOR (DORSAL) VIEW
(Right hand)

Distal phalanx
Middle phalanx
Base
Proximal phalanx
Head (knuckle)
Interphalangeal (IP) joint
Metacarpo-phalangeal (MP) joint
Intermetacarpal (IM) joint
Carpo-metacarpal (CM) joint
Intercarpal (IC) joint
Radiocarpal (wrist) joint
Styloid process of radius
Styloid process of ulna
Articular disc

ANTERIOR (PALMAR) VIEW
(Right hand)

FINGER ADDUCTION
FINGER ABDUCTION
Sesamoid bone
WRIST ADDUCTION
WRIST ABDUCTION
Ulna
Radius

The hand is a most remarkable, highly evolved, mechanical device. Movement of the hand and wrist is made possible by the architecture of the joints among the bones. The wrist joint is formed by the distal articular surface of the radius and the distal surface of the articular disc (just distal to the ulna) with the proximal articular surfaces of the *scaphoid, lunate,* and *triquetrum* bones. Forces transmitted from a fall on the hand to the wrist pass largely through the scaphoid, lunate, and radius; thus, fractures of the scaphoid and distal radius are common.

Crossing the wrist bones between the hook of the *hamate/pisiform* and the tubercle of the *trapezium/scaphoid* bones, the thin, broad transverse carpal ligament (flexor retinaculum) creates a carpal tunnel through which pass the long flexor tendons to the fingers and thumb as well as the median nerve. Compression of the nerve there can cause numbness in the radial three fingers (thumb, index, middle) and some weakness in the thumb (carpal tunnel syndrome).

Using your own hand, note that the interphalangeal (IP) joints are limited to movements of flexion/extension. The metacarpophalangeal (MP) joints permit the added movements of finger adduction/abduction. Of the carpometacarpal (CM) joints, the first (thumb) has exceptional movement (saddle type, synovial); when moving the thumb toward the little finger in an arcing motion, note that the thumbnail rotates 90°, reflecting medial rotation of the first metacarpal on the *trapezium*.

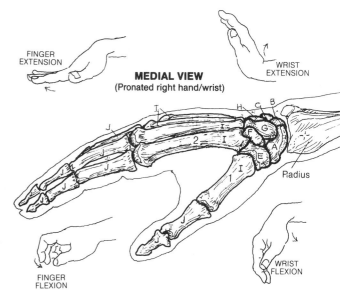

FINGER EXTENSION
WRIST EXTENSION
MEDIAL VIEW
(Pronated right hand/wrist)
Radius
FINGER FLEXION
WRIST FLEXION

CN: For all of these bones, except the carpals (F), use the same colors you used for them on Plates 25, 26, 27. Select a new, light color for F. (1) Color the arrows pointing to places where these bones can be seen or palpated on the surface of the body. (2) You may wish to test your knowledge of joints by writing their names in the spaces provided below. The answers are listed in the Appendix.

CLAVICLE A
SCAPULA B
HUMERUS C
ULNA D
RADIUS E
CARPALS F
METACARPAL G
PHALANX H

BONE
SURFACE
MARKINGS

BONE
SURFACE
MARKINGS

**POSTERIOR
VIEW**
(Right upper limb)

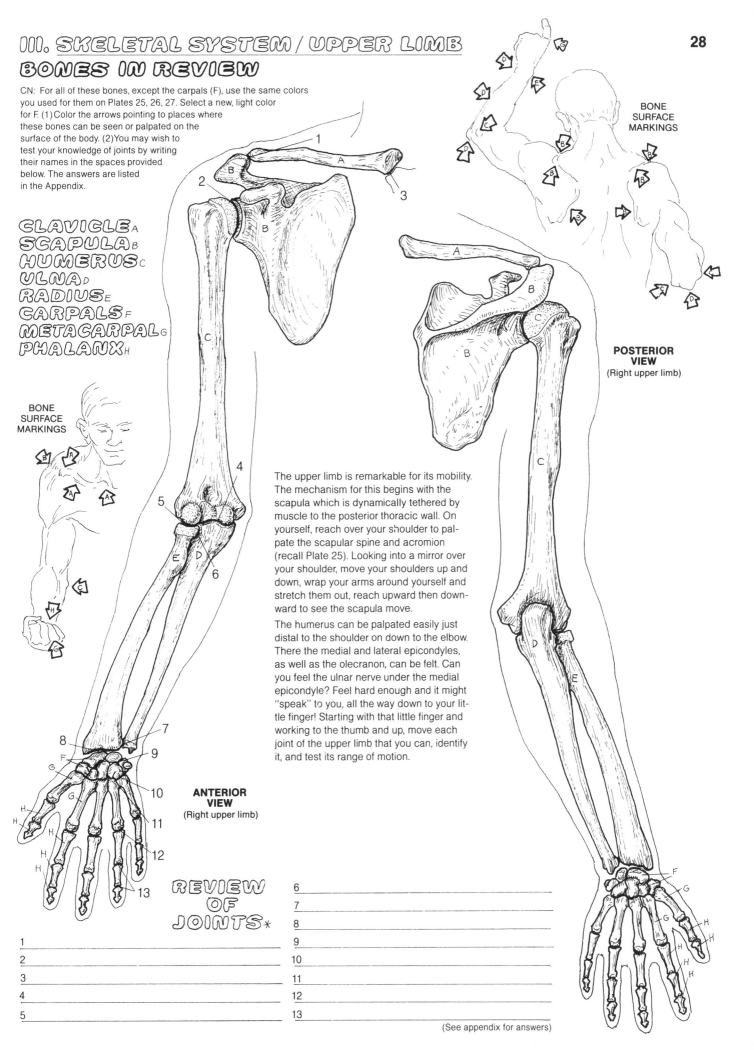

The upper limb is remarkable for its mobility. The mechanism for this begins with the scapula which is dynamically tethered by muscle to the posterior thoracic wall. On yourself, reach over your shoulder to palpate the scapular spine and acromion (recall Plate 25). Looking into a mirror over your shoulder, move your shoulders up and down, wrap your arms around yourself and stretch them out, reach upward then downward to see the scapula move.

The humerus can be palpated easily just distal to the shoulder on down to the elbow. There the medial and lateral epicondyles, as well as the olecranon, can be felt. Can you feel the ulnar nerve under the medial epicondyle? Feel hard enough and it might "speak" to you, all the way down to your little finger! Starting with that little finger and working to the thumb and up, move each joint of the upper limb that you can, identify it, and test its range of motion.

**ANTERIOR
VIEW**
(Right upper limb)

REVIEW OF JOINTS*

1 _____
2 _____
3 _____
4 _____
5 _____

6 _____
7 _____
8 _____
9 _____
10 _____
11 _____
12 _____
13 _____

(See appendix for answers)

COXAL BONE, PELVIC GIRDLE, & PELVIS

ILIUM_A
ISCHIUM_B
PUBIS_C

CN: (1) Color the two views of the coxal bone with light colors. Then color the views of the pelvic girdle. (2) Use a new color for bones of the pelvis (D) which includes the sacrum and coccyx. Then color the title *ligaments,* and all the ligaments a light gray color.

The *coxal bone* (hip bone, innominate bone, os coxa) consists of three fused bones in the adult: the *ilium*, the *ischium*, and *pubis*. The paired coxal bones constitute the *pelvic girdle*. The two somewhat-twisted coxal bones form a weightbearing arch with the sacrum and the femoral (thigh) bones, accommodating the body weight and forces imposed vertically up from the feet. The two hip bones and the sacrum constitute the *pelvis*. The pelvic inlet (superior pelvic aperture; from sacral promontory around the arcuate line at the pelvic brim) separates the true (lesser) pelvis below from the false (greater) pelvis above. The pelvic oulet (inferior pelvic aperture) is bound by the same structures as the perineum (see next plate).

The sacroiliac joint is a movable, partly synovial, partly fibrocartilaginous joint. The articular surfaces are flat but roughened. Note the larger posterior sacroiliac ligaments (compared to the anterior ligaments): they resist downward displacement of the sacrum. The sacrospinous and sacrotuberous ligaments secure the apex of the sacrum to the pelvic girdle, resisting the effects of weightbearing and gravity on the sacroiliac joint. Still, sacroiliac dysfunction is common. The iliolumbar ligaments are often involved in postural low back pain. The symphysis pubis (pubic symphysis, interpubic joint) is a partly movable, cartilaginous joint composed of a fibrocartilaginous disc interposed between cartilaginous articular surfaces.

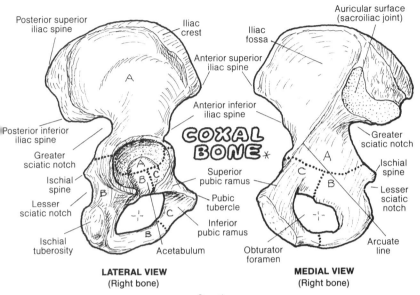

Posterior superior iliac spine
Iliac crest
Iliac fossa
Auricular surface (sacroiliac joint)
Anterior superior iliac spine
Anterior inferior iliac spine
COXAL BONE ✱
Posterior inferior iliac spine
Greater sciatic notch
Ischial spine
Lesser sciatic notch
Greater sciatic notch
Ischial spine
Lesser sciatic notch
Ischial tuberosity
Superior pubic ramus
Pubic tubercle
Inferior pubic ramus
Acetabulum
Obturator foramen
Arcuate line

LATERAL VIEW
(Right bone)

MEDIAL VIEW
(Right bone)

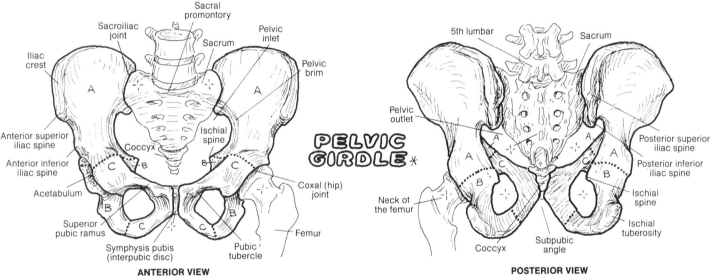

Sacral promontory
Sacroiliac joint
Sacrum
Pelvic inlet
Pelvic brim
Iliac crest
Anterior superior iliac spine
Ischial spine
Anterior inferior iliac spine
Coccyx
Acetabulum
Coxal (hip) joint
Superior pubic ramus
Symphysis pubis (interpubic disc)
Pubic tubercle
Femur

ANTERIOR VIEW

PELVIC GIRDLE ✱

5th lumbar
Sacrum
Pelvic outlet
Posterior superior iliac spine
Posterior inferior iliac spine
Ischial spine
Neck of the femur
Ischial tuberosity
Coccyx
Subpubic angle

POSTERIOR VIEW

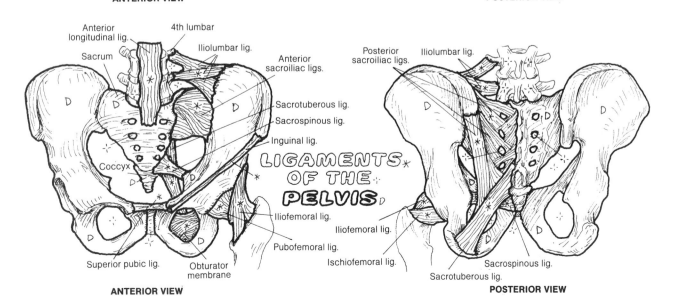

Anterior longitudinal lig.
4th lumbar
Sacrum
Iliolumbar lig.
Anterior sacroiliac ligs.
Sacrotuberous lig.
Sacrospinous lig.
Inguinal lig.
Coccyx
Iliofemoral lig.
Pubofemoral lig.
Superior pubic lig.
Obturator membrane

ANTERIOR VIEW

LIGAMENTS OF THE PELVIS_D

Posterior sacroiliac ligs.
Iliolumbar lig.
Iliofemoral lig.
Ischiofemoral lig.
Sacrospinous lig.
Sacrotuberous lig.

POSTERIOR VIEW

THIGH & LEG BONES
FEMUR_A TIBIA_B
FIBULA_C
PATELLA_D

CN: Do not use the color used for the ilium on Plate 29. Use light colors and a bright color for F. (1) Color the two large views of the lower limb. (2) Next color the femur and the six directional arrows for the hip joint. (3) Color the extension/flexion views of the knee joint. (4) Color the two views of the major ligaments and the menisci of the knee joint.

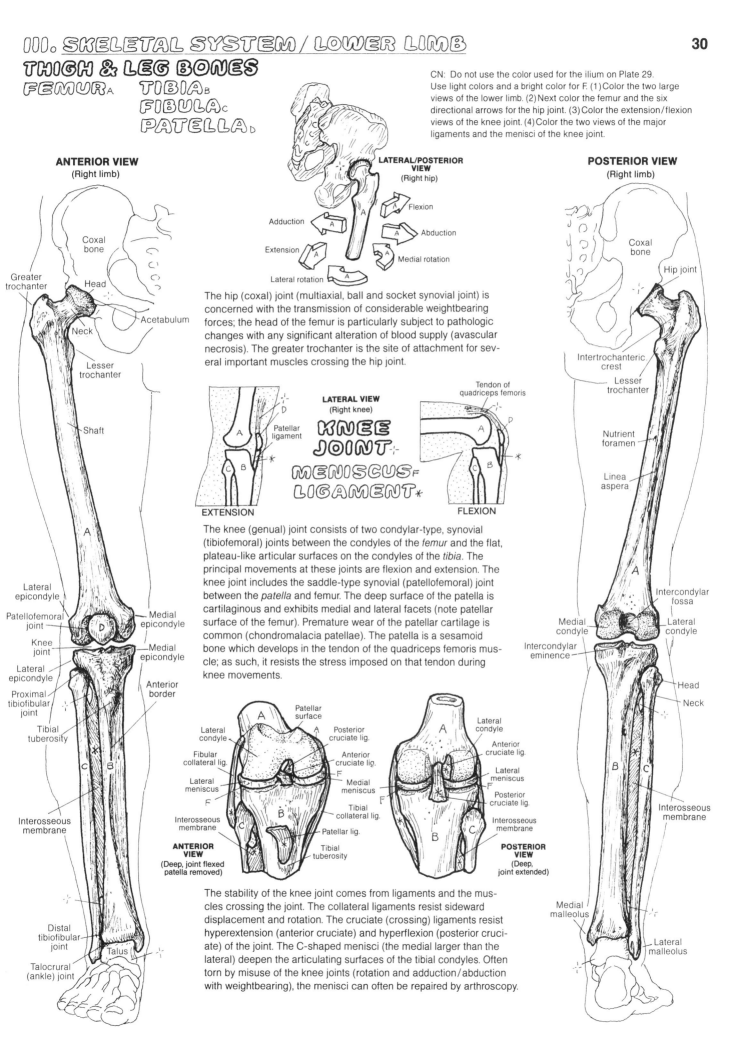

ANTERIOR VIEW
(Right limb)

- Coxal bone
- Greater trochanter
- Head
- Neck
- Acetabulum
- Lesser trochanter
- Shaft
- Lateral epicondyle
- Patellofemoral joint
- Knee joint
- Lateral epicondyle
- Proximal tibiofibular joint
- Tibial tuberosity
- Medial epicondyle
- Medial epicondyle
- Anterior border
- Interosseous membrane
- Distal tibiofibular joint
- Talus
- Talocrural (ankle) joint

LATERAL/POSTERIOR VIEW
(Right hip)

- Flexion
- Adduction
- Abduction
- Extension
- Medial rotation
- Lateral rotation

The hip (coxal) joint (multiaxial, ball and socket synovial joint) is concerned with the transmission of considerable weightbearing forces; the head of the femur is particularly subject to pathologic changes with any significant alteration of blood supply (avascular necrosis). The greater trochanter is the site of attachment for several important muscles crossing the hip joint.

LATERAL VIEW
(Right knee)

KNEE JOINT
MENISCUS_F
LIGAMENT_*

- Patellar ligament
- Tendon of quadriceps femoris

EXTENSION

FLEXION

The knee (genual) joint consists of two condylar-type, synovial (tibiofemoral) joints between the condyles of the *femur* and the flat, plateau-like articular surfaces on the condyles of the *tibia*. The principal movements at these joints are flexion and extension. The knee joint includes the saddle-type synovial (patellofemoral) joint between the *patella* and femur. The deep surface of the patella is cartilaginous and exhibits medial and lateral facets (note patellar surface of the femur). Premature wear of the patellar cartilage is common (chondromalacia patellae). The patella is a sesamoid bone which develops in the tendon of the quadriceps femoris muscle; as such, it resists the stress imposed on that tendon during knee movements.

POSTERIOR VIEW
(Right limb)

- Coxal bone
- Hip joint
- Intertrochanteric crest
- Lesser trochanter
- Nutrient foramen
- Linea aspera
- Intercondylar fossa
- Medial condyle
- Lateral condyle
- Intercondylar eminence
- Head
- Neck
- Interosseous membrane
- Medial malleolus
- Lateral malleolus

ANTERIOR VIEW
(Deep, joint flexed patella removed)

- Patellar surface
- Lateral condyle
- Fibular collateral lig.
- Lateral meniscus
- Interosseous membrane
- Posterior cruciate lig.
- Anterior cruciate lig.
- Medial meniscus
- Tibial collateral lig.
- Patellar lig.
- Tibial tuberosity

POSTERIOR VIEW
(Deep, joint extended)

- Lateral condyle
- Anterior cruciate lig.
- Lateral meniscus
- Posterior cruciate lig.
- Interosseous membrane

The stability of the knee joint comes from ligaments and the muscles crossing the joint. The collateral ligaments resist sideward displacement and rotation. The cruciate (crossing) ligaments resist hyperextension (anterior cruciate) and hyperflexion (posterior cruciate) of the joint. The C-shaped menisci (the medial larger than the lateral) deepen the articulating surfaces of the tibial condyles. Often torn by misuse of the knee joints (rotation and adduction/abduction with weightbearing), the menisci can often be repaired by arthroscopy.

ANKLE & FOOT BONES

TARSALS: (7)*
    TALUS_A CALCANEUS_B
    CUBOID_C NAVICULAR_D
    CUNEIFORMS (3)_E

**METATARSALS (5)_F
PHALANGES (14)_G
LIGAMENTS***

CN: Use different colors from those used for the ilium on Plate 29 and for the femur, tibia, fibula, and patella on Plate 30. (1) Begin with the talus (A); color that bone wherever it appears on the plate. Follow that procedure with each of the other bones. (2) Color gray all of the ligaments.

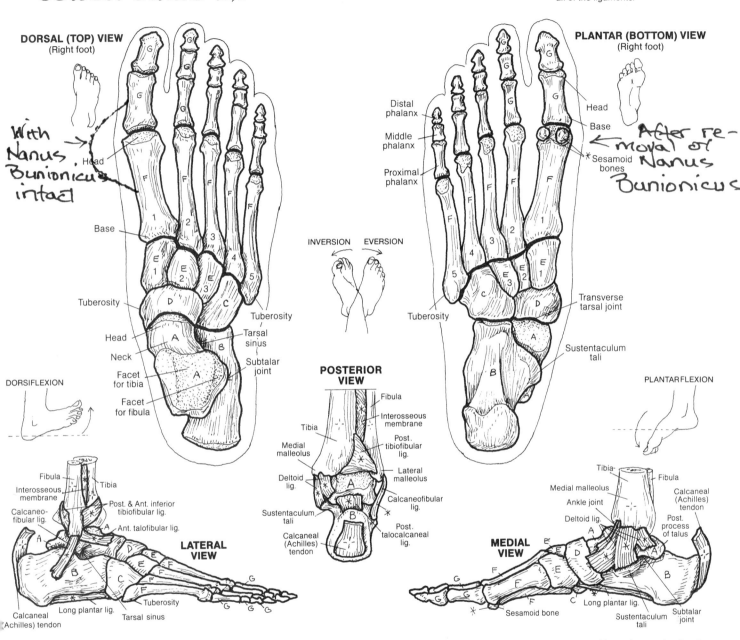

DORSAL (TOP) VIEW
(Right foot)

With Nanus Bunionicus intact →

Head

Base

1 2 3 4 5

Tuberosity

Head

Neck

Facet for tibia

Facet for fibula

Tuberosity

Tarsal sinus

Subtalar joint

DORSIFLEXION

PLANTAR (BOTTOM) VIEW
(Right foot)

Head

Base

*Sesamoid bones

← *After removal of Nanus Bunionicus*

Distal phalanx

Middle phalanx

Proximal phalanx

Tuberosity

Transverse tarsal joint

Sustentaculum tali

PLANTARFLEXION

INVERSION   EVERSION

POSTERIOR VIEW

Fibula

Interosseous membrane

Tibia

Post. tibiofibular lig.

Medial malleolus

Lateral malleolus

Deltoid lig.

Calcaneofibular lig.

Sustentaculum tali

Post. talocalcaneal lig.

Calcaneal (Achilles) tendon

LATERAL VIEW

Fibula

Tibia

Interosseous membrane

Post. & Ant. inferior tibiofibular lig.

Calcaneofibular lig.

Ant. talofibular lig.

Calcaneal (Achilles) tendon

Long plantar lig.

Tuberosity

Tarsal sinus

MEDIAL VIEW

Tibia

Fibula

Medial malleolus

Ankle joint

Deltoid lig.

Calcaneal (Achilles) tendon

Post. process of talus

Sesamoid bone

Long plantar lig.

Sustentaculum tali

Subtalar joint

The foot is a mobile, weightbearing structure. The ankle joint (hinge-type synovial joint) between tibia, fibula, and the *talus* forms a mortise, permitting only flexion (plantar flexion) and extension (dorsiflexion) here. With excessive rotation of this joint, characteristic fractures and torn ligaments occur. The foot can adjust to walking/running on tilted surfaces by virtue of the subtalar (talocalcaneal) and transverse tarsal (talocalcaneonavicular and calcaneocuboid) joints. Here inversion and eversion movements occur. The ankle has strong medial ligamentous (deltoid ligaments) and weaker lateral

ligamentous support. The relatively high frequency of inversion sprains (tearing the lateral ligaments) over eversion sprains seems to reflect this fact.

The bony architecture of the foot includes a number of arches that are reinforced and maintained by ligaments and influenced by muscles. The *medial longitudinal arch* transmits the force of body weight to the ground when standing and to the great toe in locomotion, creating a giant lever that gives spring to the gait. Both longitudinal arches function in absorbing shock loads and balancing the body.

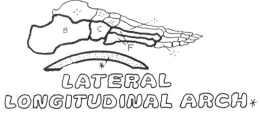

LATERAL LONGITUDINAL ARCH*

TRANSVERSE ARCH*

MEDIAL LONGITUDINAL ARCH*

CN: Use the same colors for these bones that you used for them on Plates 29-31. In the case of the coxal bone (A), use the color given to the ilium on Plate 29; for the tarsal bones (F), use any one of the tarsal colors. (1) Color the bones of the lower limb, their surface markings, and the corresponding bones on the hind limb of the dog. (2) Color the names and bones of the upper limb and the forelimb of the dog. The clavicle of the dog is not shown in this view.

LOWER LIMB: *
COXAL A
FEMUR B
PATELLA C
TIBIA D
FIBULA E
TARSAL F
METATARSAL G
PHALANX H

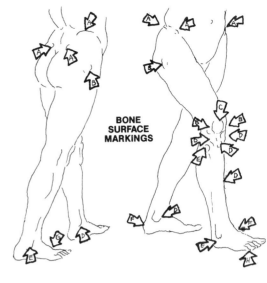

BONE SURFACE MARKINGS

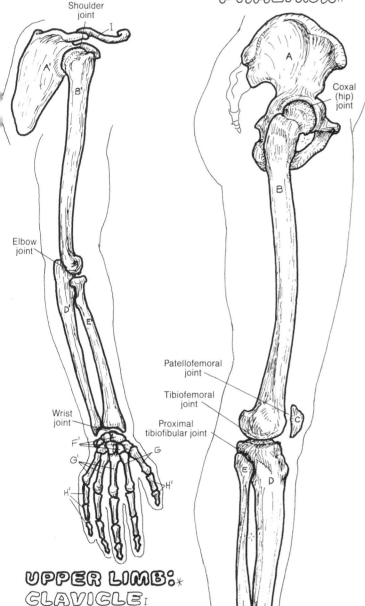

The structure of a part reflects an adaptation for function. The truth of this statement is borne out in comparing the bones of the upper and lower limbs in a biped (human) with those of a quadruped. The pectoral girdle provides a basis for mobility; the more sturdy pelvic girdle provides stability in both locomotion and weight bearing. The limb bones of the lower limb are large and solid, consistent with weight-bearing; the related joints are structurally secure, except the knee, which gives up stability for flexibility. In the upper limb, the bones are lighter, and the joints are more flexible and capable of greater ranges of motion (compare shoulder with hip, elbow with knee, wrist with ankle). Although forearm and leg each have two bones, there is little functional correlation between those pairs of bones. The foot is clearly adapted for locomotion and weight bearing, the hand (especially the thumb) for mobility and dexterity.

The quadruped (in this case, the domestic dog) uses both forelimbs and hindlimbs for supporting body weight and locomotion. The girdle (coxal/scapular) bones are adapted for locomotion, and are not as differentiated structurally or functionally as they are in humans. The canine scapula has much less scapulothoracic motion than the human scapula; the canine coxal bones do not carry a disproportionate weight, as does the human pelvic girdle. The animal bears weight on the heads of the metacarpals and metatarsals, a condition that is particularly suitable for acceleration.

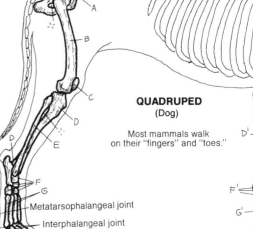

QUADRUPED (Dog)

Most mammals walk on their "fingers" and "toes."

UPPER LIMB: *
CLAVICLE I
SCAPULA A'
HUMERUS B'
ULNA D'
RADIUS E'
CARPAL F'
METACARPAL G'
PHALANX H'

CLASSIFICATION OF JOINTS

Bones are connected at joints (articulations), and all bone movements occur at joints. Joints are structurally classified as fibrous, cartilaginous, or synovial. They are functionally classified as immovable, partly movable, or freely movable. The most secure joints are immovable; the most vulnerable are freely movable. The architecture of freely movable joints determines their directions and ranges of motion.

FIBROUS JOINT*
IMMOVABLE_A /PARTLY MOVABLE_A'

Fibrous joints, where bone is connected to bone by fibrous tissue, are immovable or partly movable. Sutures are *immovable* fibrous joints; so are teeth and their sockets. Syndesmoses, here represented by the interosseous ligament of the forearm, are *partly movable* fibrous joints.

CARTILAGINOUS JOINT*
IMMOVABLE_B / PARTLY MOVABLE_B'

Cartilaginous joints, where bone is connected to bone by cartilaginous or fibrocartilaginous tissues, are immovable or partly movable. The epiphyseal growth plates are *immovable* cartilaginous joints, replaced by bone at skeletal maturity. The intervertebral discs are *partly movable* fibrocartilaginous joints.

SYNOVIAL JOINT (FREELY MOVABLE)*
ARTICULATING BONES_C
ARTICULAR CARTILAGE_D
SYNOVIAL MEMBRANE_E
SYNOVIAL CAVITY (FLUID)_F
JOINT CAPSULE_G
BURSA CAPSULE_G'
COLLATERAL LIGAMENT_H*

Synovial joints are all *freely movable* within in the limitations of the bony architecture. *Articular bones* are capped with *articular cartilage* at the joint interface. The joint cavity is lined internally with vascular *synovial membrane* (except over the articular cartilage) and secretes a nutrient, lubricating synovial *fluid*. The fibrous, sensitive *joint capsule* is reinforced by *ligaments*. A cushion of synovial membrane reinforced by dense irregular connective tissue can be found interposed between bone and a moving structure (tendon, muscle). Such a device (*bursa*) facilitates friction-free movement.

CN: Use a light blue for D, and dark color for F, and gray for H. (1) Do not color the bones in the upper half of the plate. (2) Below, color the arrows pointing to the location of the joints as well as the joint representations.

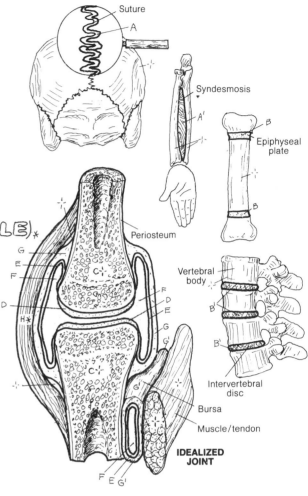

Suture
Syndesmosis
Epiphyseal plate
Periosteum
Vertebral body
Intervertebral disc
Bursa
Muscle/tendon
IDEALIZED JOINT

TYPES OF SYNOVIAL JOINTS:*

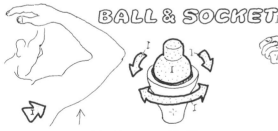

BALL & SOCKET_I

The *ball and socket* joint is best seen at the hip and shoulder joints. Movements in all directions are permitted.

HINGE_J

A *hinge* joint permits movement in only one plane: flexion/extension. It can be seen at the ankle, interphalangeal, and elbow (humeroulnar) joints.

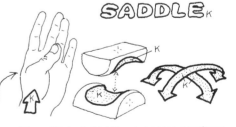

SADDLE_K

The *saddle* (sellar) joint has two concave surfaces articulating with one another. The carpometacarpal joint of the thumb is the best example of this joint which permits all movements but rotation.

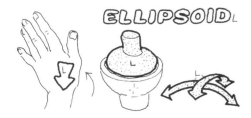

ELLIPSOID_L

The *ellipsoid* (condyloid, condylar) joint is a reduced ball and socket configuration in which significant rotation is largely excluded, e.g., the bicondylar knee and temporomandibular joints, and radiocarpal (wrist) joints.

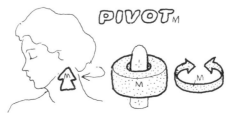

PIVOT_M

A ring of bone (C1 vertebra) rotating about an axle of bone (odontoid process of C2 vertebra) is a *pivot* joint (atlantoaxial joint). Also the rounded humeral capitulum and the radial head (radiohumeral joint).

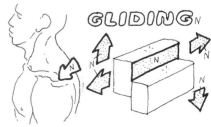

GLIDING_N

A *gliding* joint consists of generally flat surfaces gliding across one another during movement, such as the facet joints of the vertebrae, acromioclavicular, and intercarpal/intertarsal joints.

IV. ARTICULAR SYSTEM
TERMS OF MOVEMENTS

CN: Color the arrows pointing to the joints demonstrating the various movements of body. Note that inversion (K) and eversion (L) occur among bones of the foot, not at the ankle.

EXTENSION_A
DORSIFLEXION_B
FLEXION_C
PLANTARFLEXION_D
ADDUCTION_E
ABDUCTION_F
CIRCUMDUCTION_G
ROTATION_H
SUPINATION_I
PRONATION_J
INVERSION_K
EVERSION_L

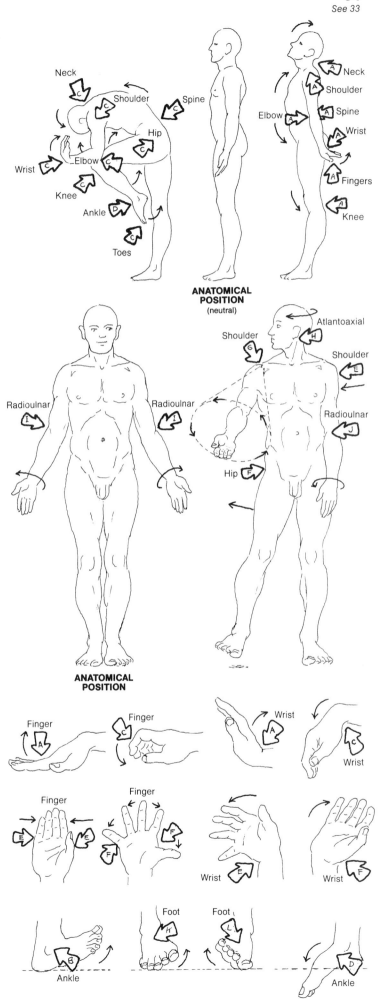

Movements of bones occur at joints. Terms of movement are therefore applicable to joints, not bones (flexion of the humerus is to break it!). Ranges of motion are limited by the bony architecture of a joint, related ligaments, and the muscles crossing that joint. It is from the anatomical position that specific directions of movement can be clearly delineated and ranges of motion measured.

Extension of a joint is to generally straighten it. In the anatomical position, most joints are in relaxed extension (neutral). In relation to the anatomical position, movements of extension are directed in the sagittal plane. Extreme, even abnormal extension is called hyperextension. At the ankle and wrist joints, extension is termed *dorsiflexion*.

Flexion of a joint is to bend it or decrease the angle between the bones of the joint. Movements of flexion are directed in the sagittal plane. At the ankle joint, flexion is also called *plantar flexion*.

Adduction of a joint moves a bone toward the midline of the body (or in the case of the fingers or toes, toward the midline of the hand or foot). In relation to the anatomical position, movements of adduction are directed medially in the coronal plane.

Abduction of a joint moves a bone away from the midline of the body (or hand or foot). Movements of abduction are directed laterally in the coronal plane.

Circumduction is a circular movement permitted at ball and socket, condylar, and saddle joints. It consists of the movements of flexion, abduction, extension, and adduction performed in sequence.

Rotation of a joint is to turn the moving bone about its axis. Rotation toward the body is internal or medial rotation; rotation away from the body is external or lateral rotation.

Supination is an external rotation of the radiohumeral joint. In the foot, it is the combined movements of inversion, adduction around a vertical axis, and plantar flexion.

Pronation is an internal rotation of the radiohumeral joint. In the foot, it is the combined movements of eversion, abduction around a vertical axis, and dorsiflexion. The joints involved in both supination and pronation are the tarsal and ankle joints.

Inversion turns the sole of the foot inward so that the medial border of the foot is elevated.

Eversion turns the sole of the foot outward so that its lateral border is elevated. Both inversion and eversion occur at subtalar (talocalcaneal) and transverse tarsal joints.

CN: Use light colors for A-E. (1) Begin with the muscle belly and tendons in the upper illustration. (2) When coloring the narrow borders of the endomysium (C) in the enlarged section, it is recommended that you also color over the muscle fiber ends (D) with the very light endomysium color, and then go back over the fiber ends with a darker color (D). Do not color the neurovascular bundle, or the cut ends of blood vessels and capillaries. (3) Color the lower illustration.

SKELETAL MUSCLE:∗
BELLY ᴀ
FASCIA:
EPIMYSIUM ᴀ'
PERIMYSIUM ʙ
ENDOMYSIUM c
MUSCLE FIBER (CELL) ᴅ
TENDON ᴇ

A named skeletal muscle (e.g., biceps brachii), surrounded by a layer of deep *fascia* (*epimysium*), consists of fascicles or bundles of muscle cells enveloped in thin fibrous tissue (*perimysium*). Each muscle cell is surrounded by a thin sheath of fibrous tissue (*endomysium*). Each of these fibrous layers is important to muscle structure and function, providing support for nerves and vessels (neurovascular bundles), ensuring uniform distribution of muscle tension during contraction, and maintaining the elasticity of muscle, permitting it to recoil to its resting length following stretching. It is the merging of these fibrous layers at the ends of the muscle fibers that form the *tendons* which integrate the muscle to its attachment site(s), such as periosteum or another tendon. Broad, flat tendons are called *aponeuroses*. The mass of the fasciae-enveloped contractile cells is called the *belly* of the muscle. It is the muscle belly that shortens during muscle contraction. The belly may be shaped one of a number of ways depending on its tendinous arrangement and attachments. Skeletal muscles are named in relation to their attachments (e.g., hyoglossus), shape (e.g., trapezius), number of heads (e.g., quadriceps), function (e.g., adductor magnus), and position (e.g., brachialis).

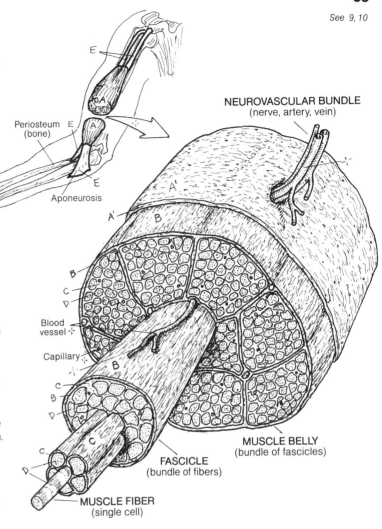

Periosteum (bone)

Aponeurosis

NEUROVASCULAR BUNDLE (nerve, artery, vein)

Blood vessel

Capillary

MUSCLE BELLY (bundle of fascicles)

FASCICLE (bundle of fibers)

MUSCLE FIBER (single cell)

MECHANICS OF MOVEMENT:∗
FULCRUM ꜰ (JOINT) ꜰ'
EFFORT ᴀ (MUSCLE) ᴀ
RESISTANCE ɢ (WEIGHT) ɢ'

1ST CLASS LEVER∗

In a *1st class lever*, the joint lies between the muscle and the load. This is the most efficient class of lever. By flexing the neck and posturing the head forward and downward, the load (G¹) is appreciably increased, and the muscular effort (A) to hold that posture may induce muscle pain and stiffness/tightness (overuse).

2ND CLASS LEVER∗

In a *2nd class lever*, the load lies between the joint and the pulling muscle. This lever system operates in lifting a wheelbarrow (the wheel is the fulcrum) as well as lifting a 75 kg (165 lb) body onto the metatarsal heads at the metatarsophalangeal joints. This is a relatively easy task for the strong calf (triceps surae) muscles; but try standing on the heads of your middle phalanges (increasing the distance F¹-G¹)!

3RD CLASS LEVER∗

In a *3rd class lever*, the muscle lies between the joint and the load and has a poor mechanical advantage here. Consider the difference in muscular effort required to carry a 45 kg (100 lb) bag of cement in your hands with flexed elbows (elbow joint: 3rd class lever) and carrying your 75 kg (165 lb) body on the heads of your metatarsals (2nd class lever at the metatarsophalangeal joints). It is all a matter of leverage.

Skeletal muscles employ simple machines, such as levers, to increase the efficiency of their contractile work about a joint. Mechanically, the degree of *muscular effort* required to overcome *resistance* to movement at a *joint* (*fulcrum*) depends upon (1) the force of that resistance (*weight*, G); (2) the relative distances from the anatomical fulcrum to the anatomical sites of muscular effort (F¹-A); and the anatomical sites of resistance (F¹-G¹). The position of the joint relative to the site of muscle pull and the site of imposed load determines the class of the lever system in use.

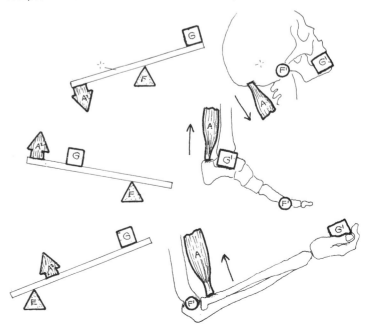

CN: Use very light colors for A and E, and a dark color for F.
(1) Begin with the skeletal muscle lifting the heel of the foot and complete the motor unit and the enlarged view of the neuromuscular junction. (2) Color carefully the motor units and related titles at the bottom of the plate: only the discharging motor units (in dark outline) are to be colored. Note that the word "partial" is not colored under the example of partial contraction.

SKELETAL MUSCLE A
MUSCLE CELL A'
MOTOR END PLATE B
MOTOR NERVE C
AXON C'
AXON BRANCH D
AXON TERMINAL E

NEUROMUSCULAR JUNCTION F
AXON TERMINAL E
MOTOR END PLATE B

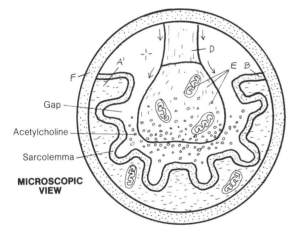

Gap
Acetylcholine
Sarcolemma

MICROSCOPIC VIEW

MOTOR UNIT: *
AXON C'
AXON BRANCH D
NEUROMUSCULAR JUNCTION F
MUSCLE CELL A'

An axon of a single motor neuron, its axon branches, and the skeletal muscle cells with which they form neuromuscular junctions constitute a *motor unit*. Within any given skeletal muscle, the number of muscle cells innervated by a single motor neuron largely determines the specificity of contraction of that muscle; the fewer the number of muscle cells in each motor unit, the more selective and refined the degree of contraction of that skeletal muscle.

Given the fact of "all or none" contraction by individual skeletal muscle cells, grades of contraction of a skeletal muscle are made possible by activating a number of motor units and not activating others. In *maximal contraction* of a skeletal muscle, all motor units are discharged. In a *partial contraction*, only some of the motor units are activated. A *resting muscle* discharges no motor units. Gluteus maximus consists of skeletal muscle cells having a nerve to muscle ratio of 1:1000 or more. There is no possibility of controlled, refined contractions from this muscle. The facial muscles, on the other hand, have a much lower nerve to muscle ratio, closer to 1:10. Here small numbers of muscle cells can be contracted by implementing one or a few motor units, generating very fine control on the muscular effect (facial expression) desired.

Skeletal muscle consists of innumerable muscle cells (fibers). Skeletal muscle requires an intact nerve (innervation) to shorten (contract). Such a nerve is called a *motor nerve*, and consists of numerous *axons* of motor neurons. A motor neuron (recall Plate 11) is dedicated solely to stimulating muscle fibers to contract. Each single *muscle cell* in a skeletal muscle is innervated by a *branch of an axon*. The microscopic site at which the axon branch attaches to the skeletal muscle cell is called the *neuromuscular junction*. Each neuromuscular junction consists of an *axon terminal* closely applied to an area of convoluted muscle cell sarcolemma called the *motor end plate*. There is a gap between the two surfaces. When a skeletal muscle cell is about to be stimulated, a chemical neurotransmitter, called acetylcholine, is released by the axon terminal into the gap. The neurotransmitter induces a change in the permeability of the sarcolemma to calcium, which initiates muscle cell contraction. A muscle cell can only contract maximally ("all or none" law).

GRADES OF CONTRACTION: *

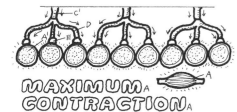

MAXIMUM A
CONTRACTION A'

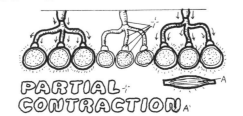

PARTIAL
CONTRACTION A'

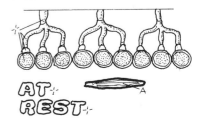

AT
REST

INTEGRATION OF MUSCLE ACTION

CN: Use a bright color for A and a light one for E. (1) Color the small arrows and the large letters of origin (O) and insertion (I) adjacent to the examples of contracted and stretched muscles. (2) In the lower illustration, color the portions of pronator teres and pronator quadratus that are outlined by dotted lines. These parts of the muscles are normally concealed by the radius in this lateral view.

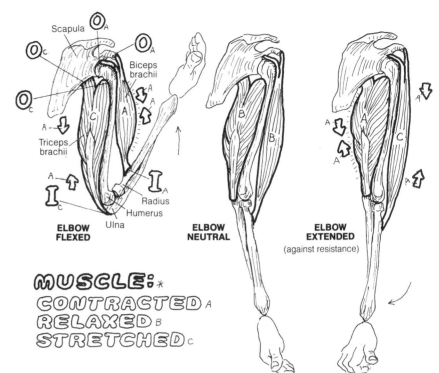

ELBOW FLEXED

ELBOW NEUTRAL

ELBOW EXTENDED (against resistance)

Scapula · Biceps brachii · Triceps brachii · Radius · Humerus · Ulna

MUSCLE:*
CONTRACTED A
RELAXED B
STRETCHED C

When a skeletal muscle shortens (contracts), a joint is moved, and two bones come closer together, isometric contraction excepted. Muscles never push; they always pull. In any given movement between two bones, one bone is generally fixed, and the other moves. The muscle attachment at the fixed bone is the *origin;* the attachment at the moving bone is the *insertion.* In complex movements where it is difficult to identify a "fixed" bone, the origin of the muscle is the more proximal attachment.

When a muscle contracts across a joint, other muscles crossing that joint are affected. No one muscle acts alone in joint movement. In flexion of the elbow joint, for example, biceps brachii (and brachialis, not shown) *contract,* while triceps brachii is *stretched.* Conversely, in elbow extension, triceps is contracted, and the biceps/brachialis muscles are stretched. In neutral, all three are *relaxed* (at rest). Tense (contracted) muscles can often be relaxed by gentle stretching.

ACTORS IN ELBOW FLEXION WITH SUPINATION vs. PRONATION *

No muscle acts alone in the movement of a joint. In the movements shown at right, various muscles are functionally integrated in the simple act of lifting an object, with the forearm supinated in the first case, and pronated in the second case.

PRIME MOVER (AGONIST) A'

The primary muscle effecting a desired joint movement is called the *prime mover* (agonist). There may be more than one; in elbow flexion with the forearm supinated, brachialis and biceps brachii are both prime movers; biceps adds significantly to the lifting power because of the added work in supinating the radius during elbow flexion. With the forearm pronated and supination resisted, the biceps loses that supinating power, and brachialis, unaffected by a pronated forearm, becomes the prime mover.

ANTAGONIST C'

Muscles which potentially or actually oppose or resist a certain movement are called *antagonists.* In the lower illustrations, triceps is the antagonist in the act of elbow flexion, even though it is being stretched and is not contracted in the case illustrated.

FIXATOR D

Fixator muscles stabilize the more proximal joints during weightbearing functions of the more distal joints. Here the trapezius muscle contracts to stabilize (immobilize) the scapula, creating a rigid platform (the scapula) for operation of the weightbearing, ipsilateral limb.

NEUTRALIZER (SYNERGIST) E

In undertaking a desired and specific movement, undesired movements are resisted by *neutralizers* (*synergists*). During flexion of the elbow with a pronated forearm, pronators of the forearm (pronator quadratus, pronator teres) contract to resist or neutralize supination of the forearm. In this action, the pronators are synergistic with the desired movement.

Globally integrated and harmonious muscle functioning makes possible painless, rhythmic, and dynamic movements, best revealed in such activities as dance, sports, and exercise. Joints affected by tense or weak interacting muscles, induced by mechanically disadvantaged posture/gait, can be subject to painful and limited movements.

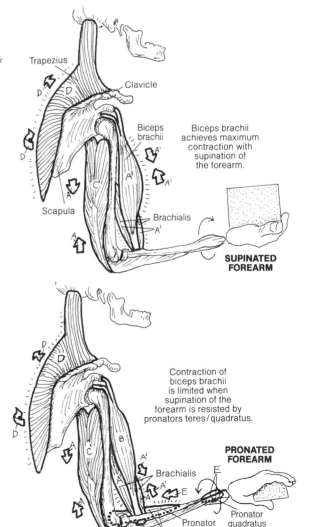

Trapezius · Clavicle · Biceps brachii · Scapula · Brachialis

Biceps brachii achieves maximum contraction with supination of the forearm.

SUPINATED FOREARM

Contraction of biceps brachii is limited when supination of the forearm is resisted by pronators teres/quadratus.

PRONATED FOREARM

Brachialis · Pronator teres · Pronator quadratus

CN: Use your lightest colors for O and Q. Use warm and cheerful colors for the muscles producing a smile (A-H). Color the muscles reflecting sadness (I-O) with greens, blues, and grays. (1) Begin with the smiling side, and color only the muscles identified by titles A-H. Also color those muscles in the profile view below. (2) Repeat the process with the sad side. Note that a portion of frontalis (I) has been cut away to reveal procerus (J). (3) Color the titles at the bottom and the related muscles on the lower view. Include the portions of the auricular muscles that disappear beneath the ear.

ORBICULARIS OCULI A
NASALIS B
LEVATOR LABII SUPERIORIS ALAEQUE NASI C
LEVATOR LABII SUPERIORIS D
LEVATOR ANGULI ORIS E
ZYGOMATICUS MAJOR F
ZYGOMATICUS MINOR G
RISORIUS H

FRONTALIS I
CORRUGATOR SUPERCILII J
ORBICULARIS ORIS K
DEPRESSOR ANGULI ORIS L
DEPRESSOR LABII INFERIORIS M
MENTALIS N
PLATYSMA O

The muscles of facial expression are generally thin, flat bands arising from a facial bone or cartilage and inserting into the dermis of the skin or the fibrous tissue enveloping the sphincter muscles of the orbit or mouth. These muscles are generally arranged into the following regional groups: (1) epicranial group (*occipitofrontalis* moving the scalp); (2) the orbital group (*orbicularis oculi, corrugator supercilii*); (3) the nasal group (*nasalis, procerus*); (4) the oral group (*orbicularis oris, zygomaticus major* and *minor*, the *levators* and the *depressors* of the lips and angles of the mouth, *risorius, buccinator*, and part of *platysma*), and (5) the group moving the ears (*auricular muscles*). The general function of each of these muscles is to move the skin wherever they insert. As you color each muscle, try contracting it on your self while looking into a mirror, and see what develops.

Orbicularis oculi and oris are sphincter muscles, tending to close the skin over the eyelids and tighten the lips, respectively. Contractions of the cheek muscle buccinator makes possible rapid changes in volume of the oral cavity, as in playing a trumpet or squirting water. The nasalis muscle has both compressor and dilator parts which influence the size of the anterior nasal openings.

GALEA APONEUROTICA Q
OCCIPITALIS R
AURICULAR MUSCLES S
PROCERUS T
BUCCINATOR P

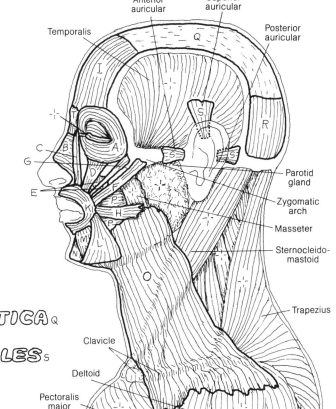

CN: Use a bright, yellowish color for the mandible (E) which appears in all the illustrations.
(1) Begin at the upper left and work your way through the various movements of the mandible.
(2) In coloring the temporomandibular joint, note that the superior and inferior joint spaces (I, I¹) are colored black.

DEEPER VIEW

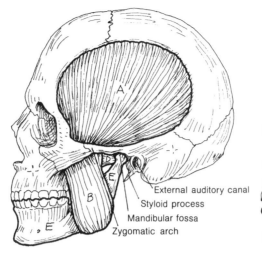
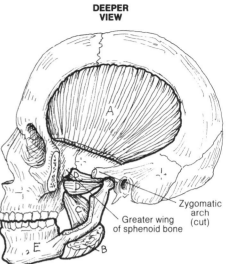

External auditory canal
Styloid process
Mandibular fossa
Zygomatic arch

Zygomatic arch (cut)
Greater wing of sphenoid bone

MUSCLES:*
TEMPORALIS A
MASSETER B
MEDIAL PTERYGOID C
LATERAL PTERYGOID D

The act of chewing is called mastication. The *muscles of mastication* move the temporomandibular joint and are largely responsible for elevation, depression, protrusion, retraction, and lateral motion of the mandible. These muscles function bilaterally to effect movements of the single bone (mandible) at two joints.

The *temporalis* and *masseter* muscles are often contracted unconsciously (clenching teeth) when stressed, giving rise to potentially severe bitemporal and preauricular headaches. The muscles can easily be palpated when contracted.

The *medial and lateral pterygoids* are in the infratemporal fossa and cannot be palpated. In the lowest drawing at right, note how the two heads of the lateral pterygoid insert on the anterior part of the articular disc. During mouth opening, the muscle pulls the articular disc forward as the mandibular head rotates forward in the fossa. On mouth closing, the muscle relaxes, permitting the disc to move posteriorly. Rapid, forced jaw opening may disrupt the articular disc.

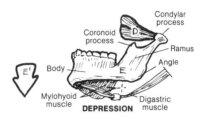

Condylar process
Coronoid process
Ramus
Body
Angle
Mylohyoid muscle
Digastric muscle
DEPRESSION

PROTRUSION

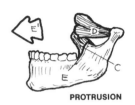

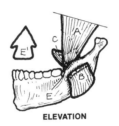

Posterior fibers

MANDIBLE MOVEMENT E E'

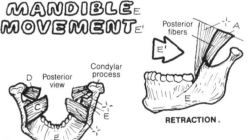

ELEVATION

Posterior view
Condylar process
LATERAL

RETRACTION

TEMPOROMANDIBULAR*
(CRANIOMANDIBULAR)
JOINT:*

JOINT CAPSULE F
ARTICULAR DISC
(MENISCUS) G
RETRODISCAL PAD H
SUP. JOINT SPACE I.
INF. JOINT SPACE I'.

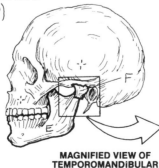

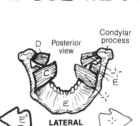

MAGNIFIED VIEW OF TEMPOROMANDIBULAR JOINT (Sagittal section)

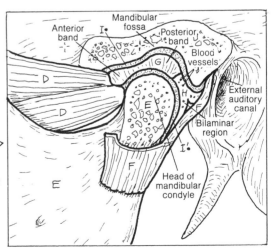

Mandibular fossa
Anterior band
Posterior band
Blood vessels
External auditory canal
Bilaminar region
Head of mandibular condyle

The mandible has two condylar processes (left and right) each of which articulates with a mandibular fossa of the temporal bone. Two *temporomandibular joints* are connected by one bone, hence the alternate term: craniomandibular joint. Intervening between the mandibular fossa and the head of the condylar process is a fibrocartilaginous, movable oval plate called the *articular disc* or meniscus. The disc embodies two (anterior and posterior) avascular fibrous bands, their long axes directed in the coronal plane. The two bands are separated by a thinner intermediate zone. With closed mouth, the mandibular head rests against the posterior band; with full opening (35–50 mm between upper and lower incisors), the head lies against the anterior band. Posterior to the disc is the *retrodiscal pad*, a

two-layered (bilaminar) region of loose fibrous, vascular, sensitive connective tissue from which the disc gets its nutrition. Medially and laterally, the disc is attached to the condyle of the mandible. The mandibular head, the disc, and the fossa are enclosed by a joint capsule. The disc divides the joint space into two compartments, one above and one below the disc. The disc may be structurally incomplete from birth, even perforated. It frequently tends to fray with aging. Disc displacement may be responsible for clicking and limited range of jaw motion.

ANTERIOR & LATERAL MUSCLES

CN: Except for B and E, use your lightest colors throughout the plate. (1) Begin with the diagrams of the triangles of the neck and the sternocleidomastoid (A, B, C). Color over all the muscles within the triangles. (2) Then work top and bottom illustrations simultaneously, coloring each muscle in as many views as you can find it. Note the relationship between muscle name and attachment.

The neck is a complex tubular region of muscles, viscera, vessels, and nerves surrounding the cervical vertebrae. The muscles of the neck are arranged in superficial and deep groups. Here we concentrate on superficial muscles. The superficial posterior and posterolateral muscle of the neck is trapezius (Plate 46). The deep posterior muscles are covered in Plate 41. The most superficial anterior muscle of the neck is platysma (Plate 38). The anterior and lateral muscle groups are divided into triangular areas by the *sternocleidomastoid* muscle.

ANTERIOR TRIANGLE A
SUPRAHYOID MUSCLES: (D)
STYLOHYOID D¹ DIGASTRIC D²
MYLOHYOID D³ HYOGLOSSUS D⁴
GENIOHYOID D⁵
HYOID BONE E
INFRAHYOID MUSCLES: (F)
STERNOHYOID F¹ OMOHYOID F²
THYROHYOID F³ STERNOTHYROID F⁴

The anterior region of the neck is divided in the midline; each half forms an *anterior triangle*. The borders of the anterior triangle of superficial neck muscles are clearly illustrated. The *hyoid bone*, suspended from the styloid processes of the skull by the stylohyoid ligaments, divides each anterior triangle into upper *suprahyoid* and lower *infrahyoid* regions.

The *suprahyoid muscles* arise from the tongue (glossus), mandible (mylo-, genio-, anterior digastric), and skull (stylo-, posterior digastric) and insert on the hyoid bone. They elevate the hyoid bone, influencing the movements of the floor of the mouth and the tongue, especially during swallowing. With a fixed hyoid, the suprahyoid muscles, especially the digastrics, depress the mandible.

The *infrahyoid muscles* generally arise from the sternum, thyroid cartilage of the larynx, or the scapula (omo-) and insert on the hyoid bone. These muscles partially resist elevation of the hyoid bone during swallowing. *Thyrohyoid* elevates the larynx during production of high-pitched sounds; sternohyoid depresses the larynx to assist in production of low-pitched sounds.

POSTERIOR TRIANGLE C
SEMISPINALIS CAPITIS C¹
SPLENIUS CAPITIS C²
LEVATOR SCAPULAE C³
SCALENUS: ANT. C⁴ MED. C⁵ POST. C⁶

The *posterior triangle* consists of an array of muscles covered by a layer of deep (investing) cervical fascia just under the skin between sternocleidomastoid and trapezius. The borders of the triangle are clearly illustrated. Muscles of this region arise from the skull and cervical vertebrae; they descend to and insert upon the upper two ribs (*scalenes*), the upper scapula (*omohyoid, levator scapulae*), and the cervical/thoracic vertebral spines (*splenius capitis, semispinalis capitis*). Visualizing their attachments, these muscles' function becomes clear.

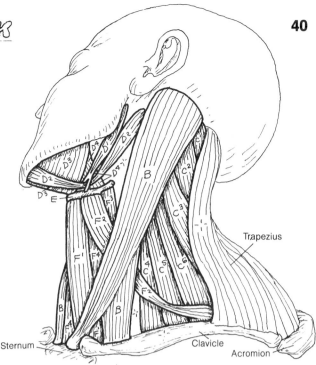

Trapezius
Sternum
Clavicle
Acromion

STERNOCLEIDOMASTOID B

The *sternocleidomastoid* muscle, acting unilaterally, tilts the head laterally on the same side while simultaneously rotating the head and pulling the back of the head downward, lifting the chin, and rotates the front of the head to the opposite side. Both muscles acting together move the head forward (anteriorly) while extending the upper cervical vertebrae, lifting the chin upward.

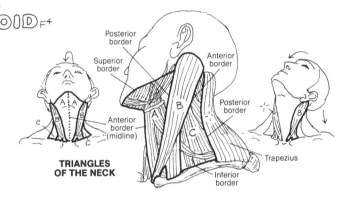

Posterior border
Superior border
Anterior border
Anterior border (midline)
Posterior border
Trapezius
Inferior border

TRIANGLES OF THE NECK

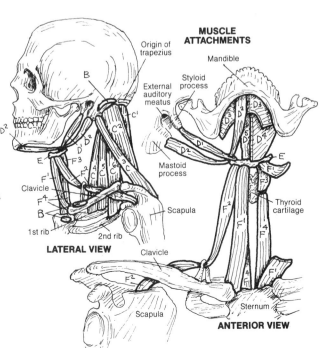

Origin of trapezius
MUSCLE ATTACHMENTS
Mandible
Styloid process
External auditory meatus
Mastoid process
Clavicle
Scapula
Thyroid cartilage
1st rib
2nd rib
Clavicle
LATERAL VIEW
Scapula
Sternum
ANTERIOR VIEW

DEEP MUSCLES OF BACK & POSTERIOR NECK

CN: Use your lightest colors on the B and C groups. Note that splenius (A) and semispinalis (C¹) represent more than one muscle; the muscle subsets are identified. (1) After coloring the muscles of the back and posterior neck, color the lower right diagram which describes the location and function of the deep movers of the spine.

The *deep muscles of the back and posterior neck* extend, rotate, or laterally flex one or more of the 24 paired facet joints and the 22 intervertebral disc joints of the vertebral column. The long muscles move several motion segments (recall Plate 21) with one contraction, while the short muscles can move one or two motion segments at a time (see intrinsic movers).

COVERING MUSCLE:∗
SPLENIUS A

The *splenius* muscles extend and rotate the neck and head in concert with the opposite sternocleidomastoid muscle. Splenius capitis covers the deeper muscles of the upper spine.

VERTICAL MUSCLES:∗
ERECTOR SPINAE B
SPINALIS B¹
LONGISSIMUS B²
ILIOCOSTALIS B³

The *erector spinae* group are the principal extensors of the vertebral motion segments. Oriented vertically along the longitudinal axis of the back, they are thick, quadrilateral muscles in the lumbar region, splitting into smaller, thinner separate bundles attaching to the ribs (*iliocostalis*), and upper vertebrae and head (*longissimus, spinalis*). Erector spinae arises from the lower thoracic and lumbar spines, the sacrum, ilium, and intervening ligaments.

OBLIQUE MUSCLES:∗
TRANSVERSOSPINALIS GROUP:(C)
SEMISPINALIS C¹
MULTIFIDUS C²
ROTATORES C³

The *transversospinalis* group extends the motion segments of the back, and rotates the thoracic and cervical vertebral joints. These muscles generally run from the transverse processes of one vertebra to the spine of the vertebra above, spanning three or more vertebrae. The *semispinales* are the largest muscles of this group, reaching from mid-thorax to the posterior skull; the *multifidi* consist of deep fasciculi spanning 1–3 motion segments from sacrum to C2; the *rotatores* are well defined only in the thoracic region.

DEEPEST MUSCLES:∗
INTERTRANSVERSARII D
INTERSPINALIS E
SUBOCCIPITAL MUSCLES F

These small, deep-lying muscles cross the joints of only one motion segment. They are collectively major postural muscles. Electromyographic evidence has shown that these short muscles remain in sustained contraction for long periods of time during movement and standing/sitting postures. They are most prominent in the cervical and lumbar regions. The small muscles set deep in the posterior, *suboccipital* region (deep to semispinalis and erector spinae) rotate and extend the joints between the skull and C1 and C2 vertebrae.

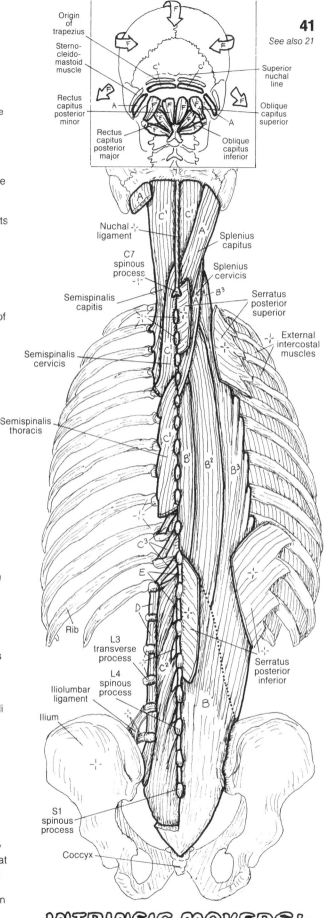

INTRINSIC MOVERS:∗
EXTENSOR E
ROTATOR C³
LATERAL FLEXOR D

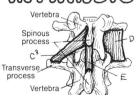

MUSCLES OF THORAX & POSTERIOR ABDOMINAL WALL

CN: Use blue for E and red for G. (1) You may wish to darken the underside of the diaphragm (A) in the anterior view. Do not confuse the arcuate ligaments with the 12th rib. (2) In the cross sectional view above, color the broken lines which represent transparent, membranous portions of the intercostal muscles.

THORAX MUSCLES:*

THORACIC DIAPHRAGM A
EXTERNAL INTERCOSTAL B
INTERNAL INTERCOSTAL C
INNERMOST INTERCOSTAL D

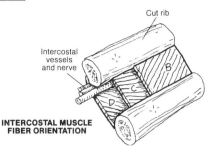

Cut rib

Intercostal vessels and nerve

INTERCOSTAL MUSCLE FIBER ORIENTATION

The *thoracic diaphragm* is a broad, thin muscle spanning the thoracoabdominal cavity; its origin, much of which is illustrated here, includes the lower six ribs (not shown).

The left and right halves of the diaphragm insert into each other (central tendon). The diaphragm is responsible for 75% of the respiratory air flow. Openings (hiatuses) in the diaphragm provide passage for major transiting structures.

The *intercostal muscles* alter the dimensions of the thoracic cavity by collectively moving the ribs, resulting in 25% of the total respiratory effort. The specific function of each of these muscles, with respect to fiber orientation, is not understood. The innermost intercostals are an inconstant layer, and here include the transversus thoracis and subcostal muscles.

INFERIOR VENA CAVA E
ESOPHAGUS F
AORTA G

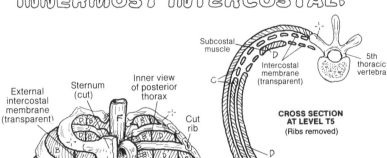

Subcostal muscle

Intercostal membrane (transparent)

5th thoracic vertebra

CROSS SECTION AT LEVEL T5
(Ribs removed)

Transversus thoracis muscle

Sternum

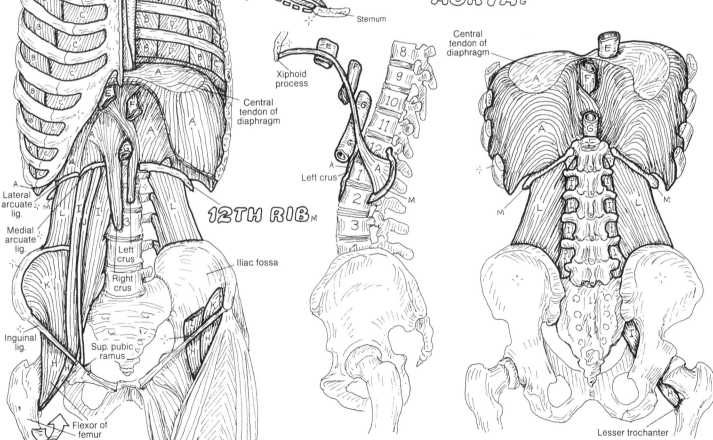

External intercostal membrane (transparent)

Sternum (cut)

Inner view of posterior thorax

Cut rib

Central tendon of diaphragm

Xiphoid process

Central tendon of diaphragm

Left crus

12TH RIB M

Lateral arcuate lig.

Medial arcuate lig.

Inguinal lig.

Left crus

Right crus

Iliac fossa

Sup. pubic ramus

Flexor of femur

Lesser trochanter

ANTERIOR VIEW

LATERAL VIEW

POSTERIOR VIEW

POSTERIOR ABDOMINAL WALL MUSCLES:*

ILIOPSOAS H
PSOAS MAJOR I MINOR J
ILIACUS K
QUADRATUS LUMBORUM L

The tendons of *iliacus* and *psoas major* converge to a single insertion (*iliopsoas*). Iliopsoas, a strong flexor of the hip joint, is a powerful flexor of the lumbar vertebrae; a weak psoas may contribute to low back pain. *Quadratus lumborum* is an extensor of the lumbar vertebrae (bilaterally) and a lateral flexor unilaterally. It functions in respiration by securing the 12th rib. Immediately anterior to these muscles is the retroperitoneum (see Plate 109).

MUSCLES OF ANTERIOR ABDOMINAL WALL & INGUINAL REGION

CN: Use a dark color for J and bright ones for B and I. (1) Color the 3 layers of the abdominal wall. (2) Color the sheath of the rectus abdominis in gray. Color the two locational arrows in this and the upper illustration. (3) Beginning with J and K, and followed by H, color the coverings of the spermatic cord.

ANT. ABDOMINAL WALL:*
TRANSVERSUS ABDOMINIS_A
RECTUS ABDOMINIS_B
INTERNAL OBLIQUE_C
EXTERNAL OBLIQUE_D

The anterior abdominal wall consists of three layers of flat muscles, the tendons (aponeuroses) of which interlace in the midline, and a vertically oriented pair of segmented muscles which are ensheathed incompletely by the aponeuroses of the three flat muscles (*sheath of the rectus abdominis*). The flat muscles arise from the lateral aspect of the torso (inguinal ligament, iliac crest, thoracolumbar fascia, lower costal cartilages, ribs). The lowest fibers of *external oblique* roll inwardly to form the *inguinal ligament*. These three muscles act to compress the abdominal contents during expiration, urination, and defecation. They assist in maintaining pressure on the curve of the low back, resisting "sway back" (excess lumbar lordosis) and extension of the low back.

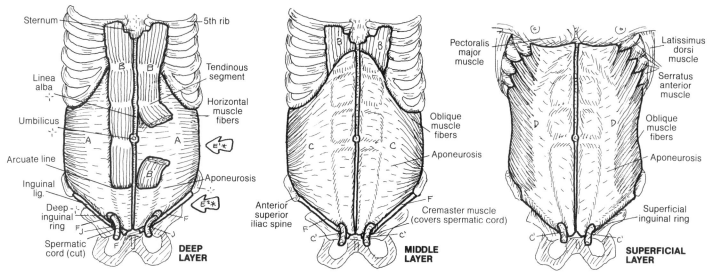

Each segmented *rectus abdominis* muscle arises from the pubic crest and tubercles and inserts on the lower costal cartilages and xiphoid process (sternum). They are flexors of the vertebral column. The *sheath of the rectus* varies in its extent, running from deep to superficial from below upward, as illustrated. Below the arcuate line, no muscle contributes to its posterior layer (E²*); in the middle, all three flat muscles contribute equally to the sheath (E¹*); above, the anterior sheath is formed from external oblique; posteriorly, the rectus contacts the costal cartilages.

The *inguinal region* is the lower medial portion of the abdominal wall. Herein exists the inguinal canal with its inner (deep inguinal ring) and outer (superficial inguinal ring) openings, and the resident *spermatic cord* (ductus deferens and testicular vessels and nerves) in the male and the round ligament of the uterus in the female. In its apparent "migration" from the abdominal cavity through the inguinal canal into the scrotum during the last months of fetal development, the testis and spermatic cord push through (not perforate) and carry with them each layer of abdominal wall they traverse, much as a finger might push through four layers of latex to form a four-layered finger glove. Each of these layers constitutes a *covering of the cord*. The inguinal canal is a site of relative structural weakness in both males and females, and the abdominal wall here is subject to protrusions (inguinal hernia) from intra-abdominal fat and intestines, indirectly through the deep ring or directly through the wall near the superficial ring.

SHEATH OF RECTUS ABDOMINIS_E*

INGUINAL REGION:*
INGUINAL LIG._F
CREMASTER MUS._C'
PYRAMIDALIS MUS._G
PERITONEUM_H
TRANSVERSALIS FASCIA_I
SPERMATIC CORD_J
TESTIS/EPIDIDYMIS_K

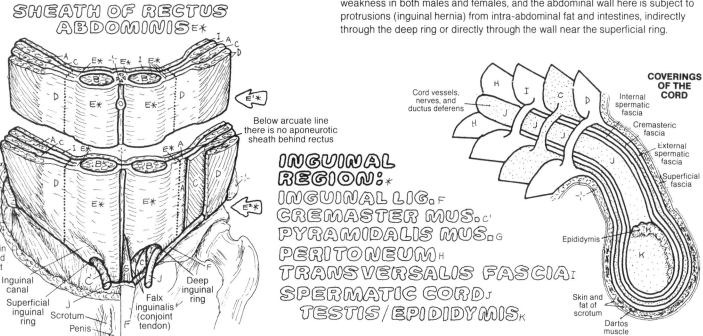

PELVIC DIAPHRAGM (FLOOR):*¹
LEVATOR ANI:⋰
LEVATOR PROSTATAE/VAGINAE_A
PUBORECTALIS_B
PUBOCOCCYGEUS_C
ILIOCOCCYGEUS_D
COCCYGEUS_E

PELVIC WALL:*²
OBTURATOR INTERNUS_F
PIRIFORMIS_G
SACROTUBEROUS LIGAMENT_H*
SACROSPINOUS LIGAMENT_I*
TENDINOUS ARCH_J

CN: Use bright colors for A and J, and gray for H and I.
(1) Begin with the illustration of the pelvic floor muscles, labeled "pelvic diaphragm," just below the large illustration at mid-right. Then go to that large illustration and color the same muscles. Continue with all drawings showing pelvic floor muscles.
(2) Color the pelvic wall muscles and the ligaments in the "pelvic wall" diagram. Then color these muscles/ligaments in the large illustration at right, followed by the rest of the drawings showing pelvic wall muscles/ligaments.

The muscles of the pelvis form the pelvic floor in the pelvic outlet (*coccygeus* and the *levator ani*) and the pelvic wall (*obturator internus* and *piriformis*). The fascia-covered pelvic floor muscles constitute the pelvic diaphragm, separating pelvic viscera from the perineal structures inferiorly. The pelvic wall includes the *sacrotuberous* and *sacrospinous ligaments*.

The levator ani on each side arises from the pubic bone and ischial spine and the intervening *tendinous arch*, droops downward as it passes toward the midline, and inserts on the anococcygeal ligament and the coccyx with the contralateral levator ani. The muscle essentially has four parts (A, B, C, and D). Coccygeus is the posterior muscle of the pelvic floor, on the same plane as iliococcygeus and immediately posterior to it. The pelvic diaphragm counters abdominal pressure, and with the thoracic diaphragm, assists in micturition, defecation, and childbirth. It is an important support mechanism for the uterus, resisting prolapse.

The obturator internus, a lateral rotator of the hip joint, arises, in part, from the margins of the obturator foramen on the pelvic side. It passes downward and posterolaterally past the obturator foramen to and through the lesser sciatic foramen, inserting on the medial surface of the greater trochanter of the femur. Its covering fascia forms the tendinous arch from which levator ani arises in part. Piriformis, a lateral rotator of the hip joint, takes a course similar to obturator internus to arrive at the greater trochanter.

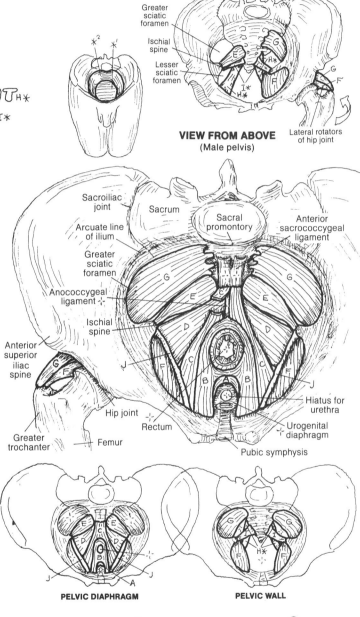

VIEW FROM ABOVE
(Male pelvis)

ANTERIOR VIEW
(Coronal section)

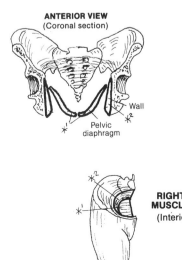

PELVIC DIAPHRAGM

PELVIC WALL

RIGHT PELVIC WALL MUSCLES/LIGAMENTS
(Interior view/female)

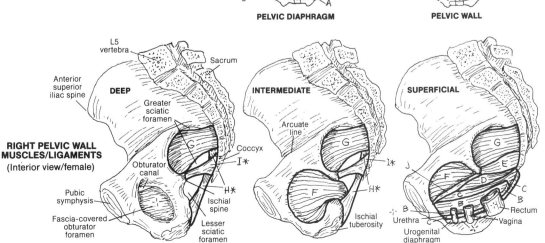

DEEP

INTERMEDIATE

SUPERFICIAL

MUSCLES OF THE PERINEUM

CN: (1)Color gray the title *perineum* (*¹) and the two related diamond-shaped areas at the top of the plate. Color the elements forming the boundaries of the peri- neum, as seen from below. (2)Color gray the titles *urogenital triangle* and *anal triangle,* and their respective triangles. (3)Color the lower views simultaneously.

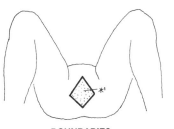

PERINEUM:*¹ (BOUNDARIES):
SYMPHYSIS PUBIS ᴀ
COCCYX ʙ
ISCHIAL TUBEROSITY ᴄ
SACROTUBEROUS LIGAMENT ᴅ
ISCHIOPUBIC RAMUS ᴇ

The 3-dimensional *perineum* is the region below the pelvic cavity situated within the pelvic outlet. It is bordered by structures A–E. Its floor (inferior border) is skin and fascia; its roof is the pelvic diaphragm. The diamond-shaped perineum is bisected at the *ischial tuberosities* into the urogenital and anal triangles.

UROGENITAL TRIANGLE:*²
ISCHIOCAVERNOSUS M. ꜰ
BULBOSPONGIOSUS M. ɢ
SUP. TRANSVERSE PERINEAL M. ʜ
UROGENITAL DIAPHRAGM ɪ

The *urogenital triangle* contains the penis, scrotum, and related structures in the male and the clitoris, urethral and vaginal orifices, and related structures in the female. It includes: (1) the anterior recess of the ischiorectal fossa, (2) the *urogenital diaphragm* composed of the deep transverse perineal muscle and sphincter urethrae, and (3) the superficial perineal space, containing the *ischiocavernosus* and *bulbospongiosus muscles*. These muscles ensheathe the roots of the erectile bodies of the penis / clitoris, aiding in their erection. *Bulbospongiosus* also contracts rhythmically during the ejaculation of semen. The *superficial transverse perineal* muscles stabilize the fibrous perineal body which helps anchor the perineal structures.

ANAL TRIANGLE:*³
LEVATOR ANI M. ᴊ
EXTERNAL SPHINCTER ANI M. ᴋ
ANOCOCCYGEAL LIGAMENT ʟ

The *anal triangle* includes the fat-filled ischiorectal fossae and the anal opening (anus) and its *external sphincter ani muscle*. The inferior fibers of the *anococcygeal ligament* give attachment to the anal sphincter.

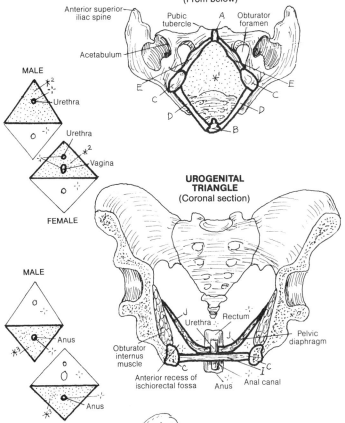

BOUNDARIES OF PERINEUM (From below)

UROGENITAL TRIANGLE (Coronal section)

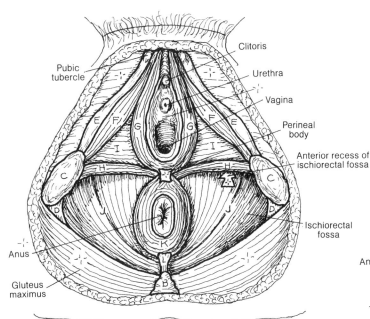

FEMALE PERINEUM

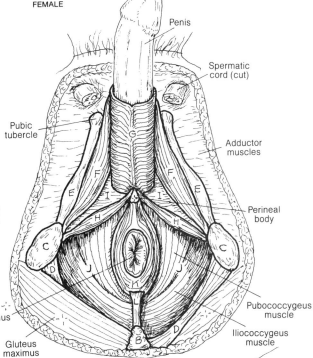

MALE PERINEUM

MUSCLES OF SCAPULAR STABILIZATION
TRAPEZIUS_A RHOMBOID MAJOR_B RHOMBOID MINOR_B' LEVATOR SCAPULAE_C SERRATUS ANTERIOR_D PECTORALIS MINOR_E

CN: (1) Color the six muscles of scapular stabilization. Note that the two rhomboids receive the same color (B). In the two main views, color gray the nuchal ligament and its title. (2) Color the attachment site diagrams at upper right. (3) In the illustrations below describing scapular movement, note that the three regions of trapezius (A) play different roles. Color gray the scapulae, the arrows, and the movement titles.

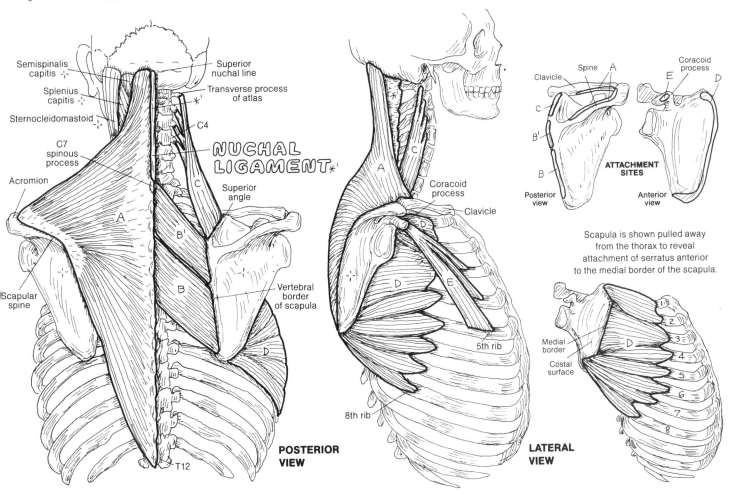

The scapula lies on the posterior thorax, roughly from T 2 to T 8. It has no direct bony attachment with the axial skeleton. Enveloped by muscle, it glides over the fascial-covered thorax during upper limb movement (scapulothoracic motion). Bursae have been reported between the thorax and the scapula; so has bursitis. The scapula is dynamically moored to the axial skeleton by muscles attaching the scapula to the axial skeleton. These *muscles of scapular stabilization* make possible considerable scapular mobility and, therefore, shoulder/arm mobility.

Note the roles of these 6 muscles in scapular movement, and note how the shoulder and arm are affected. *Pectoralis minor* assists *serratus anterior* in protraction of the scapula such as in pushing against a wall; it also helps in depression of the shoulder and downward rotation of the scapula. Consider the power resident in serratus anterior and trapezius in pushing or swinging a bat. Note the especially broad sites of attachment of the *trapezius* muscle. Trapezius commonly manifests significant tension with hard work—mental and physical. A brief massage of this muscle often brings quick relief.

MOVEMENTS OF THE SCAPULA*

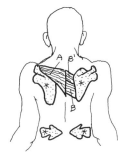
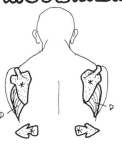
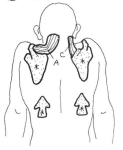
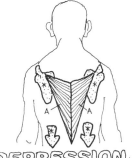
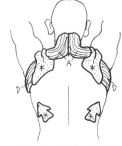
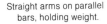

RETRACTION*
Military posture ("squaring the shoulders")

PROTRACTION*
Pushing forward with outstretched arms and hands.

ELEVATION*
Shrugging the shoulders or protecting the head.

DEPRESSION*
Straight arms on parallel bars, holding weight.

UPWARD ROT.*
Lifting or reaching over head.

CN: (1) In addition to the four muscles, color the arrows and titles describing their actions. (2) Color the muscular attachment sites and the diagram of the function of the cuff muscles at mid-right. (3) Do not color the problem spot numerals in the lower illustration. They are there to identify locations discussed in the text.

SUPRASPINATUS_A
INFRASPINATUS_B
TERES MINOR_C
SUBSCAPULARIS_D

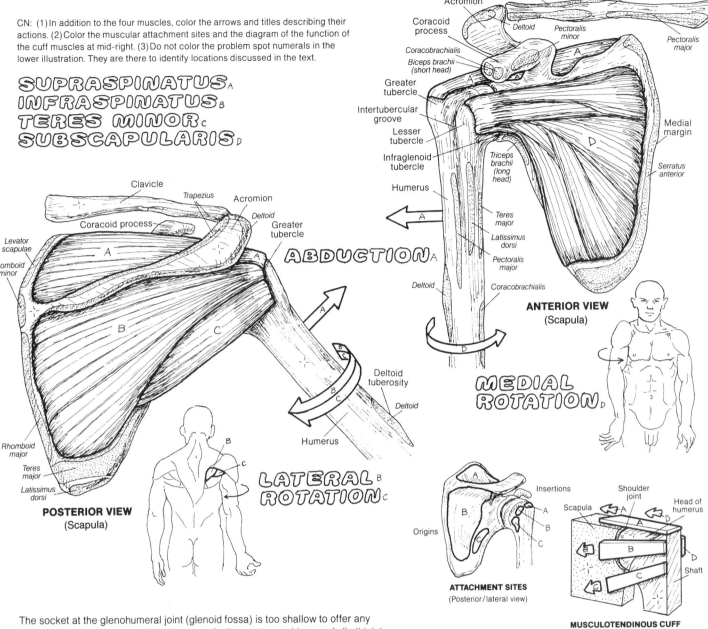

ABDUCTION_A

MEDIAL ROTATION_D

LATERAL_B ROTATION_C

POSTERIOR VIEW
(Scapula)

ANTERIOR VIEW
(Scapula)

ATTACHMENT SITES
(Posterior / lateral view)

MUSCULOTENDINOUS CUFF
(Posterior / lateral view)

The socket at the glenohumeral joint (glenoid fossa) is too shallow to offer any bony security for the head of the humerus. As ligaments would severely limit joint movement, muscle tension must be employed to pull the humeral head in to the shallow scapular socket during shoulder movements. Four muscles fulfill this function: *supraspinatus*, *infraspinatus*, *teres minor*, and *subscapularis* ("SITS muscles"). These muscles form a musculotendinous ("rotator") cuff around the head of the humerus, enforcing joint security. Especially effective during robust shoulder movements, they permit the major movers of the joint to work without risking joint dislocation.

The SITS muscles have come to be known as the "rotator cuff" muscles, in spite of the fact that one of them, supraspinatus, is an abductor of the shoulder joint and not a rotator. Independent of their collective function, all of these muscles are important movers of the joint.

The shoulder joint and rotator cuff muscles are subject to overuse and early degeneration. A common problem arises from repeated contact (impingement) at the undersurface of the acromion (1), the coracoacromial ligament (2), and the underside of the distal clavicle with the subacromial bursa (3), supraspinatus tendon (4), and shoulder joint capsule (5). Progressive rubbing contacts can induce bursal irritation and inflammation (bursitis) and a tear in the supraspinatus tendon ("rotator cuff rupture"). Bony spurs can form on the underside of the acromion, or prior acromioclavicular dislocation can put the distal clavicle closer to the supraspinatus tendon, irritating the latter (supraspinatus tendinitis).

PROBLEM SPOTS IN THE SHOULDER REGION
(Anterior view)

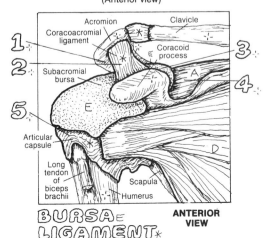

BURSA_E LIGAMENT_*

ANTERIOR VIEW

MOVERS OF SHOULDER JOINT

DELTOID_A PECTORALIS MAJOR_B LATISSIMUS DORSI_C TERES MAJOR_D CORACOBRACHIALIS_E BICEPS BRACHII_F TRICEPS BRACHII (LONG HEAD)_G

CN: (1) Begin with both posterior views; note that the biceps and triceps are not shown on the lateral view.
(2) When coloring the muscles below, note the actions of different parts of the deltoid (A) and pectoralis major (B).

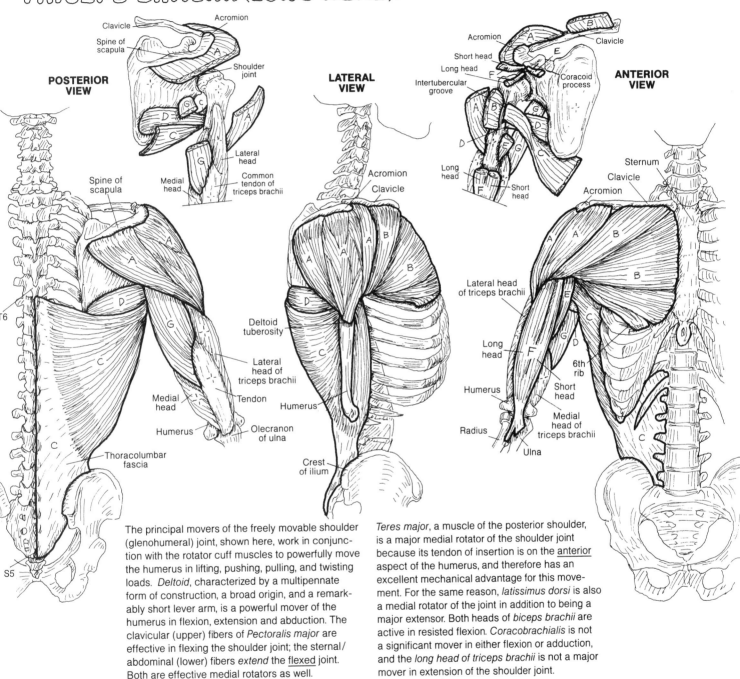

POSTERIOR VIEW

LATERAL VIEW

ANTERIOR VIEW

The principal movers of the freely movable shoulder (glenohumeral) joint, shown here, work in conjunction with the rotator cuff muscles to powerfully move the humerus in lifting, pushing, pulling, and twisting loads. *Deltoid*, characterized by a multipennate form of construction, a broad origin, and a remarkably short lever arm, is a powerful mover of the humerus in flexion, extension and abduction. The clavicular (upper) fibers of *Pectoralis major* are effective in flexing the shoulder joint; the sternal/abdominal (lower) fibers *extend* the flexed joint. Both are effective medial rotators as well.

Teres major, a muscle of the posterior shoulder, is a major medial rotator of the shoulder joint because its tendon of insertion is on the anterior aspect of the humerus, and therefore has an excellent mechanical advantage for this movement. For the same reason, *latissimus dorsi* is also a medial rotator of the joint in addition to being a major extensor. Both heads of *biceps brachii* are active in resisted flexion. *Coracobrachialis* is not a significant mover in either flexion or adduction, and the *long head of triceps brachii* is not a major mover in extension of the shoulder joint.

MOVEMENTS OF THE HUMERUS AT THE SHOULDER JOINT*

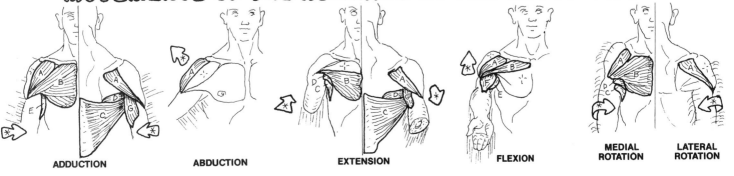

ADDUCTION   **ABDUCTION**   **EXTENSION**   **FLEXION**   **MEDIAL ROTATION**   **LATERAL ROTATION**

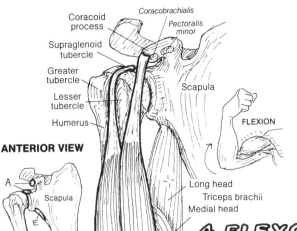

ANTERIOR VIEW

Coracoid process
Supraglenoid tubercle
Greater tubercle
Lesser tubercle
Humerus
Coracobrachialis
Pectoralis minor
Scapula
FLEXION
Long head
Triceps brachii
Medial head
Scapula
A
E
Humerus
Long head
Short head
C
B
D
G
Radial tuberosity
A
D
Ulna
Radius
C
H
Bicipital aponeurosis (cut)
Radius
Ulna
Interosseous ligament
Styloid process

CN: Use the same colors for biceps brachii (A) and triceps brachii (E) as you did for those muscles on Plate 48. (1) Color the four flexors and their attachment sites on the drawings to their left. Do the same for the extensors on the right. (2) Color the supinators and pronators below, the arrows demonstrating their actions, and their attachment sites at upper left.

4 FLEXORS:*
BICEPS BRACHII(A)
BRACHIALIS(B)
BRACHIORADIALIS(C)
PRONATOR TERES(D)

Medial epicondyle

The principal *flexors* of the elbow joint are *brachialis* and *biceps brachii*, of which the former has the best mechanical advantage. Yet it's the bulge of a contracted biceps that gets all the visual attention! The tendon of biceps inserts at the tuberosity of the radius, making the muscle a supinator of the forearm as well. With the limb supinated, the biceps works to fulfill flexion of the elbow and supination of the elbow. Take away the supinating function (flexing the pronated elbow), and the appearance of biceps is disappointing (in most of us!). Note the additional attachment of the bicipital aponeurosis into the deep fascia of the common flexor group (not shown) in the forearm. *Brachioradialis* is active in flexion of the elbow and in rapid extension where it counters the centrifugal force produced by that movement. *Pronator teres* assists in elbow flexion as well as pronation.

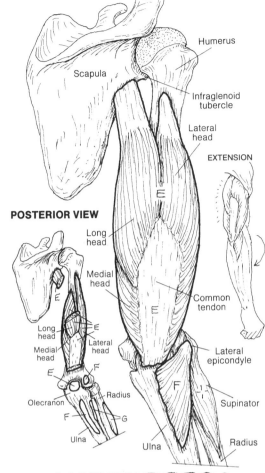

POSTERIOR VIEW

Scapula
Humerus
Infraglenoid tubercle
Lateral head
EXTENSION
Long head
Medial head
E
Common tendon
Long head
Lateral head
Medial head
E
F
Lateral epicondyle
Olecranon
Radius
F
G
Ulna
Supinator
Radius
F
Ulna

2 EXTENSORS:*
TRICEPS BRACHII(E)
ANCONEUS(F)

The principal extensor of the elbow joint is the three-headed *triceps brachii* with its massive tendon of insertion. The smaller *anconeus* assists in this function. Triceps is a powerful antagonist to the elbow flexors.

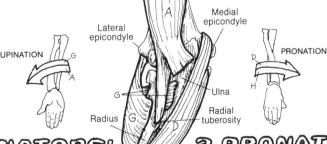

Lateral epicondyle
Medial epicondyle
A
SUPINATION
G
A
Ulna
Radial tuberosity
Radius
G
G
PRONATION
D
H
Interosseous ligament
H

2 SUPINATORS:*
BICEPS BRACHII(A)
SUPINATOR(G)

Biceps brachii is the more powerful supinator of the elbow, but *supinator* is important in maintaining supination. Supinator arises from the lateral aspect of the elbow, passing obliquely downward and forward to a rather broad insertion on the upper lateral and anterior surface of the radius. A bundle of fibers from the upper lateral ulna passes behind the radius to join the lateral fibers of supinator.

2 PRONATORS:*
PRONATOR TERES(D)
PRONATOR QUADRATUS(H)

Pronator quadratus is the principal pronator of the elbow joint, superior in its mechanical advantage to *pronator teres*. Pronating the forearm (palm down) involves medial rotation of the radius. Since only the radius can rotate in the forearm, the pronators clearly cross the radius on the anterior side of the forearm and their origin is ulnar.

ANTERIOR VIEW

MOVERS OF WRIST & HAND JOINTS

FLEXORS: *¹

DEEP LAYER
FLEX. DIGITORUM PROFUNDUS ᴀ
FLEX. POLLICIS LONGUS ʙ

INTERMEDIATE LAYER
FLEX. DIGITORUM SUPERFICIALIS ᴄ

SUPERFICIAL LAYER
FLEX. CARPI ULNARIS ᴅ
PALMARIS LONGUS ᴇ
FLEX. CARPI RADIALIS ꜰ

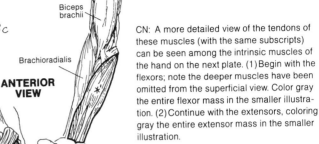

Biceps brachii

Brachioradialis

ANTERIOR VIEW

CN: A more detailed view of the tendons of these muscles (with the same subscripts) can be seen among the intrinsic muscles of the hand on the next plate. (1) Begin with the flexors; note the deeper muscles have been omitted from the superficial view. Color gray the entire flexor mass in the smaller illustration. (2) Continue with the extensors, coloring gray the entire extensor mass in the smaller illustration.

Humerus
Radius
Interosseous membrane
Supinator
Medial epicondyle
Ulna
Pronator teres

C
A
B
Pronator quadratus

Ulna
Carpals
Metacarpal
Palmar aponeurosis (cut)

DEEP **INTERMEDIATE** **SUPERFICIAL**

The *flexors of the wrist* (carpus) *and fingers* (digits) take up most of the anterior compartment of the forearm, arising as a group from the medial epicondyle, the upper radius and ulna, and the intervening interosseous membrane. The deep layer of muscles in the anterior forearm (*flexor pollicis longus* or FPL in the radial half, *flexor digitorum profundus* or FDP in the ulnar half) lie in contact with the radius and ulna. The superficial layer of muscles (wrist flexors: the *"carpi" muscles* and *palmaris longus*) is seen just under the skin and thin superficial fascia. The intermediate layer *(flexor digitorum superficialis*, FDP) exists between the superficial and deep groups. In the anterior (palmar) fingers, note how the tendons of FDS, which insert on the sides of the middle phalanges, split at the level of the proximal phalanges, permitting the deeper (posterior) tendons of FDP to pass on through to the bases of the distal phalanges.

EXTENSORS: *²

DEEP LAYER
EXT. INDICIS ɢ
EXT. POLLICIS LONGUS ʜ
EXT. POLLICIS BREVIS ɪ

SUPERFICIAL LAYER
EXT. CARPI ULNARIS ᴊ
EXT. DIGITI MINIMI ᴋ
EXT. DIGITORUM ʟ
EXT. CARPI RADIALIS LONGUS ᴍ
EXT. CARPI RADIALIS BREVIS ɴ
ABDUCTOR POLLICIS LONGUS ᴏ

The *extensors of the wrist and fingers* arise from the lateral epicondyle and upper parts of the bones and interosseous membrane of the forearm, forming an extensor compartment on the posterior side of the forearm. The wrist extensors insert on the distal carpal bones or metacarpals, while the finger extensors form an expansion of tendon over the middle and distal phalanges to which the small intrinsic muscles of the hand insert. The wrist extensor muscles are critical to hand function: grasp a finger of one hand with your fingers and an extended wrist of the other; now try it with wrist fully flexed. Note the power of the hand exists only with an extended wrist.

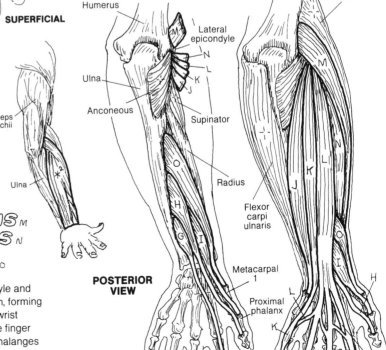

Triceps brachii

Ulna

POSTERIOR VIEW

Humerus
Lateral epicondyle
Ulna
Anconeous
Supinator
Radius
Flexor carpi ulnaris

Brachioradialis

Metacarpal 1
Proximal phalanx
Distal phalanx
Index finger

DEEP **SUPERFICIAL**

MOVERS OF HAND JOINTS (INTRINSICS)

CN: The extrinsic muscles which move the wrist and finger joints were covered on Plate 50, and their tendons are shown in dark line and labeled here for identification and study, *but not for coloring*. If possible, use different colors on this plate. (1) Color the muscles of the two anterior views, as well as the flexor retinaculum (gray). (2) Color the posterior view. (3) In the illustration of finger abduction (at the bottom) note that the little finger is not moved by the dorsal interosseous (U).

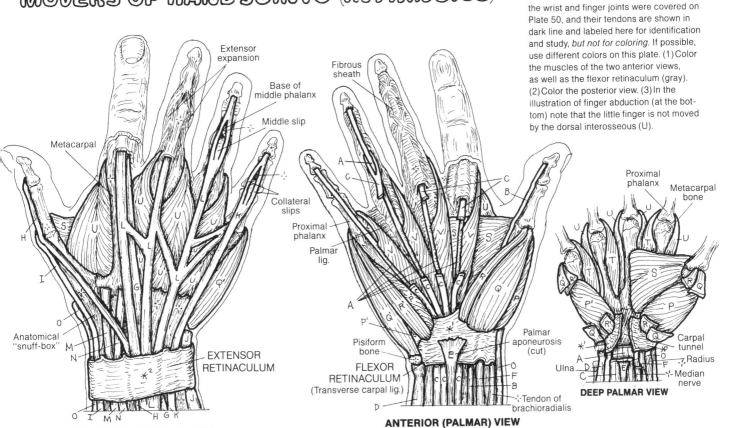

POSTERIOR (DORSAL) VIEW

- Extensor expansion
- Base of middle phalanx
- Middle slip
- Metacarpal
- Collateral slips
- Anatomical "snuff-box"
- EXTENSOR RETINACULUM

ANTERIOR (PALMAR) VIEW

- Fibrous sheath
- Proximal phalanx
- Palmar lig.
- Pisiform bone
- FLEXOR RETINACULUM (Transverse carpal lig.)
- Palmar aponeurosis (cut)
- Tendon of brachioradialis

DEEP PALMAR VIEW

- Proximal phalanx
- Metacarpal bone
- Carpal tunnel
- Radius
- Median nerve
- Ulna

THENAR EMINENCE *'
OPPONENS POLLICIS P
ABDUCTOR POLLICIS BREVIS Q
FLEXOR POLLICIS BREVIS R

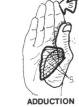

Note the palpable bulge of muscle (*thenar eminence*) just proximal to the thumb on your own hand. Integrated with the action of the other thumb movers, these three muscles make possible complex movements of the thumb. The thenar muscles arise/insert in the same general area as one another; however, their different orientation orders different functions.

HYPOTHENAR EMINENCE *²
OPPONENS DIGITI MINIMI P'
ABDUCTOR DIGITI MINIMI Q'
FLEXOR DIGITI MINIMI BREVIS R'

These muscles move the 5th digit; they are complementary to the thenar muscles in attachment and function. The function of opposition is basic to some of the complex grasping functions of the hand.

DEEP MUSCLES: *
ADDUCTOR POLLICIS S
PALMAR INTEROSSEUS T
DORSAL INTEROSSEUS U
LUMBRICAL V

Adductor pollicis, in concert with the first dorsal interosseous muscle provides great strength in grasping an object between thumb and index finger . . . try it. The *interossei* and *lumbrical muscles* insert into expanded finger extensor tendons (extensor expansion; see posterior view) forming a complex mechanism for flexing the metacarpophalangeal joints and extending the interphalangeal joints. By their phalangeal insertions, the interossei abduct/adduct certain digits.

ACTIONS OF INTRINSIC MUSCLES: ON THE THUMB *

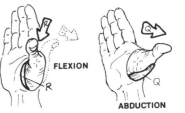

FLEXION   **ABDUCTION**   **ADDUCTION**

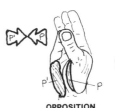

OPPOSITION   **CIRCUMDUCTION**

ON THE FINGERS *

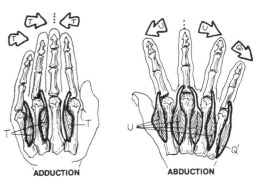

ADDUCTION   **ABDUCTION**

CN: Color all the muscles in a single group of each view before going on to the next group.

The muscles shown are the superficial muscles of the upper limb, many of which can be felt or seen on yourself. Try to find the outline and bulk of as many of these muscles as you can as you color each group.

MUSCLES ACTING ON THE SCAPULA A
MUSCLES ACTING ON THE SHOULDER JOINT B
FLEXORS OF THE ELBOW JOINT C
EXTENSORS OF THE ELBOW JOINT D
FLEXORS OF WRIST, HAND, & FINGERS E
EXTENSORS OF WRIST, HAND, & FINGERS F
FOREARM MUSCLES ACTING ON THE THUMB G
THENAR & HYPOTHENAR MUSCLES H
OTHER MUSCLES ACTING ON THUMB & FINGERS I

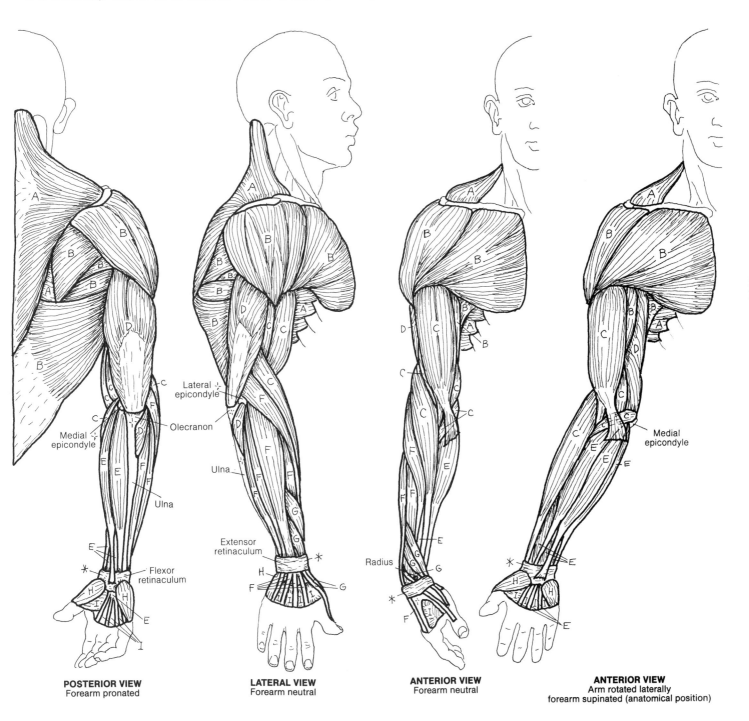

POSTERIOR VIEW
Forearm pronated

LATERAL VIEW
Forearm neutral

ANTERIOR VIEW
Forearm neutral

ANTERIOR VIEW
Arm rotated laterally
forearm supinated (anatomical position)

CN: Note in the two superficial views that the upper part of the iliotibial tract (*¹, title in the left, lower corner), normally covering gluteus medius (B), has been cut away. (1) Color each muscle in all views, including the directional arrows, before going on to the next one. The site of origin of the piriformis muscle (E) on the anterior sacrum cannot be seen at lower right. The origin of the obturator internus (F) on the right cannot be seen, but the origin of the muscle on the contralateral side can be colored. See plate 44 for additional views of these muscles.

GLUTEUS MAXIMUS A
GLUTEUS MEDIUS B
GLUTEUS MINIMUS C
TENSOR FASCIAE LATAE D

6 DEEP, LATERAL ROTATORS: *
PIRIFORMIS E
OBTURATOR INTERNUS F
OBTURATOR EXTERNUS G
QUADRATUS FEMORIS H
GEMELLUS SUPERIOR I
GEMELLUS INFERIOR J

The gluteal muscles are arranged in three layers: the most superficial is *gluteus maximus*. The large sciatic nerve runs deep to it, as every student nurse has learned well. Its thickness varies. Gluteus maximus extends the hip joint during running and walking up-hill, but does not act in relaxed walking. The inter-mediately placed, more lateral *gluteus medius* is a major abductor of the hip joint and an important sta-bilizer (leveler) of the pelvis when the opposite lower limb is lifted off the ground.

The deepest layer of gluteal muscles is the *gluteus minimus* and the *lateral rotators* of the hip joint. They cover up/fill the greater and lesser sciatic notches. These muscles generally insert at the posterior aspect of the greater trochanter of the femur. The gluteal muscles (less gluteus maximus) correspond to some degree with the rotator cuff of the shoulder joint: lateral rotators posteriorly, abductor (gluteus medius) superiorly, medial rotators (gluteus medius and minimus, tensor fasciae latae) anteriorly.

ILIOTIBIAL TRACT *¹

The *iliotibial tract*, a thickening of the deep fascia (fascia lata) of the thigh, runs from ilium to tibia and helps stabilize the knee joint laterally. The muscle *tensor fasciae latae*, a frequently visible and palpable flexor and medial rotator of the hip joint, inserts into this fibrous band, tensing it.

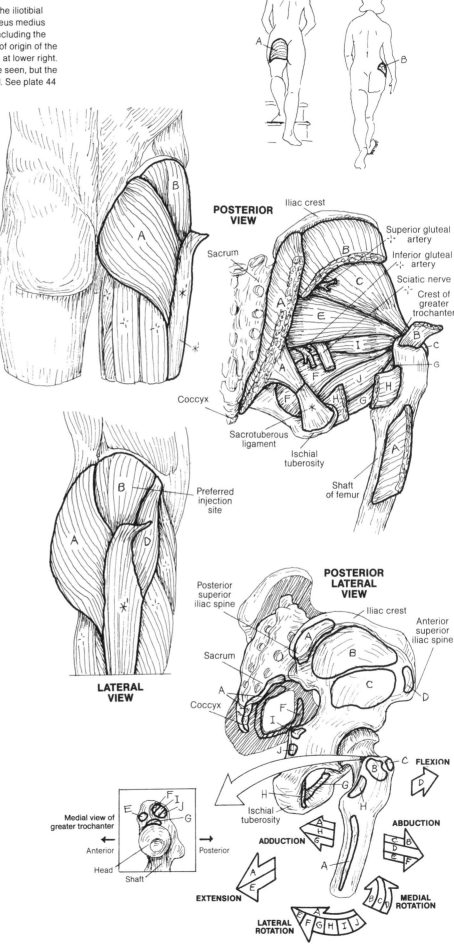

MUSCLES OF POSTERIOR THIGH

HAMSTRINGS:* SEMIMEMBRANOSUS_A SEMITENDINOSUS_B BICEPS FEMORIS_C

CN: (1) Color each hamstring muscle in the deep view before going on to the superficial. Then color the diagrams of flexion and extension.
(2) Color gray the outline of the muscles in the drawings at upper right.

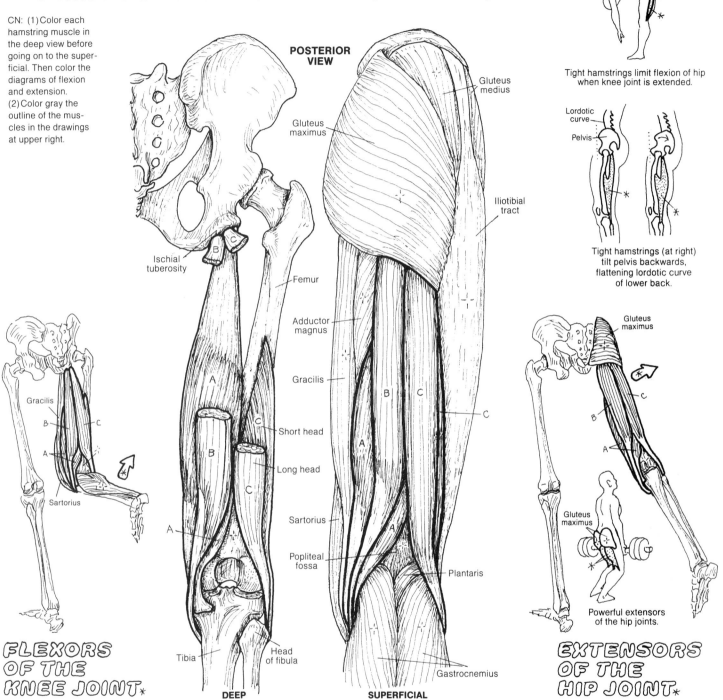

POSTERIOR VIEW

Gluteus medius
Gluteus maximus
Iliotibial tract
Ischial tuberosity
Femur
Adductor magnus
Gracilis
Short head
Long head
Sartorius
Popliteal fossa
Plantaris
Gastrocnemius
Tibia
Head of fibula

DEEP   **SUPERFICIAL**

Gracilis
Sartorius

FLEXORS OF THE KNEE JOINT*

Tight hamstrings limit flexion of hip when knee joint is extended.

Lordotic curve
Pelvis

Tight hamstrings (at right) tilt pelvis backwards, flattening lordotic curve of lower back.

Gluteus maximus
Gluteus maximus

Powerful extensors of the hip joints.

EXTENSORS OF THE HIP JOINT*

The *muscles of the posterior thigh* (called hamstrings after a procedure for cutting the tendons of these muscles in certain domestic animals) are equally effective at both extension of the hip joint and flexion of the knee joint; contraction of antagonists can isolate one or the other joint movement. Unlike the hip extensor gluteus maximus, the hamstrings are active during normal walking. In relaxed standing, the hamstrings (and gluteus maximus) are inactive. In knee flexion, the hamstrings act in concert with sartorius, gracilis, and gastrocnemius.

Reduced hamstring stretch ("tight hamstrings") limits hip flexion with the knee extended; flexion of the knee permits increased hip flexion. Try this on yourself. Tight hamstrings, by their ischial origin, pull the posterior pelvis down, lengthening the erector spinae muscles, and flattening the lumbar lordosis, potentially contributing to limitation of lumbar movement and back pain. The long tendons of the hamstrings can be easily felt just above the partially flexed knee.

PECTINEUS A
ADDUCTOR BREVIS B
ADDUCTOR LONGUS C
ADDUCTOR MAGNUS D
GRACILIS E
OBTURATOR EXTERNUS F

CN: Color one muscle at a time in the five main views before going to the next one. Note that the attachment sites on the posterior surface of the femur are represented by dotted lines.

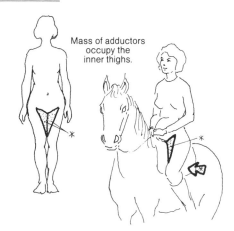

Mass of adductors occupy the inner thighs.

ANTERIOR VIEW

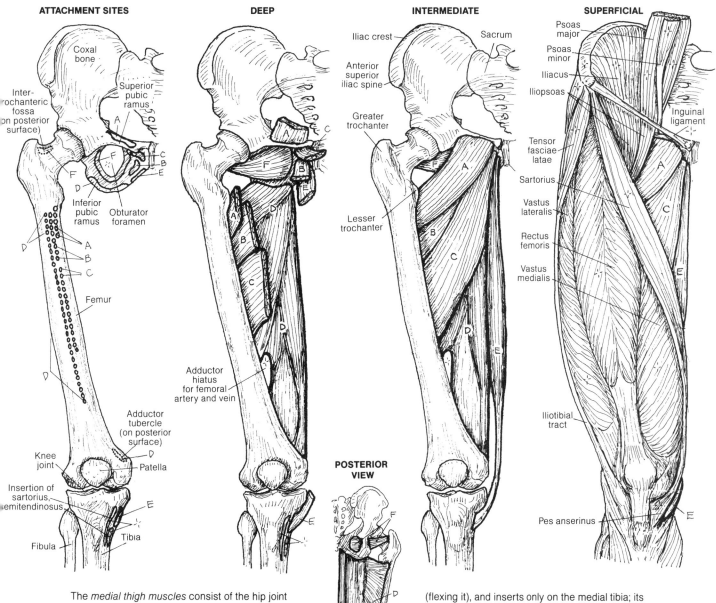

ATTACHMENT SITES

Coxal bone
Inter-trochanteric fossa (on posterior surface)
Superior pubic ramus
Inferior pubic ramus
Obturator foramen
Femur
Adductor tubercle (on posterior surface)
Knee joint
Patella
Insertion of sartorius, semitendinosus
Fibula
Tibia

DEEP

Adductor hiatus for femoral artery and vein

POSTERIOR VIEW

Linea aspera
Adductor hiatus

INTERMEDIATE

Iliac crest
Sacrum
Anterior superior iliac spine
Greater trochanter
Lesser trochanter

SUPERFICIAL

Psoas major
Psoas minor
Iliacus
Iliopsoas
Inguinal ligament
Tensor fasciae latae
Sartorius
Vastus lateralis
Rectus femoris
Vastus medialis
Iliotibial tract
Pes anserinus

The *medial thigh muscles* consist of the hip joint *adductors* (A through E) and *obturator externus*, a lateral rotator of that joint. The latter was colored on Plate 53 as one of the deep gluteal muscles, as its tendon passes into that region. However, it is compartmentalized by fasciae in the medial thigh, covers the external surface of the obturator foramen in the deep upper medial thigh, and receives the same innervation as the adductors. The *gracilis* is the longest of the adductor group, crosses the medial knee (flexing it), and inserts only on the medial tibia; its tendon joins the tendons of sartorius and semitendinosus to form an insertion shaped like a goose's foot (hence called the pes anserinus). The *adductor magnus* is the most massive of the group (see posterior view). In its lower half, adductor magnus fibers give way to the passage of the femoral vessels (adductor hiatus). All the adductors, except gracilis, insert on the vertical rough line (linea aspera) on the posterior surface of the femur.

MUSCLES OF ANTERIOR THIGH

SARTORIUS ᴀ
QUADRICEPS FEMORIS:
RECTUS FEMORIS ʙ
VASTUS LATERALIS c
VASTUS INTERMEDIUS ᴅ
VASTUS MEDIALIS ᴇ
ILIOPSOAS ꜰ

PATELLAR LIGAMENT ɢ✱

CN: The patellar ligament (G) is colored gray but the patella is left uncolored.
(1) Begin with the deep view of the thigh and then complete the superficial view.
(2) On the far left, color the portions of the quadriceps that are antagonists to the hamstring group. (3) Complete the action diagrams along the right margin.

FLEXORS OF THE HIP JOINT ✱

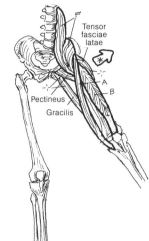

Tensor fasciae latae
Pectineus
Gracilis

ANTERIOR VIEW

DEEP

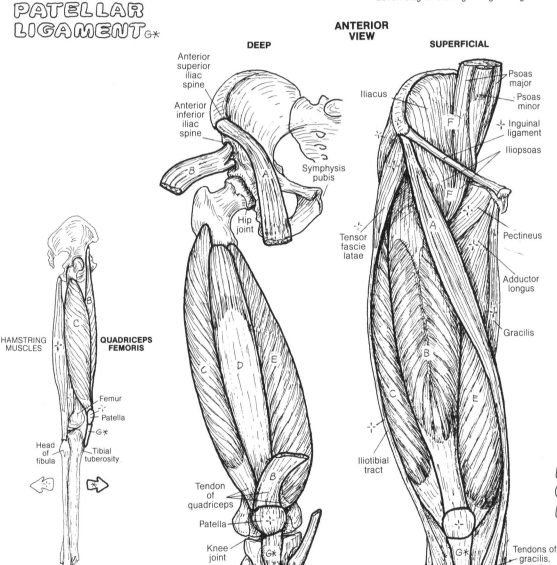

Anterior superior iliac spine
Anterior inferior iliac spine
Symphysis pubis
Hip joint

SUPERFICIAL

Iliacus
Psoas major
Psoas minor
Inguinal ligament
Iliopsoas
Tensor fascie latae
Pectineus
Adductor longus
Gracilis
Iliotibial tract
Tendons of gracilis, semi-tendinosus

HAMSTRING MUSCLES
QUADRICEPS FEMORIS
Femur
Patella
G✱
Head of fibula
Tibial tuberosity

Tendon of quadriceps
Patella
Knee joint
Tibial tuberosity

LATERAL VIEW

FLEXOR OF THE KNEE JOINT ✱

EXTENSORS OF THE KNEE JOINT ✱

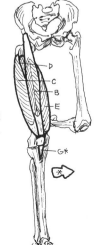

The *sartorius* ("tailor's" muscle; so-called because of the role of this muscle in enabling a crossed-legs sitting posture) is a flexor and lateral rotator of the hip joint, and a flexor of the knee joint, as you can infer from its illustrated attachments. The *quadriceps femoris* muscle arises from four heads. The *vastus medialis* and *lateralis* arise from the linea aspera on the posterior aspect of the femur; the *vastus intermedius* arises from the anterior femoral shaft. All four converge on to the superior aspect (base) of the patella to form the patellar tendon. Some tendon fibers continue over the patellar surface to join the ligament below. At the inferior aspect (apex) of the patella, the tendinous fibers continue to the tibial tuberosity.

The tendon between the patella and the tibial tuberosity is called the *patellar ligament*. *Rectus femoris* is a strong hip joint flexor, and is the only member of quadriceps to cross that joint. Quadriceps femoris is the only knee extensor. The significance of its role becomes crystal clear to those having experienced a knee injury; the muscles tend to atrophy and weaken rapidly with disuse, and "quad" exercises are essential to maintain structural stability of the joint.

The *iliopsoas* is the most powerful flexor of the hip, having a broad thick muscle belly and attaching at the lesser trochanter at the proximal end of the femoral shaft. Recall Plate 42 for its posterior abdominal origin.

CN: Begin with the attachment sites of the anterior leg. Note that the muscles A, B, and C arise from the interosseous ligament as well as the tibia and the fibula. Do not color the attachment sites of the thigh muscles (in small italics). Attachment sites on the plantar surface of the foot are shown at upper right.

The muscles of the leg are arranged into anterior-lateral, lateral, and posterior compartments. The bony ridge (anterior margin) of the tibia creates two oblique surfaces the anterolateral of which relates to the anterior leg muscles; the anteromedial surface is bony (ouch!) and devoid of muscle. The lateral compartment muscles largely arise from the fibula and the interosseous ligament between tibia and fibula.

ANTERIOR LEG:*
TIBIALIS ANTERIOR_A
EXTENSOR DIGITORUM LONGUS_B
EXTENSOR HALLUCIS LONGUS_C
PERONEUS TERTIUS_D

All of the anterior leg muscles are dorsiflexors (extensors) of the ankle; *extensors hallucis and digitorum longus* are toe extensors; *tibialis anterior* is an invertor of the subtalar joints as well, and *peroneus tertius* (the 5th tendon of extensor digitorum) is an evertor. Due to rotation of the lower limb during embryonic development, these extensors are anterior to the bones in the anatomical position (unlike the upper limb wrist extensors). Tibialis anterior is particularly helpful in lifting the foot up during the swing phase of walking to avoid striking the toes.

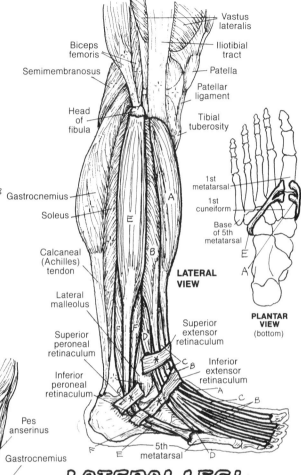

LATERAL VIEW

PLANTAR VIEW (bottom)

LATERAL LEG:*
PERONEUS LONGUS_E
PERONEUS BREVIS_F

The *peroneal muscles* are principally evertors of the foot, and are especially active during plantar flexion, as in walking on the toes or pushing off with the great toe. Peroneus tertius arises in the peroneal compartment but is actually part of extensor digitorum.

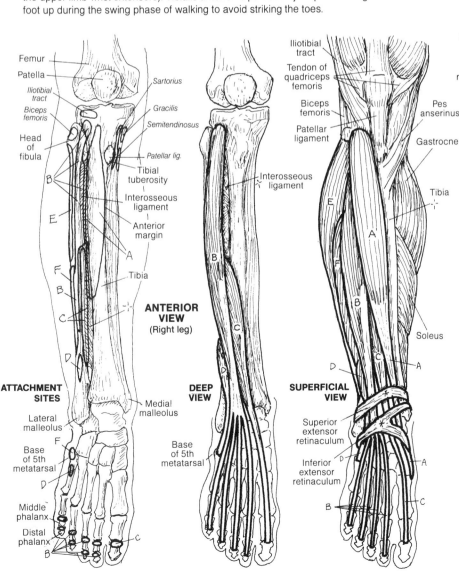

ATTACHMENT SITES

ANTERIOR VIEW (Right leg)

DEEP VIEW

SUPERFICIAL VIEW

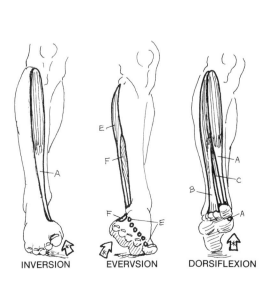

INVERSION   **EVERVSION**   **DORSIFLEXION**

TIBIALIS POSTERIOR G
FLEXOR DIGITORUM LONGUS H
FLEXOR HALLUCIS LONGUS I
POPLITEUS J
PLANTARIS K
SOLEUS L
GASTROCNEMIUS M

CN: The muscles to be colored on this plate are labeled G-M; any other letter label found here (A–F from Pl. 57; N–Y from Pl. 59) is for identification only, and those muscles should be left uncolored. You may repeat colors used for muscles on Plate 57 on this and/or the next plate. (1) Color one muscle at a time in each of the posterior views. Note that the plantaris (K), the soleus (L), and the gastrocnemius (M) all insert into the same tendon (tendocalcaneus) which receives the color M. (2) Color the upper and lower medial views.

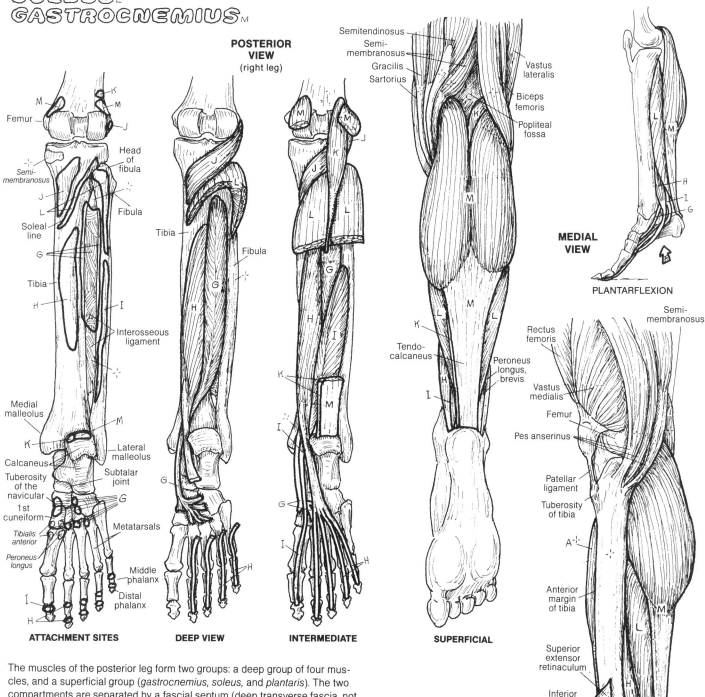

POSTERIOR VIEW (right leg)

ATTACHMENT SITES

DEEP VIEW

INTERMEDIATE

SUPERFICIAL

MEDIAL VIEW

PLANTARFLEXION

The muscles of the posterior leg form two groups: a deep group of four muscles, and a superficial group (*gastrocnemius, soleus,* and *plantaris*). The two compartments are separated by a fascial septum (deep transverse fascia, not shown). The fascial compartments are fairly non-expandable; muscle swelling secondary to vascular insufficiency may result in serious muscle compression/muscle death (compartment syndrome) without fascial decompression.

The major calf muscle is *gastrocnemius* which flexes the knee and, with its two fellows, plantarflexes the ankle joint. In knee flexion it is aided by *popliteus* which also rotates the tibia medially. The other deep flexors plantarflex the ankle joint (both toe and great toe *flexors* and *tibialis posterior*), flex the toes (the flexors), and invert the foot (tibialis posterior).

CN: Feel free to use the colors used for the letter labels on plates 57 and 58. Those letters are presented here for identification, and the muscles they refer to are not meant to be colored. Also note that plantar surface attachment sites for those extrinsic foot muscles have been omitted in the illustration of the fourth layer but can be found on the two preceding plates. (1) Begin with the fourth layer and complete each illustration before going on to the next.

The dorsal intrinsic muscles of the foot (those that arise and insert within the dorsum of the foot) are limited to two *small extensors* of the toes shown at right, most of the extensor function being derived from extrinsic extensors.

The intrinsic muscles of the plantar region of the foot are shown here in four layers. The *plantar interossei*, wedged between the metatarsal bones, constitute the deepest (4th) layer. They adduct toes 3-5, flex the metatarsophalangeal (MP) joints of these toes, and contribute to extension of the interphalangeal (IP) joints of these toes through the mechanism of the extensor expansion. The *dorsal interossei* abduct toes 3-5, and facilitate the other actions of the plantar interossei.

The third layer of muscles acts on the great toe (hallux) and 5th digit (digiti minimi). The second layer includes the *quadratus plantae*, inserting into the lateral border of the common tendon (H) of flexor digitorum longus (FDL). It assists that muscle in flexion of the toes. The *lumbricals* arise from the individual tendons of FDL and insert into the medial aspect of the extensor expansion (dorsal aspect). They flex the MP joints, and extend the IP joints of toes 2-5 via the extensor expansion.

The superficial (first) layer consists of the *abductors* of the 1st and 5th digits, and the *flexor digitorum brevis*. The plantar muscles are covered by the thickened deep fascia of the sole, the plantar aponeurosis, extending from calcaneus to the fibrous sheath of the flexor tendons.

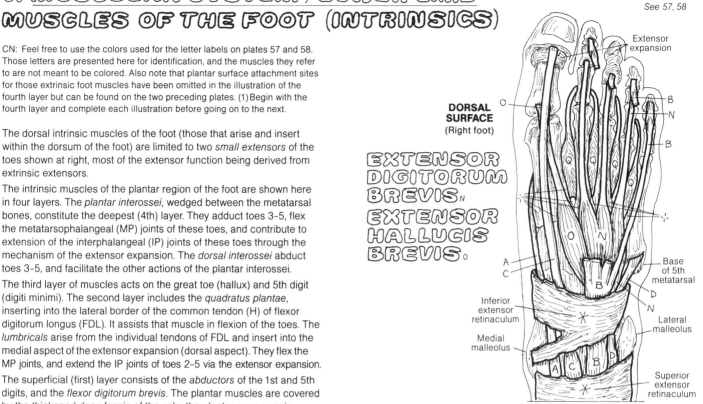

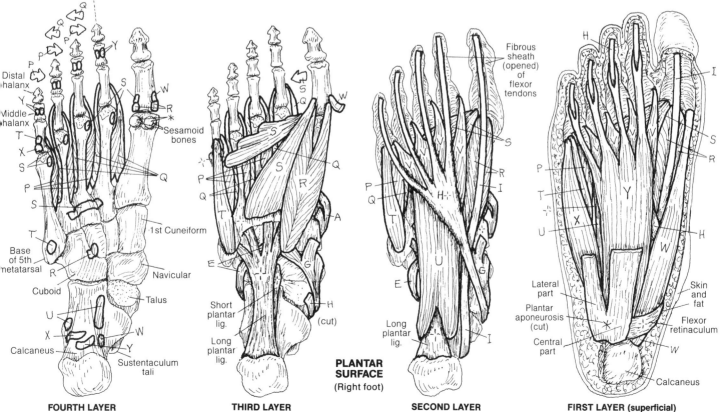

FOURTH LAYER

THIRD LAYER

SECOND LAYER

FIRST LAYER (superficial)

SUMMARY OF MUSCLE GROUPS

ABDUCTORS OF THE HIP_A
ADDUCTORS OF THE HIP_B
FLEXORS OF THE HIP_C
EXTENSORS OF THE HIP_D
FLEXORS OF THE KNEE_E
EXTENSORS OF THE KNEE_F
PLANTAR FLEXORS OF ANKLE & FOOT_G
DORSAL FLEXORS OF ANKLE & FOOT_H
INVERTORS OF THE ANKLE_I
EVERTORS OF THE ANKLE_J
INTRINSIC MUSCLES OF THE FOOT_K

CN: Color one group of muscles in as many views as it appears before going on to the next. Most of the intrinsic muscles of the foot (K) are on the plantar surface and are not shown here.

Some of the muscles shown here cross two joints and/or have more than one function; the primary functions of the muscles are labeled here, as in Plate 52 (summary of upper limb muscles).

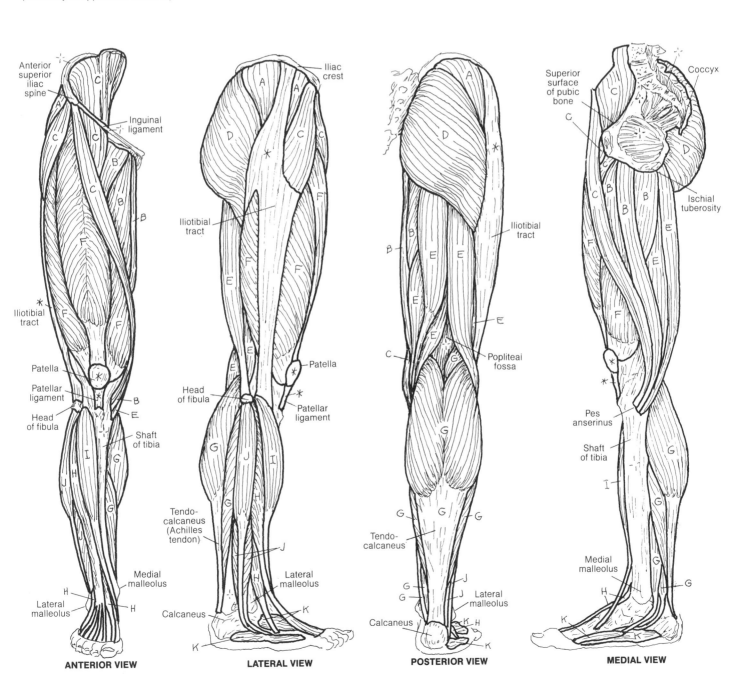

ANTERIOR VIEW **LATERAL VIEW** **POSTERIOR VIEW** **MEDIAL VIEW**

FUNCTIONAL OVERVIEW

FLEXOR_A
EXTENSOR_B
ABDUCTOR_C
ADDUCTOR_D
ROTATOR_E
SCAPULAR
STABILIZER_F
EVERTOR_G
INVERTOR_H

CN: Use light colors throughout (especially for A and B). Deeper muscles are not included in the large illustrations. (1) Color all of the muscle groups in the anterior view before going on to the posterior view at right. Only the muscles on one side of the figure have been labeled. As you color the muscle, also color its opposite. (2) Color the small diagram below.

Upon coloring these functional groups, note the spatial relationship of adductors to abductors and evertors to invertors. Take particular note of the extensors and flexors. Recall that extension of weightbearing joints is an anti-gravity function, and extensor muscles of these joints tend to keep the standing body vertically straight. Note the line of gravity and its relationship to the vertebral, hip, knee, and ankle joints. The center of gravity of an average human being standing with perfect posture is just anterior to the motion segment of S1–S2. Flexion of the neck and torso moves the center of gravity forward, loading the posterior cervical, thoracic, and lumbar paraspinal (extensor) muscles. The actors moving the vertebral, hip, knee, and ankle joints make possible erect standing and walking/running posture.

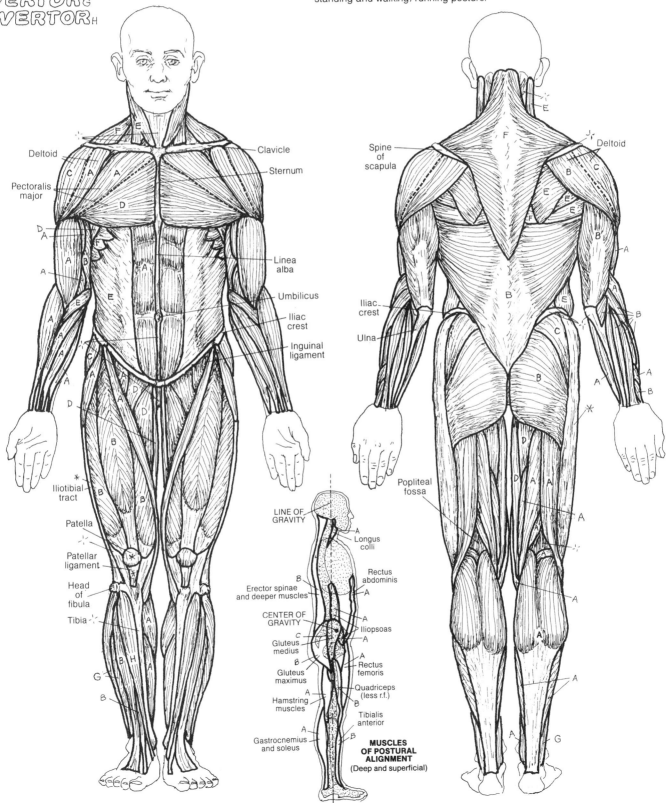

Deltoid
Pectoralis major
Clavicle
Sternum
Linea alba
Umbilicus
Iliac crest
Inguinal ligament
Iliotibial tract
Patella
Patellar ligament
Head of fibula
Tibia

Spine of scapula
Deltoid
Iliac crest
Ulna
Popliteal fossa

LINE OF GRAVITY
Longus colli
Rectus abdominis
Erector spinae and deeper muscles
CENTER OF GRAVITY
Iliopsoas
Gluteus medius
Rectus femoris
Gluteus maximus
Quadriceps (less r.f.)
Hamstring muscles
Tibialis anterior
Gastrocnemius and soleus

MUSCLES OF POSTURAL ALIGNMENT
(Deep and superficial)

CN: Color these subscripts: DP = dark purple, LP = light purple, O = orange, P = pink, PB = pale blue, PO = pale orange, R = red, and T = tan. These colors match the stains usually employed to observe these cells. (1) Color the test tube (use any color for B and C). (2) In the eosinophil and basophil (bottom left), first color the cytoplasm and then stipple in the rest.

ERYTHROCYTES R
(RED BLOOD CORPUSCLES)

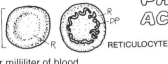

6-8 μm   RETICULOCYTE

Erythrocytes (RBCs; approximately 4.5-6.2 million per milliliter of blood in men; 4-5.5 million/ml in women) are formed in the bone marrow where they lose their nucleus and most of their organelles prior to their release into the peripheral blood. Recently released immature erythrocytes (reticulocytes) may retain some ribosomal RNA in their cytoplasm; these granules appear dark purple and reticulated when stained. Normally making up about 1% of the RBC population, reticulocytes may increase in number during chronic oxygen lack (e.g., as at prolonged high altitude). The circulating RBC (without nucleus or organelles, it is truly a corpuscle and not a cell) is a non-rigid, bi-concave-shaped, membrane-lined sac of hemoglobin, a large iron-containing protein. Hemoglobin (12-16 grams/decaliter of blood in women; 14-18 g/dL in men) has a powerful affinity for oxygen, and is the principal carrier of oxygen in the body; plasma is the other carrier. Erythrocytes pick up oxygen in the lungs and release it in the capillaries to the tissues/cells. RBCs circulate for about 120 days until defective, and are then taken out of the blood and broken down by cells of the spleen.

THROMBOCYTES LP
(PLATELETS)

Thrombocytes or platelets (150,000-400,000/ml of blood; 2-5 μm in diameter) are small bits of cytoplasm from giant cells (megakaryocytes) of the bone marrow. Circulating in the blood for a lifetime of 10 days or so, platelets adhere to injured endothelium and play a significant role in limiting hemorrhage (aggregation of platelets, blood coagulation/clotting, and clot removal).

PLASMA (55%) T
WATER (90%) A
PROTEINS (8%) B
ACIDS, SALTS (2%) C

FORMED ELEMENTS OF THE BLOOD (45%) *
ERYTHROCYTES (99%) R
THROMBOCYTES (0.6-1.0%) LP
LEUKOCYTES (0.2%) D

All constituents of the blood that can be observed as discrete structures with the aid of the light microscope are called *formed elements of the blood.* The rest of the blood is a protein-rich fluid called *plasma.* When blood is allowed to clot, the cells disintegrate (hemolysis) and a thick yellow fluid emerges called serum. Serum is basically plasma less clotting elements. If whole blood is centrifuged in a test tube, the RBCs will settle to the bottom, the *leukocyte fraction* will form a buffy coat on top of that, and the *plasma,* being the lightest, will take up the upper 55% of the total volume. Packed RBC's in a test tube constitute a *hematocrit* (40-52% of the blood volume in men; 37-47% in women). The difference in blood values between men and women is probably related to iron storage and metabolism differences (men store up to 50% more iron than women). A low hematocrit may be associated with anemia or hemorrhage.

LEUKOCYTES D
(WHITE BLOOD CELLS)

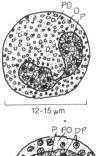

Cytoplasm
Nucleus
Granules
LP
P

SEGMENTED (MATURE)

LP
P

BAND (IMMATURE)

GRANULAR *
NEUTROPHIL

Segmented neutrophils (55-75% of the WBC population) arise in the bone marrow and live short lives in the blood and connective tissues (hours-4 days). Immature forms (*band neutrophils,* 1-5%) may be seen in the blood; their numbers often increase in acute infections. Neutrophils rapidly engulf foreign elements/cellular debris; strong enzymes in specific granules and lysosomes destroy them (phagocytosis).

EOSINOPHIL

PO O P

12-15 μm

Eosinophils (1-3% of WBCs) exhibit colorful granules when properly stained. Eosinophils are phagocytic in immune reactions. They are involved in the late-onset phase of asthmatic attacks (subsequent bronchial constriction), possibly enhancing cell injury by increasing cell membrane permeability in the bronchial mucosa to allergic substances. They also appear to limit the expression of mast cell degranulation (histamine release and effects) during allergic reactions.

BASOPHIL

P PO DP

Basophils (0-1% of WBCs) contain dark-staining granules. Basophils are known to degranulate in allergic reactions, releasing histamine, serotonin, and heparin. Such degranulation induces contraction of smooth muscle, increases vascular permeability (enhancing the effects of inflammation), and slows down movement of white blood cells in inflammation.

NONGRANULAR
LYMPHOCYTE

PB P

6-18 μm

Lymphocytes (20-45% of WBCs) arise from the bone marrow, and reside in the blood as well as the lymphoid tissues (lymph nodes, thymus, spleen, and so on). Lymphocytes generally consist of about 20% B cells (short-lived cells from the bone marrow and concerned with humoral immunity, transformation into plasma cells and the secretion of antibodies or immunoglobulins) and 70% T cells (long-lived cells from the thymus; may be cytotoxic, helper, or suppressor cells associated with cell-mediated immunity). Lymphocytes with neither B or T surface antigens (less than 5%) are called natural killer cells.

MONOCYTE

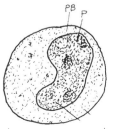

PB P

12-20 μm

Monocytes (2-8% of WBCs) arise in the bone marrow, mature in the blood (about eight hours), then leave the circulation to enter the extracellular spaces as macrophages. They are critical to the functioning of the immune system, as they present antigen to the immune cells, secrete substances in immune reactions, and destroy antigens (see glossary). They phagocytose cellular and related debris in wound healing, bone formation, and multiple other cellular activities where breakdown occurs.

CN: Use blue for A, purple for B, red for C, and very light colors for D and E. (1)Color the titles for systemic and pulmonary circulation; the two figures; and the borders bracketing the large illustration. Also color purple (representing the transitional state between oxygenation and deoxygenation) the two capillaries, demonstrating the difference between capillary function in the lungs versus the body. (2)Begin in the right atrium of the heart and color the flow of deoxygenated blood (A) into the lungs. After coloring the pulmonary capillary network (B), color the oxygenated blood (C) that re-enters the heart and is pumped into and through the systemic circuit.

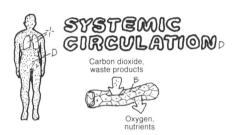

SYSTEMIC CIRCULATION D

Carbon dioxide, waste products

Oxygen, nutrients

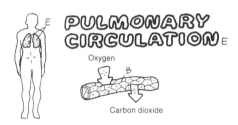

PULMONARY CIRCULATION E

Oxygen

Carbon dioxide

DEOXYGENATED BLOOD A
CAPILLARY BLOOD B
OXYGENATED BLOOD C

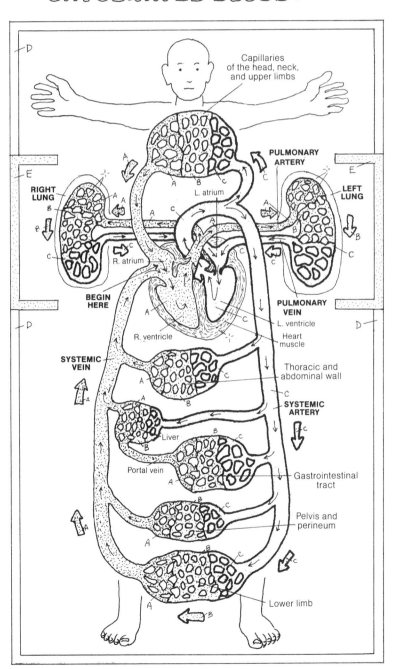

Circulation of blood begins with the heart which pumps blood into arteries and receives blood from veins. *Arteries* conduct blood *away* from the heart regardless of the amount of oxygen (oxygenation) in that blood. *Veins* conduct blood toward the heart, regardless of the degree of oxygenation of the blood. *Capillaries* are networks of extremely thin-walled vessels throughout the body tissues that permit the exchange of gases and nutrients between the vessel interior (vascular space) and the area external to the vessel (extracellular space). Capillaries receive blood from small arteries and conduct blood to small veins.

There are two circuits of blood flow: (1) the pulmonary circuit, which conveys *deoxygenated blood* from the right side of the heart to the lungs and freshly *oxygenated blood* back to the left side of the heart, and (2) the systemic circuit, which conveys oxygenated blood from the left heart to the body tissues and returns deoxygenated blood to the right heart. The color red is used universally for *oxygenated blood*; the color blue is used for *deoxygenated blood*.

Clearly, not all arterial blood is oxygenated (in the pulmonary circulation, arteries conduct deoxygenated blood to the lungs), and not all venous blood is deoxygenated (pulmonary veins conduct oxygenated blood to the heart).

Capillary blood is mixed; it is largely oxygenated on the arterial side of the capillary bed, and it is largely deoxygenated on the venous side, as a consequence of delivering oxygen to and picking up carbon dioxide from the tissues it supplies.

One capillary network generally exists between an artery and a vein; an exception is the portal circulation characterized by two capillary sets between artery and vein. The vein between the two networks is the portal vein. Such can be seen between the gastrointestinal tract and the liver.

BLOOD VESSELS

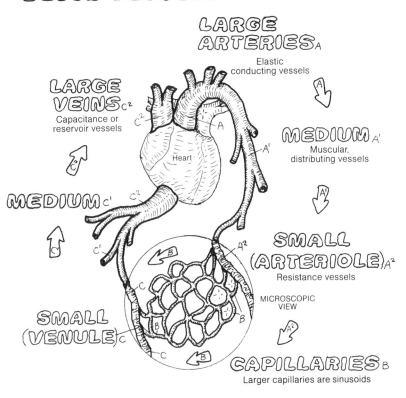

LARGE ARTERIESₐ
Elastic conducting vessels

LARGE VEINS C²
Capacitance or reservoir vessels

MEDIUM A'
Muscular, distributing vessels

MEDIUM C'

Heart

SMALL (ARTERIOLE) A²
Resistance vessels

MICROSCOPIC VIEW

SMALL (VENULE) C

CAPILLARIES B
Larger capillaries are sinusoids

CN: Use red for A, purple for B, blue for C, and very light colors for D, F, and H. (1) Complete the upper left diagram, beginning with the large arteries. (2) Color the blood vessels and their titles at the bottom of the plate. Note that the vas and nervus vasorum in the fibrous tissue layer (H) are not colored. (3) In the diagram of venous valve action, the blood in both vein and artery is colored gray.

Large arteries (elastic or conducting arteries), such as the aorta or common carotid, contain multiple layers of elastic tissue. They are roughly the size of a finger. *Medium arteries* (muscular, distributing arteries), averaging the size of a pencil, are generally named (e.g., brachial). Diminutive branches of medium arteries are called small arteries (*arterioles*); unnamed, they control the flow of blood into capillary beds (resistance vessels). *Capillaries* are unnamed simple endothelial tubes supported by thin fibrous tissue. Microscopic in dimension, some capillaries are larger (sinusoids) or more specialized than others.

Veins get progressively larger as they get closer to the heart. Veins have tributaries; except in portal circulations, they do not have branches. *Venules* (small veins) are formed by the merging of capillaries and are basically of the same construction. Venules merge to form *medium veins* and these are the tributaries of *large veins* (capacitance or reservoir vessels). Certain specialized large veins, as in the skull, are called sinuses. The walls of these veins are thinner than those of their arterial counterparts, and their lumens are generally larger. Large veins can stretch significantly, becoming virtual reservoirs of blood.

All vessels demonstrate a simple squamous epithelial (*endothelial*) lining (tunica interna) supported by a thin layer of fibrous tissue (not shown). Most medium veins of the neck and extremities have a series of small pockets formed from the endothelial layer. These valves are paired and point in the direction of blood flow. Though offering no resistance to blood flow, they will bend into and close off the lumen of the vein when the flow of blood is reversed. Valves resist gravity-induced blood pooling, especially in the lower limb vessels. Venous flow here is enhanced by the contraction of skeletal muscles the bulges of which give an anti-gravity boost to the movement of blood. The *internal elastic lamina* is a discrete layer only in medium-sized arteries and assists in maintaining blood pressure; this tissue is more diffuse in other vessels. The *tunica media* consists of concentrically arranged smooth muscle fibers. It is well developed in medium arteries, least developed in veins. Medium arteries use this layer in distributing blood from one field to another. In arterioles, reduced to only one or two layers, the smooth muscle can literally block blood flow into capillary fields. The *external elastic lamina* exists as a discrete layer only in muscular arteries. The *tunica externa* (adventitia) is fibrous tissue contiguous with the fascial layer in which the vessel is located; within this tunic much smaller nutrient vessels (vasa vasorum) and motor nerves (nervi vasorum) are found.

STRUCTURE:

TUNICA INTERNA:✳
ENDOTHELIUM D
INTERNAL ELASTIC LAMINA E
TUNICA MEDIA:✳
SMOOTH MUSCLE F
EXTERNAL ELASTIC LAMINA G
TUNICA EXTERNA:✳
FIBROUS TISSUE H

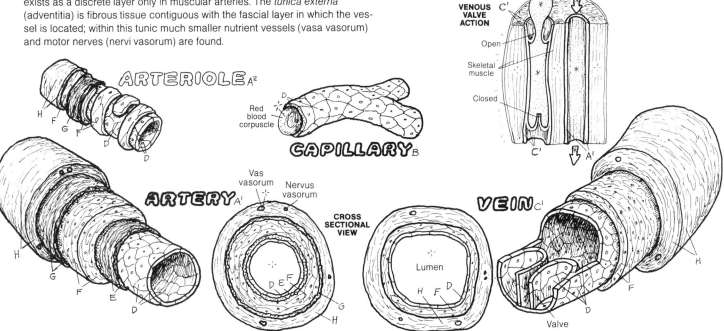

ARTERIOLE A²

CAPILLARY B
Red blood corpuscle

VENOUS VALVE ACTION
Open
Skeletal muscle
Closed

ARTERY A'
Vas vasorum
Nervus vasorum
CROSS SECTIONAL VIEW

VEIN C'
Lumen
Valve

MEDIASTINUM REGIONS:*

SUPERIOR A
INFERIOR: ÷
  ANTERIOR B
  MIDDLE C
  POSTERIOR D

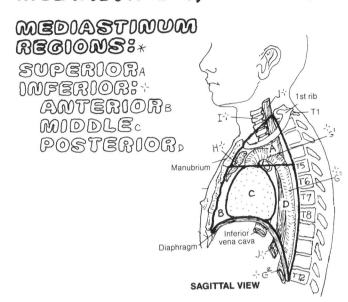

SAGITTAL VIEW

CN: Use blue for F, red for G, and your lightest colors for A–D. (1) Begin with the regions of the mediastinum at upper left and color over all the structures within the dark outline. (2) Color the major structures within the mediastinum in the anterior view. Note that the lungs, not being in the mediastinum, remain uncolored. Note that the thymus, which can be seen in the sagittal view, has been deleted here to show the great vessels covered by it. (3) Finally, color the walls of the heart and layers of pericardium at lower left. The pericardial cavity has been greatly exaggerated for coloring. It is normally only a potential space.

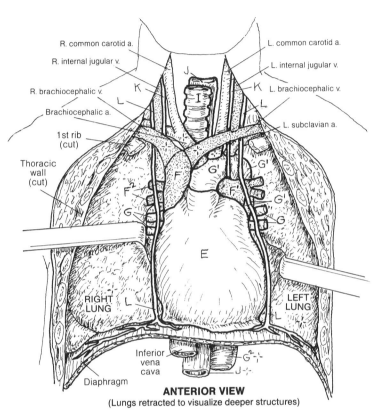

ANTERIOR VIEW
(Lungs retracted to visualize deeper structures)

STRUCTURES:*

PERICARDIUM-LINED HEART E
GREAT VESSELS: ÷
  SUPERIOR VENA CAVA F
  PULMONARY TRUNK F¹
  PULMONARY ARTERY F²
  PULMONARY VEIN G
  AORTIC ARCH G¹
  THORACIC AORTA G² ÷

THYMUS H ÷
TRACHEA I
ESOPHAGUS J
VAGUS NERVE K
PHRENIC NERVE L

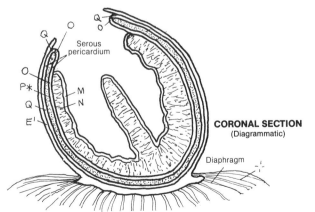

CORONAL SECTION
(Diagrammatic)

WALLS OF THE HEART/PERICARDIUM:*

ENDOCARDIUM M
MYOCARDIUM N
VISCERAL PERICARDIUM O
PERICARDIAL CAVITY P*
PARIETAL PERICARDIUM Q
FIBROUS PERICARDIUM E¹

The mediastinum (median septum or partition) is a highly populated region between and excluding the lungs. A variety of passageways, nerves, and vessels enter, pass through, and exit the mediastinum. For descriptive purposes, the mediastinum is divided into the subdivisions (regions) illustrated. The *superior mediastinum* is remarkable for the array of *great vessels* of the *heart*, and the *trachea, esophagus, vagus* and *phrenic nerves*. At the level of the T4–T5 vertebrae (superior/inferior mediastinal border), the trachea bifurcates into the main bronchi (see Plate 95) posterior to the great vessels, and the aorta makes its *arch*. The *posterior mediastinum* includes the inferior continuation of the esophagus embraced by a fine network of vagal nerve fibers, the thoracic duct (see Plate 83), and the descending (*thoracic*) aorta. The floor of the mediastinum is the diaphragm penetrated by the thoracic aorta, esophagus, and inferior vena cava.

The heart wall consists of an inner layer of simple squamous epithelium (*endocardium*) overlying a variably thick *myocardium* (cardiac muscle). External to the myocardium is a three-layered sac (*pericardium*). The innermost layer of this sac is the *visceral pericardium* (epicardium), clothing the heart. At the origin of the aortic arch, this layer turns (reflects) outward to become the *parietal pericardium* (imagine a fist clutching the edges around the opening of a paper bag; now push the fist into the closed bag still clutching the edges; as you do so, note that your fist becomes surrounded by two layers of the paper bag, yet is not inside the bag itself). The relationship of your fist to the two layers of the bag is the relationship of the heart to the visceral and parietal pericardium. The cavity of the bag is empty—the fist is not in the bag (if you did it right!). Similarly, the *pericardial cavity* between the two pericardial layers is empty as well, except for serous fluid that makes for friction-free movement of the heart in its sac.

The *fibrous pericardium* is the outer surface of the parietal pericardium; it is fibrous and fatty, and is strongly attached to the sternum, the great vessels, and the diaphragm. It keeps the twisting, contracting, squeezing heart within the middle mediastinum.

CHAMBERS OF THE HEART

CN: Use blue for A, red for H, and your lightest colors for B, C, I, and J.
(1) Begin with the four chambers of the heart, and follow the direction of blood flow as you color your way down the list of titles. Also color the directional arrows blue (dotted) and red; their titles are at lower right. (2) Color the circulation chart below, beginning with numeral one (1) in the right atrium. Color the arrows accordingly, along with the four numerals. Do not color the chambers or the vessels.

SUPERIOR VENA CAVA_A
INFERIOR VENA CAVA_{A'}
RIGHT ATRIUM_B
RIGHT VENTRICLE_C
  A-V TRICUSPID VALVE_D
    CHORDAE TENDINEAE_E
    PAPILLARY MUSCLE_F
PULMONARY TRUNK_{A²}
  PUL. SEMILUNAR VALVE_G
  PUL. ARTERY_{A³}
PULMONARY VEIN_H
LEFT ATRIUM_I
LEFT VENTRICLE_J
  A-V BICUSPID (MITRAL) VALVE_{D'}
    CHORDAE TENDINEAE_{E'}
    PAPILLARY MUSCLE_{F'}
ASCENDING AORTA_{H'}
  AORTIC SEMILUNAR VALVE_{G'}
  AORTIC ARCH_{H²}
  THORACIC AORTA_{H³}

The heart is the muscular pump of the blood vascular system. It contains four cavities (chambers): two on the right side (pulmonary heart), two on the left (systemic heart). The pulmonary "heart" includes the right atrium and right ventricle. The thin-walled *right atrium* receives *deoxygenated blood* from the *superior* and the *inferior vena cava*, and from the coronary sinus (draining the heart vessels). The thin-walled *left atrium* receives *oxygenated blood* from pulmonary veins. Atrial blood is pumped at a pressure of about 5 mm Hg into the *right and left ventricles* simultaneously through the atrioventricular orifices, guarded by the 3-cusp *tricuspid valve* on the right and the 2-cusp *bicuspid valve* on the left. The cusps are like panels of a parachute, secured to the *papillary muscles* in the ventricles by tendinous *chordae tendineae*. These muscles contract with the ventricular muscles, tensing the cords, and resisting cusp over-flap as ventricular blood bulges into them during ventricular contraction (systole). The ventricles, significantly more muscular than their fellow atria, pump deoxygenated blood to the lungs via the *pulmonary trunk* at a pressure of about 25 mm Hg (right ventricle), and into the *ascending aorta* at a pressure of about 120 mm Hg (left ventricle) simultaneously. This pressure difference is reflected in the thicker walls of the left ventricle compared to the right. The pocket-like *pulmonary and aortic semilunar valves* guard the trunk and aorta, respectively. As blood falls back toward the ventricle from the trunk/aorta during the resting phase (diastole), these pockets fill, closing off their respective orifices, and preventing reflux into the ventricles.

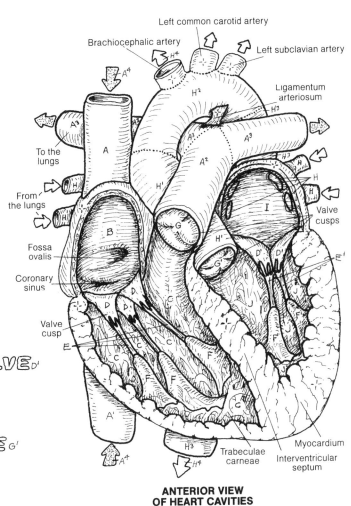

ANTERIOR VIEW OF HEART CAVITIES

CIRCULATION THROUGH THE HEART*

OXYGENATED BLOOD ➡ H⁴
DEOXYGENATED BLOOD ➡ A⁴

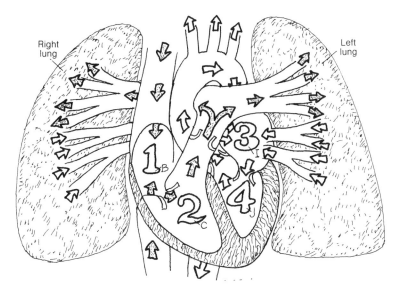

CARDIAC CONDUCTION SYSTEM & THE ECG

CN: Use blue for C and red for D. (1)Begin with the conduction system in the large illustration. (2)Color the waves of the ECG chart and the corresponding drawings illustrating simultaneous events in the heart. The T wave receives both blue and red as it represents the refilling of the atria with deoxygenated (right) and oxygenated blood (left). Be sure to color the large letters as well.

CONDUCTION SYSTEM: *
SA (SINOATRIAL) NODE A
INTERNODAL TRACT A'
AV (ATRIOVENTRICULAR) NODE B
AV BUNDLE B'
PURKINJE PLEXUS B²

Contraction of cardiac muscles occurs without motor nerves. Groups of specialized muscle cells conduct electrochemical impulses among the cardiac muscles. These cells constitute the cardiac conduction system. They are more excitable than their cardiac muscle counterparts; thus they are capable of controlling and synchronizing cardiac muscle contraction to permit the movement of blood from atria to ventricles and from ventricles to the pulmonary trunk/ascending aorta. The most excitable cells of the system are in the *sinoatrial (SA) node*. It is called the "pacemaker" as it sets the pace (or rate) of impulse conduction; the artificial replacement for a diseased, dysfunctional SA node is also called the pacemaker. Impulses travel the *internodal tracts* through the atrial wall to the *atrioventricular (AV) node* located in the lower interatrial septum; then through the *atrioventricular (AV) bundle* in the interventricular septum to and through the *Purkinje plexus* supplying the ventricular walls. The SA node fires about 45 times a minute. Each cycle of impulse conduction initiates a sequence of coordinated pumping actions by the heart cavities. Contraction of the ventricular muscles constitutes the heart beat, the rate of which is mediated by nerves of the autonomic nervous system.

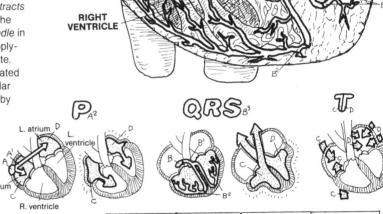

DEOXYGENATED BLOOD C
OXYGENATED BLOOD D

ELECTROCARDIOGRAM (ECG): *
P WAVE A²
QRS WAVE (COMPLEX) B³
T WAVE D C

The activity of the conduction system generates electrical activity about the heart, and this activity (specifically voltage changes) can be monitored, assessed, and measured by electrocardiograph machines. The recorded data (various waves of varying voltage over time) from such machines constitute the electrocardiogram (ECG). To gain these data, electrodes are placed on a number of body points on the skin. When the SA node fires, it causes the baseline voltage to drop (depolarize) over both atria, and this is reflected in the ECG by an upward deflection of the baseline voltage (a line produced on an oscilloscope or on a moving paper tape). This upward deflection is called the *P wave*, which causes the right and left atrial musculature to contract. The AV node takes about 0.1 second to excite the AV bundle, which then conducts the impulses to the Purkinje plexus. The electrical events here constitute the *QRS complex* of waves. As a result, the ventricles depolarize and the ventricular muscles contract. The *T wave* represents a period of recovery of baseline voltage (repolarization), following which the atria fill with blood from the vena cavae and the pulmonary veins. The ST segment is an especially important indicator of heart muscle function/dysfunction in certain electrode configurations. The shape and direction of wave deflection are dependent upon the spatial relationship of the electrodes (leads) on the body surface.

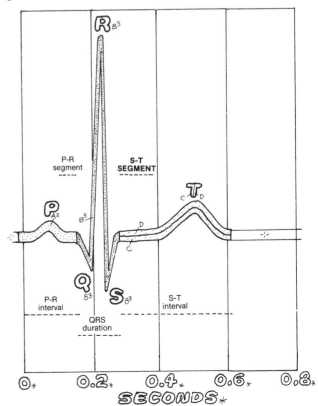

CORONARY ARTERIES & CARDIAC VEINS

CN: Use your brightest colors for A, D, and L. (1) When coloring the arteries, include the broken lines which represent vessels on the posterior surface of the heart. (2) Do the same with the veins. (3) Color the artery in front of the plaque in the circled view; color the vessel after the plaque a lighter shade of the same color or do not color it at all.

CORONARY ARTERIES: *
RIGHT CORONARY ARTERY A
MUSCULAR BRANCH A'
MARGINAL BRANCH B
POSTERIOR INTERVENTRICULAR (DESCENDING) BRANCH C
LEFT CORONARY ARTERY D
ANTERIOR INTERVENTRICULAR (DESCENDING) BRANCH E
MUSCULAR BRANCH E'
CIRCUMFLEX BRANCH F

The *coronary arteries* form an upside down crown (*L. corona*) about or just deep to the surface of the heart. The arteries lie in grooves or sulci, often covered over by the epicardium and sometimes the myocardium as well.

Both left and right arteries arise from small openings (aortic sinuses) just above the two aortic semilunar valve cusps. Generally, the left coronary artery is somewhat larger than the right; the flow rate through the left is greater in most people than that through during the cardiac cycle. There may be considerable variation in the anastomotic pattern of the left and right arterial branches. These branches terminate in multitudes of arterioles supplying the vast capillary network among the muscle fibers. The apparent multiple communications among the left and right coronary arteries notwithstanding, varying degrees of vascular insufficiency occur when there is significant obstruction of one or both coronary arteries. There is some extra-coronary arterial supply to the heart from the epicardial vessels (branches of internal thoracic arteries) and aortic vasa vasorum.

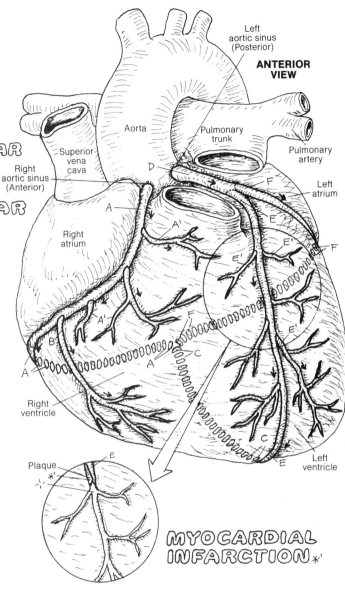

ANTERIOR VIEW

Aorta, Pulmonary trunk, Pulmonary artery, Left atrium, Left aortic sinus (Posterior), Superior vena cava, Right aortic sinus (Anterior), Right atrium, Right ventricle, Left ventricle, Plaque

MYOCARDIAL INFARCTION *'

Insufficient oxygenation (hypoxia) of the myocardium occurs with significantly reduced blood flow to the muscle (ischemia), often inducing sharp, crushing chest pain (angina pectoris). About 75% or more obstruction by atherosclerotic plaque (thrombi) can cause myocardial damage (myocardial infarction). Significant plaque in the artery to the sinoatrial node (not shown) can cause arrhythmias in the cardiac conduction system, often necessitating the installation of an artificial pacemaker in the chest.

CARDIAC VEINS: *
GREAT CARDIAC V. G
MIDDLE CARDIAC V. H
MARGINAL V. I
ANTERIOR CARDIAC V. J
SMALL CARDIAC V. K
CORONARY SINUS L

The *cardiac veins* travel with the coronary arteries, but incompletely. Vast anastomoses of veins occur throughout the myocardium; most drain into the right atrium by way of the *coronary sinus*. The *anterior cardiac veins* conduct blood directly into the right atrium. Other small veins may drain directly into the right atrium as well. Some deep (arteriosinusoidal) veins drain directly into the atria and ventricles. Extracardiac venous drainage can also occur through the vasa vasorum of the vena cavae.

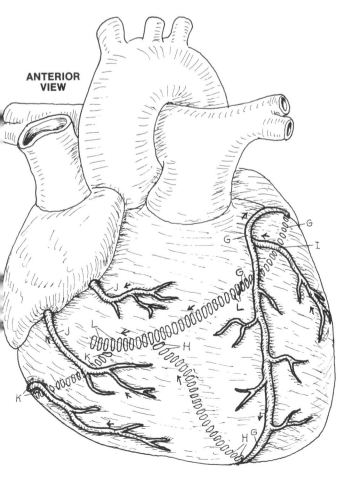

ANTERIOR VIEW

CN: Use red for A and dark or bright colors for B and L. (1) Begin with the brachiocephalic (A) and the right subclavian (B) and its branches. Color the broken lines which represent deeper vessels. (2) Do the same for the right common carotid (L) and its branches. (3) Color the arrows pointing to the four sites where the arterial pulse may be palpated.

BRACHIOCEPHALIC A

RIGHT SUBCLAVIAN B
INTERNAL THORACIC c
VERTEBRAL D
THYROCERVICAL TRUNK E
  INFERIOR THYROID F
  SUPRASCAPULAR G
  TRANSVERSE CERVICAL H
COSTOCERVICAL TRUNK I
  DEEP CERVICAL J
  HIGHEST INTERCOSTAL K

RIGHT COMMON CAROTID L
INTERNAL CAROTID M
  OPHTHALMIC N
EXTERNAL CAROTID O
  SUPERIOR THYROID P
  LINGUAL Q
  FACIAL R
  OCCIPITAL S
  MAXILLARY T
    ALVEOLAR BRANCHES: U
     INF. U SUP. U'
    MIDDLE MENINGEAL V
  POSTERIOR AURICULAR W
  SUPERFICIAL TEMPORAL X
  TRANSVERSE FACIAL Y

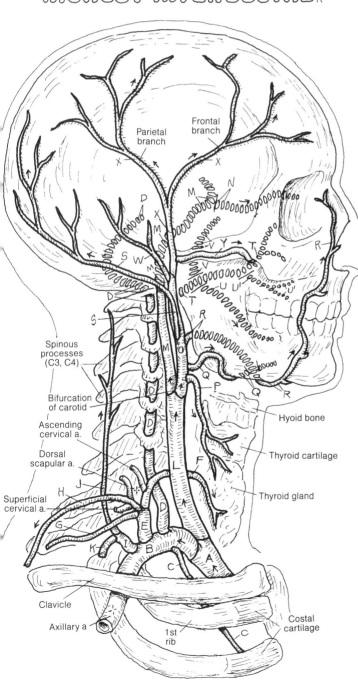

Parietal branch

Frontal branch

Spinous processes (C3, C4)

Bifurcation of carotid

Ascending cervical a.

Dorsal scapular a.

Superficial cervical a.

Hyoid bone

Thyroid cartilage

Thyroid gland

Clavicle

Axillary a

1st rib

Costal cartilage

The *subclavian artery* is the major source of blood to the upper limb, and contributes vessels to the lateral and posterior neck and shoulder. On the right, the artery springs from the brachiocephalic; on the left, the artery comes directly off the aortic arch as does the common carotid (Plate 73). The *vertebral artery* dives deep into the neck to find and enter the transverse foramen of the 6th cervical vertebra. It supplies vessels to the spinal cord, brain stem, and cerebellum. The *thyrocervical trunk* arises just medial to the anterior scalene muscle (see Plate 40) and immediately gives off its branches, the destinations of which are obvious by name. The subclavian artery ends and the axillary artery begins at the lateral border of the first rib.

The *common carotid artery* ascends the neck ensheathed with the internal jugular vein and vagus nerve (not shown). Between the hyoid bone and the upper thyroid cartilage, the artery bifurcates into *internal and external carotid arteries*. The internal carotid passes into the skull, gives off the *ophthalmic artery* to the orbital region and joins the circulus arteriosus (Plate 70). The external carotid artery and its branches supply all of the visceral, musculoskeletal, and dental structures of the head and neck less the brain and orbit. The external carotid divides into *maxillary* and *superficial temporal arteries*. The maxillary artery is a major source of blood to the deep skull cavities, the orbit, teeth, the muscles of mastication, and the dura mater (*middle meningeal artery*). The middle meningeal artery on the dura mater immediately deep to the temporal bone is a potential site of rupture with a hard fall on the side of the head (epidural hematoma).

PULSE SITES *

CN: Use red for A and dark or bright colors for F and G. (1) Begin with the internal carotid (A) and related arteries. Color all illustrations simultaneously as you proceed through the titles. (2) Do the same for the vertebral artery and its branches.

INTERNAL CAROTID A
ANTERIOR CEREBRAL B
ANTERIOR COMMUNICATING C
MIDDLE CEREBRAL D
POSTERIOR COMMUNICATING E

VERTEBRAL F
BASILAR G
CEREBELLAR (3) H
POSTERIOR CEREBRAL I
ANTERIOR SPINAL J

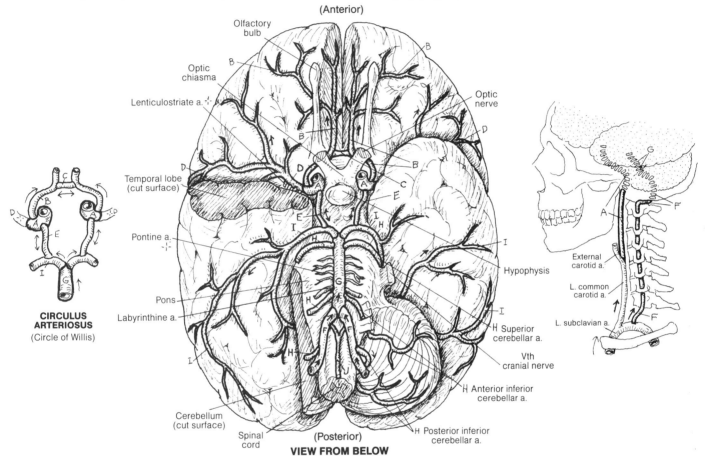

CIRCULUS ARTERIOSUS
(Circle of Willis)

(Anterior)

Olfactory bulb
Optic chiasma
Lenticulostriate a.
Temporal lobe (cut surface)
Pontine a.
Pons
Labyrinthine a.
Cerebellum (cut surface)
Spinal cord

Optic nerve
Hypophysis
H Superior cerebellar a.
Vth cranial nerve
H Anterior inferior cerebellar a.
H Posterior inferior cerebellar a.

(Posterior)
VIEW FROM BELOW

External carotid a.
L. common carotid a.
L. subclavian a.

The paired *internal carotid arteries* pass up through the carotid canals of the temporal bones, curve within the base of the skull (petrous part), travel through the cavernous sinuses (cavernous part), and divide into the *anterior* and *middle cerebral arteries* just lateral to the crossing of the optic nerves (cerebral part). Small lenticulostriate arteries come off the middle cerebral at right angles, supplying the basal ganglia. These "stroke arteries" are commonly the ruptured vessels in intracerebral hemorrhage, often resulting in at least partial paralysis of the limb muscles on the side of the body contralateral to the hemorrhage. Note the anastomosis permitting blood from the internal carotids to mix with that of the vertebrals. Note how these vessels form an irregular vascular circle (circulus arteriosus; circle of Willis). There is considerable variation in the size of the communicating vessels, casting some doubt on the functional significance of these interconnections of the arterial circle. Note the vessels serving the brain stem. You can see that cerebral hemorrhages usually involve the carotid system, while brain stem infarcts relate to the *vertebral* system. Note that a major supplier of the spinal cord (*anterior spinal artery*) comes off the vertebral arteries as well.

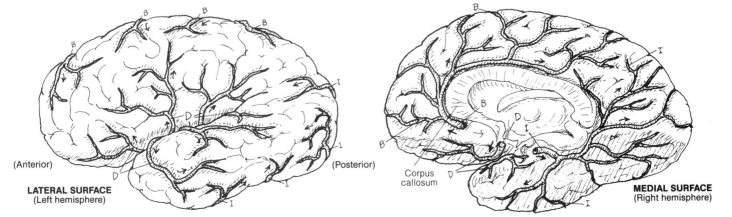

(Anterior)   (Posterior)

LATERAL SURFACE
(Left hemisphere)

Corpus callosum

MEDIAL SURFACE
(Right hemisphere)

ARTERIES & VEINS OF UPPER LIMB

Internal jugular v.

External jugular v.

PULSE POINTS*

Common carotid a.

Clavicle

Acromion

Coracoid process

Humerus

CN: Use red for A (under the title *arteries*) and blue for I (under the title *veins*). You can repeat colors, but do not use the same color for an artery and its corresponding vein. Color the arterial pulse point arrows gray. Broken lines represent veins on the posterior surface.

The arteries here are major ones; their multiple branchings are not shown to avoid visual congestion. Although the basic arterial route from the heart to the fingers is uncomplicated, the existence of collateral routes of flow should be appreciated. In the event of *subclavian* or *axillary* obstruction or ligation, blood can still flow to the distal extremities. The scapular anastomoses (from suprascapular and superficial cervical branches of the subclavian, around the scapula to the *subscapular* branch of the axillary; see Plate 69) do provide an alternate means of getting blood to the *brachial artery*. Note the collateral routes about the neck of the humerus, the elbow, and the wrist.

Middle collateral a.

Recurrent interosseous a.

Radius

Ulna

ANTERIOR VIEW
(Right limb)

◁ ARTERIES:*

BRACHIOCEPHALIC A
SUBCLAVIAN B
AXILLARY C
   SUPERIOR THORACIC D
   THORACO-ACROMIAL & BRANCHES E
   LATERAL THORACIC F
   SUBSCAPULAR G
   ANT./POST. CIRCUMFLEX HUMERAL H
BRACHIAL I
   PROFUNDA BRACHII & BRANCH J
   SUPERIOR ULNAR COLLATERAL K
   INFERIOR ULNAR COLLATERAL L
RADIAL M
   RADIAL RECURRENT N
ULNAR O
   ANT. ULNAR RECURRENT P
   POST. ULNAR RECURRENT Q
   COMMON INTEROSSEOUS R
   SUPERFICIAL PALMER ARCH S
   PALMER DIGITAL T
   DEEP PALMER ARCH U

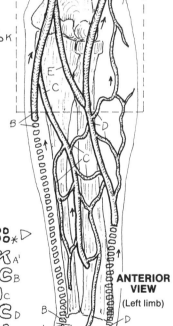

ANTERIOR VIEW
(Left limb)

VEINS:* ▷

DORSAL DIGITAL A & NETWORK A'
BASILIC B
MEDIAN V. OF FOREARM C
CEPHALIC D
MEDIAN CUBITAL E
BRACHIAL F
AXILLARY G
SUBCLAVIAN H
BRACHIOCEPHALIC I

The veins of the upper limb are quite variable in their number and pattern, as are most veins. There are generally two sets of veins in the limbs: deep and superficial. The deep set follow the arteries and are not shown in the anterior aspect of the forearm and lower arm. These deep veins often travel in pairs (venae comitantes; not shown). The superficial veins of the hand and forearm (many observable under the skin) are drained by the *basilic* and *cephalic veins* that travel in the superficial fascia. The superficial veins of the elbow are frequent sites for blood sampling and administration of intravenous medication.

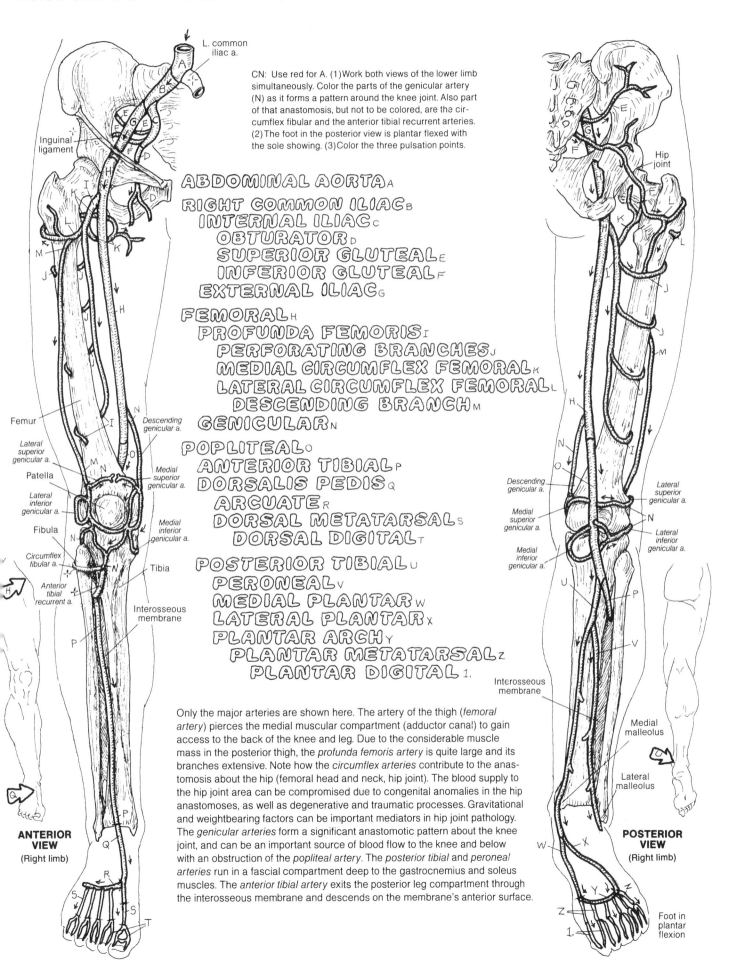

L. common iliac a.

CN: Use red for A. (1)Work both views of the lower limb simultaneously. Color the parts of the genicular artery (N) as it forms a pattern around the knee joint. Also part of that anastomosis, but not to be colored, are the circumflex fibular and the anterior tibial recurrent arteries. (2)The foot in the posterior view is plantar flexed with the sole showing. (3)Color the three pulsation points.

Inguinal ligament

ABDOMINAL AORTA A
RIGHT COMMON ILIAC B
INTERNAL ILIAC C
OBTURATOR D
SUPERIOR GLUTEAL E
INFERIOR GLUTEAL F
EXTERNAL ILIAC G

FEMORAL H
PROFUNDA FEMORIS I
PERFORATING BRANCHES J
MEDIAL CIRCUMFLEX FEMORAL K
LATERAL CIRCUMFLEX FEMORAL L
DESCENDING BRANCH M
GENICULAR N

POPLITEAL O
ANTERIOR TIBIAL P
DORSALIS PEDIS Q
ARCUATE R
DORSAL METATARSAL S
DORSAL DIGITAL T

POSTERIOR TIBIAL U
PERONEAL V
MEDIAL PLANTAR W
LATERAL PLANTAR X
PLANTAR ARCH Y
PLANTAR METATARSAL Z
PLANTAR DIGITAL 1.

Femur

Lateral superior genicular a.

Patella

Lateral inferior genicular a.

Fibula

Circumflex fibular a.

Anterior tibial recurrent a.

Descending genicular a.

Medial superior genicular a.

Medial inferior genicular a.

Tibia

Interosseous membrane

ANTERIOR VIEW
(Right limb)

Hip joint

Descending genicular a.

Medial superior genicular a.

Medial inferior genicular a.

Lateral superior genicular a.

Lateral inferior genicular a.

Interosseous membrane

Medial malleolus

Lateral malleolus

POSTERIOR VIEW
(Right limb)

Foot in plantar flexion

Only the major arteries are shown here. The artery of the thigh (*femoral artery*) pierces the medial muscular compartment (adductor canal) to gain access to the back of the knee and leg. Due to the considerable muscle mass in the posterior thigh, the *profunda femoris artery* is quite large and its branches extensive. Note how the *circumflex arteries* contribute to the anastomosis about the hip (femoral head and neck, hip joint). The blood supply to the hip joint area can be compromised due to congenital anomalies in the hip anastomoses, as well as degenerative and traumatic processes. Gravitational and weightbearing factors can be important mediators in hip joint pathology. The *genicular arteries* form a significant anastomotic pattern about the knee joint, and can be an important source of blood flow to the knee and below with an obstruction of the *popliteal artery*. The *posterior tibial* and *peroneal arteries* run in a fascial compartment deep to the gastrocnemius and soleus muscles. The *anterior tibial artery* exits the posterior leg compartment through the interosseous membrane and descends on the membrane's anterior surface.

CN: Use red for A and use light, bright colors for B, I, J, and L. (1)When coloring the posterior intercostal arteries (K) under the thoracic aortic section, color all the arteries even though only a few are labeled. (2) In the abdomen, note the location of the inferior phrenic, suprarenal, middle sacral, internal iliac, and external iliac arteries, none of which are to be colored. Also note the inferior vena cava (stippled) which echoes the shape of the abdominal aorta.

AORTIC ARCH A
CORONARY B
BRACHIOCEPHALIC C
COMMON CAROTID D
SUBCLAVIAN E
INTERNAL THORACIC F
MUSCULOPHRENIC F'
SUPERIOR EPIGASTRIC F²
COSTOCERVICAL TRUNK G
HIGHEST INTERCOSTAL H

THORACIC AORTA A'
BRONCHIAL I
ESOPHAGEAL J
POSTERIOR INTERCOSTAL K

ABDOMINAL AORTA A²
CELIAC TRUNK L
LEFT GASTRIC M
SPLENIC N
COMMON HEPATIC O
SUPERIOR MESENTERIC P
RENAL Q
TESTICULAR / OVARIAN R
LUMBAR S
INFERIOR MESENTERIC T
COMMON ILIAC U

The branches of the *aortic arch* are unpaired except for the coronary arteries/sinuses. The branches of the *thoracic aorta* are paired. The *bronchial arteries* supply the lung tissues with oxygenated blood. Branches of the *abdominal aorta* are usually described as visceral and parietal. The visceral arteries are both single and paired; note the vessels to the gastrointestinal tract are unpaired. Parietal branches of the abdominal aorta (*lumbar arteries*) are segmental and bilateral, in register with the segmental and bilateral posterior intercostals and subcostals of the thoracic aorta.

There are 9 intercostal spaces anteriorly (not shown), and 11 posteriorly (due to the floating ribs 11 and 12, not shown). These spaces are supplied by both *anterior* and *posterior intercostal arteries*. The posterior intercostals for spaces 1 and 2 are branches of the *highest* (supreme, superior) *intercostal artery* from the costocervical trunk. The posterior intercostals for spaces 3-11 come directly off the thoracic aorta. The subcostal arteries (not shown) run inferiorly to the 12th rib, posterior to the diaphragm. The paired first lumbar arteries (not shown) leave the abdominal aorta immediately below the diaphragm. The anterior intercostals arise from the *internal thoracic artery* for the first 6 intercostal spaces; then the latter artery bifurcates into *superior epigastric* and *musculophrenic* branches. The musculophrenic artery supplies the lower 3 intercostal spaces.

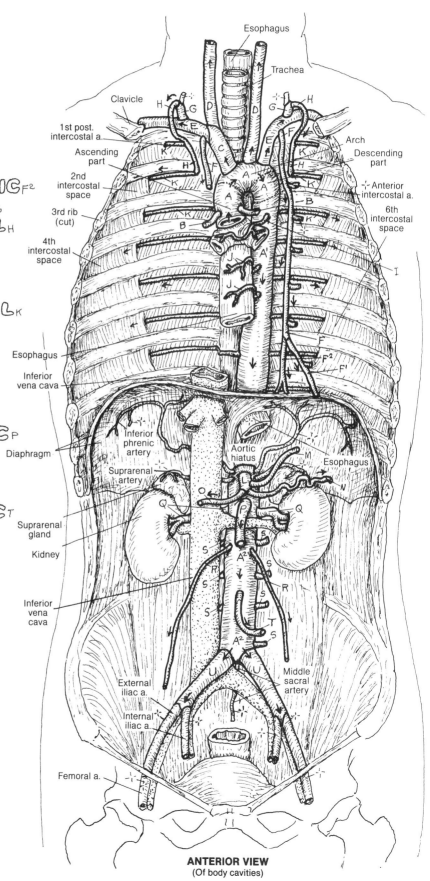

ANTERIOR VIEW
(Of body cavities)

ARTERIES TO GASTROINTESTINAL TRACT & RELATED ORGANS

CN: Use red for A, and use the same colors for celiac trunk (B), superior mesenteric artery (L), and inferior mesenteric artery (Q) that you used for those structures on Plate 73, where they had different subscripts.

(1) Begin with the large illustration. Note that the two pancreaticoduodenal arteries (H, H¹) receive the same color. (2) Color the upper illustration which summarizes the 3 sources of blood supply to the digestive system.

AORTA A
CELIAC TRUNK B
HEPATIC: COMMON C / LEFT C¹ / RIGHT C²
RIGHT GASTRIC D
GASTRODUODENAL E
R. GASTROEPIPLOIC F
L. GASTROEPIPLOIC G
PANCREATICODUODENAL (SUP.) H
CYSTIC I
LEFT GASTRIC J
SPLENIC K

The *celiac trunk*, the first single visceral artery off the abdominal aorta, is a very short vessel that divides immediately into arteries to the liver, spleen, and stomach. Only the major branches of these three arteries are shown here. Note the anastomotic pattern of arteries to the stomach. The blood supply to the pancreas can be better appreciated in Plate 129.

SUPERIOR MESENTERIC L
PANCREATICO-DUODENAL (INF.) H¹
MIDDLE COLIC M
RIGHT COLIC N
ILEO-COLIC O
BRANCHES TO SMALL INTESTINE P

The *superior mesenteric* artery supplies most of the small intestine, head of the pancreas, cecum, ascending colon, and part of the transverse colon. It travels in the common mesentery. Notice the collateral circulation between the celiac and superior mesenteric arteries in the curve of the duodenum. The superior and inferior mesenteric arteries also interconnect via a marginal artery that runs along the length of the large intestine and is fed by both these arteries. The arteries to the ileum/jejunum (cut short) run in the common mesentery.

INFERIOR MESENTERIC Q
LEFT COLIC R
SIGMOID BRANCHES S
SUPERIOR RECTAL T

The *inferior mesenteric* artery supplies the transverse colon down to the rectum. Its branches lie, for the most part, behind the peritoneum (retroperitoneal); the vessels to the sigmoid colon run in the sigmoid mesocolon. Note the anastomoses between branches of the *superior rectal artery* and those of the middle and inferior rectal arteries.

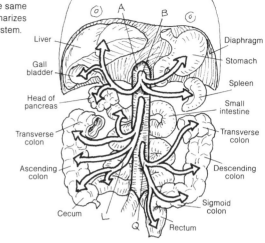

3 SOURCES OF DIGESTIVE BLOOD SUPPLY

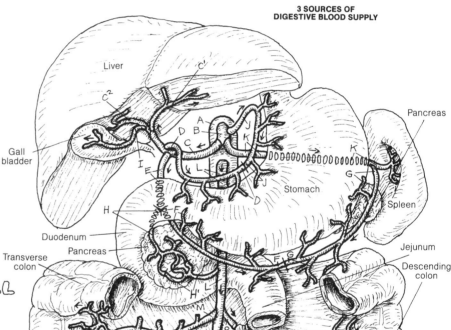

ANTERIOR VIEW
(Diagrammatic)

CN: Use a light color for A. (1)Color the medial views of both pelves simultaneously. Only the disposition of reproductive organs and their vessels varies in these two views. (2)Color both halves of the dissected perineum seen from below. The names of the male vessels can be seen in the medial view.

PELVIS: *

INTERNAL ILIAC A

POSTERIOR TRUNK: ⊹
ILIOLUMBAR B
SUPERIOR GLUTEAL C
LATERAL SACRAL D

ANTERIOR TRUNK: ⊹
UMBILICAL (FETAL) E⊹
SUP. VESICAL / A. TO VAS DEF. F
OBTURATOR G
UTERINE H
VAGINAL I
INFERIOR VESICAL J
MIDDLE RECTAL K
INFERIOR GLUTEAL L

The *internal iliac artery* supplies the pelvis and perineum, with some collaterals from the inferior mesenteric and femoral arteries. Its branches are usually organized into posterior (parietal) and anterior (visceral) divisions. The vascular pattern here is variable; the one shown is characteristic. From the posterior trunk, the *superior gluteal* passes through the greater sciatic foramen above piriformis. The *inferior gluteal* and *pudendal arteries,* from the anterior trunk, depart the pelvis through the lesser sciatic foramen below piriformis. Proximal to the formation of these latter two vessels, the anterior trunk of the internal iliac gives off four branches in both male and female: the first is the *superior vesical* (arising from the proximal part of the fetal umbilical artery; when the umbilical cord is cut, the distal part of the artery atrophies, forming the medial umbilical ligament, the remaining umbilical artery becomes the superior vesical artery, supplying the upper bladder and ductus deferens). The second is the *obturator artery* to the medial thigh region. The third is the *uterine artery*; in the male, it is the *inferior vesical artery.* The *vaginal artery* comes off the uterine artery. The arteries to the prostate and seminal vesicles (not shown) come off the *inferior vesical.* The fourth is the *middle rectal.*

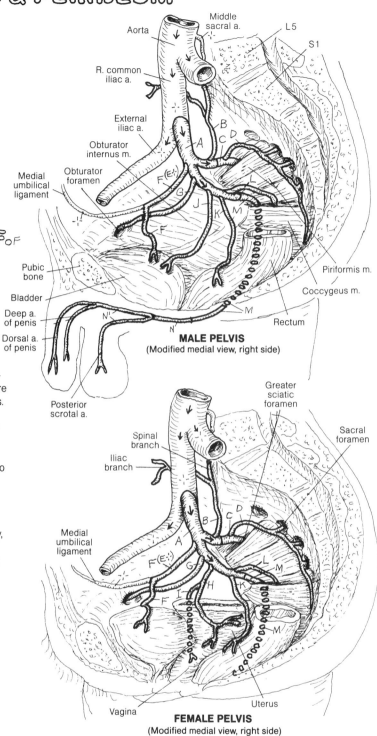

MALE PELVIS
(Modified medial view, right side)

FEMALE PELVIS
(Modified medial view, right side)

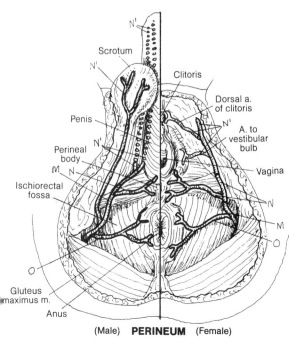

(Male) **PERINEUM** (Female)

PERINEUM: *
PUDENDAL M
PERINEAL N / BRANCHES N'
INFERIOR RECTAL O

The external genital structures are supplied by the *internal pudendal arteries* which pass through the pudendal (fascial) canal alongside the inferior pubic ramus. The arteries (and nerves of the same name) enter the perineum just medial to the ischial tuberosities. The *perineal arteries* are significant in that they provide the vascular basis for erection of the penis and clitoris in sexual stimulation.

CN: Use red for A. Using the preceding plates as reference where necessary, and after reading the text on this plate, color each artery identified by subscript and write its name in that color in the appropriate space at left. If you are using pens with tips too broad to permit legible writing, write the name of the artery in pencil and circle the subscript next to it with the related color.

See glossary in the back of the book for answers.

A _____
A¹ _____
A² _____
B _____
C _____
D _____
E _____
F _____
F¹ _____
G _____
H _____
I _____
J _____
K _____
L _____
M _____
N _____
O _____
P _____
Q _____
R _____
S _____
T _____
U _____
V _____
W _____
X _____
Y _____
Z _____
1 _____
2 _____
3 _____
4 _____
5 _____
6 _____
7 _____
8 _____
9 _____
10 _____
11 _____
12 _____
13 _____

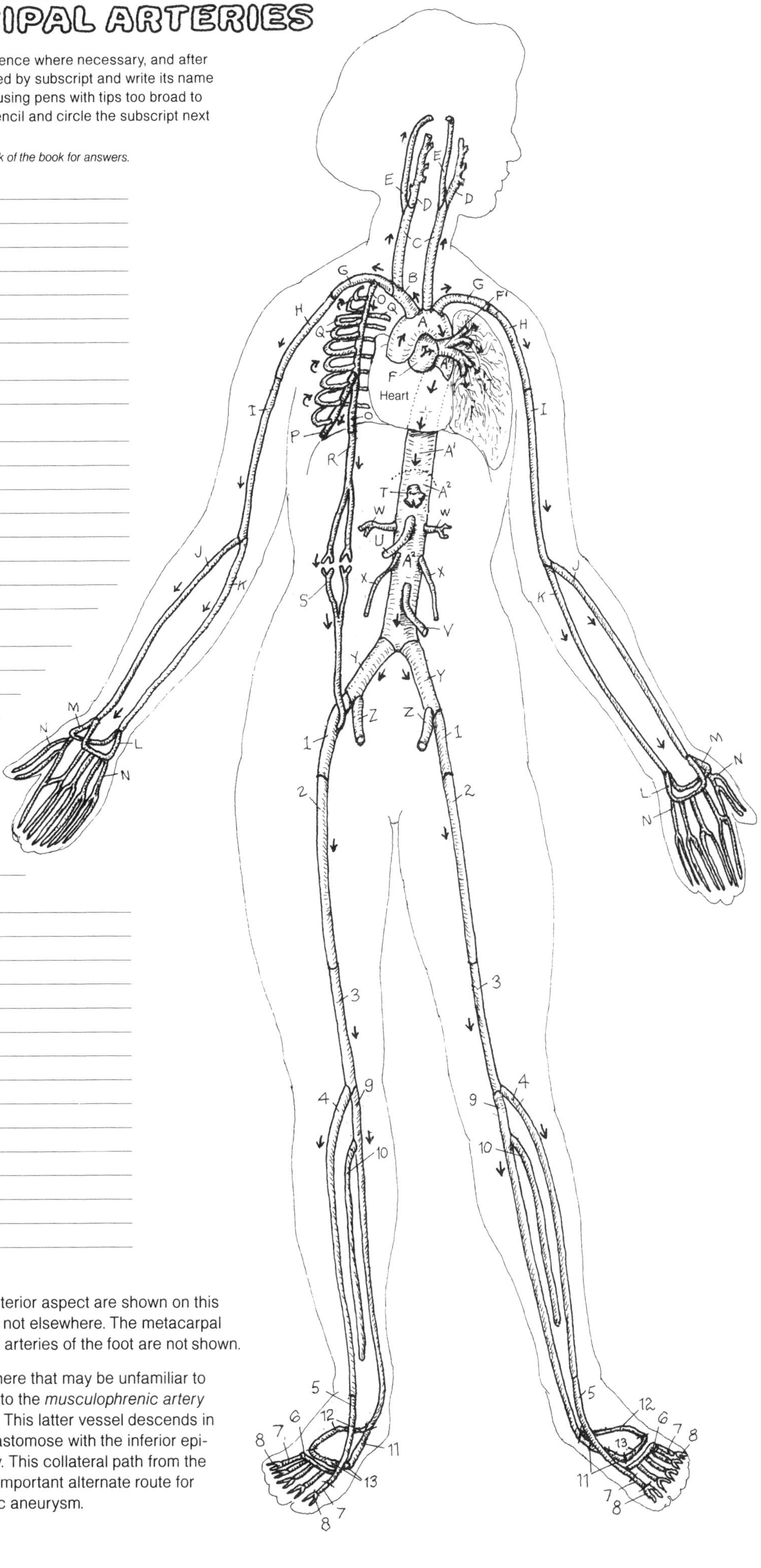

Only the major arteries clearly visible from the anterior aspect are shown on this plate. The arteries are duplicated in the limbs but not elsewhere. The metacarpal arteries, the plantar metatarsal and plantar digital arteries of the foot are not shown.

There is one pair of anastomotic vessels shown here that may be unfamiliar to you: note the *internal thoracic artery* bifurcates into the *musculophrenic artery* and the more medial *superficial epigastric artery*. This latter vessel descends in the sheath of the rectus abdominis muscle to anastomose with the inferior epigastric artery, a branch of the external iliac artery. This collateral path from the heart (indirectly) to the lower limb represents an important alternate route for blood in the event of a seriously obstructing aortic aneurysm.

VEINS OF THE HEAD & NECK

CN: Note the order of titles and their indentations. We begin with titles of tributaries, indenting them above the vein with which they merge or join. This order is in the direction of flow. It will hold for all plates on the veins. Use lighter colors for the sinuses (A–K), represented in the lateral view by broken lines. (1) Begin with the venous sinuses. When coloring the falx and tentorium gray, color lightly over the vessels contained within (A, B, D, and E). Do not color the superior cerebral veins which join the superior sagittal sinus (A). The occipital sinus (K) is shown only in the lateral view.

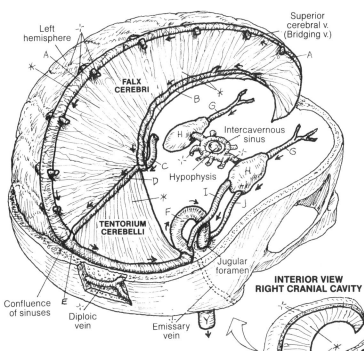

Left hemisphere

Superior cerebral v. (Bridging v.)

FALX CEREBRI

Intercavernous sinus

Hypophysis

TENTORIUM CEREBELLI

Jugular foramen

Confluence of sinuses

Diploic vein

Emissary vein

INTERIOR VIEW RIGHT CRANIAL CAVITY

SINUSES OF DURA MATER: *
SUPERIOR SAGITTAL SINUS A
INFERIOR SAGITTAL SINUS B
GREAT CEREBRAL V. C
STRAIGHT SINUS D
TRANSVERSE SINUS E
SIGMOID SINUS F
SUPERIOR OPHTHALMIC V. G
CAVERNOUS SINUS H
SUPERIOR PETROSAL SINUS I
INFERIOR PETROSAL SINUS J
OCCIPITAL SINUS K

The veins draining the brain are tributaries of large venous channels (cranial dural venous sinuses) in the dura mater. The external cerebral ("bridging") veins drain the cerebral surface and merge with the *superior sagittal sinus*. They are known to rupture when excessive inertial loads are imposed on the brain (subdural hematoma). Two internal cerebral veins form the *great cerebral vein*; these drain the deeper hemispheres (subcortical areas). The *confluence* of merging *sinuses* (*occipital, straight,* and superior sagittal) is variable. The *cavernous sinus* offers significant collateral drainage of blood from the brain. Other collateral veins include the diploic and emissary veins.

VEINS OF HEAD & NECK: *
PTERYGOID PLEXUS L
MAXILLARY M
RETROMANDIBULAR N
SUPERFICIAL TEMPORAL O
POSTERIOR AURICULAR P
ANTERIOR JUGULAR Q
EXTERNAL JUGULAR R

ANGULAR S
DEEP FACIAL T
FACIAL U
LINGUAL V
SUPERIOR THYROID W
MIDDLE THYROID X
INTERNAL JUGULAR Y

DEEP CERVICAL Z
VERTEBRAL 1.
RIGHT SUBCLAVIAN 2.
RIGHT BRACHIOCEPHALIC 3.

The *internal jugular veins* are the principal vessels draining the venous sinuses; *angular veins* (draining the *superior ophthalmic veins*) and the *pterygoid plexuses* assist. The tributaries of the internal/*external jugular veins* are variable. The internal jugular vein travels with the common/internal carotid arteries and the vagus nerve, while the *external jugular vein* can often be seen in the superficial fascia at the side of the neck.

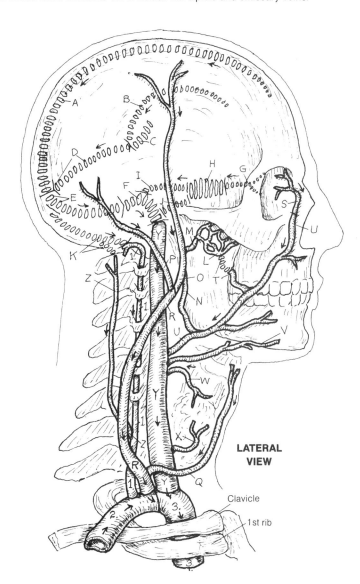

LATERAL VIEW

Clavicle

1st rib

CN: Use blue for the superior and inferior venae cavae (H, H¹). Note that a large segment of the latter has been deleted to reveal the azygos vein (N). Use bright colors for the first posterior intercostal (D) and internal thoracic (F) veins, both of which drain into the brachiocephalic.

SUPERIOR VENA CAVAL SYSTEM: *

SUPERIOR THYROID ∆
MIDDLE THYROID ʙ
INTERNAL JUGULAR c
1ST POSTERIOR INTERCOSTAL ᴅ
INFERIOR THYROID ᴇ
INTERNAL THORACIC ꜰ
R. & L. BRACHIOCEPHALIC ɢ
SUPERIOR VENA CAVA ʜ

AZYGOS SYSTEM: *

POSTERIOR INTERCOSTAL ᴅ'
SUPERIOR INTERCOSTAL ɪ
LUMBAR ᴊ
ASCENDING LUMBAR ᴋ
HEMIAZYGOS (ACCESSORY) ʟ
HEMIAZYGOS ᴍ
AZYGOS ɴ

INFERIOR VENA CAVAL SYSTEM: *

COMMON ILIAC o
TESTICULAR / OVARIAN ᴘ
RENAL ꞯ
HEPATIC ʀ
INFERIOR VENA CAVA ʜ'

The *superior vena cava* drains the head, neck, and upper limbs. In addition, it drains the intercostal and lumbar regions by way of a remarkable but irregular and variable collection of veins called the azygos system. In conjunction with the veins of the vertebral canal (vertebral venous plexus, not shown), the azygos system (*azygos, accessory hemiazygos, and hemiazygos veins*) provides a means of returning blood to the heart from the lower limb and trunk in the event of obstruction of the inferior vena cava.

The azygos and hemiazygos veins are derived from the merging of the *ascending lumbar* and subcostal veins. The anterior intercostal veins (not shown) drain into the *internal thoracic veins*. The inferior vena caval (and azygos) systems have no major tributaries draining the gastrointestinal tract, gall bladder, and pancreas. These viscera have their own venous (portal) system (see Plate 80). However, note that the liver is drained by *hepatic veins* into the inferior vena cava (IVC) just below the diaphragm. The *renal veins* enter the vena cava at right angles; note that the left *testicular/ovarian veins* merge with the left renal vein and that such is not the case on the right.

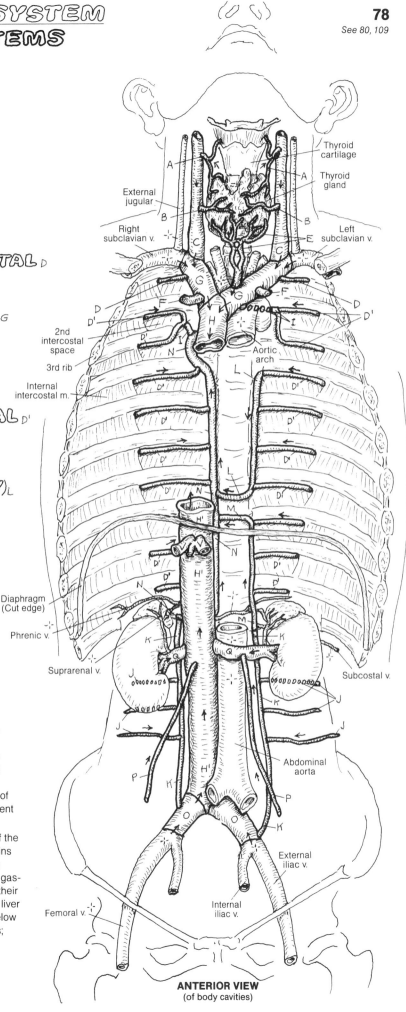

Thyroid cartilage
Thyroid gland
External jugular
Right subclavian v.
Left subclavian v.
2nd intercostal space
3rd rib
Aortic arch
Internal intercostal m.
Diaphragm (Cut edge)
Phrenic v.
Suprarenal v.
Subcostal v.
Abdominal aorta
External iliac v.
Internal iliac v.
Femoral v.

ANTERIOR VIEW
(of body cavities)

CN: Use blue for P and light colors for A-O, the deep veins. Use dark colors for the superficial veins (Q-V). (1) Begin with the deep veins and work both views simultaneously. (2) After completing the superficial veins, color the main ones in each of the small illustrations, but not the fine lines representing their tributaries.

DEEP VEINS:*
PLANTAR DIGITAL A /METATARSAL A'
DEEP PLANTAR VENOUS ARCH B
MED. C /LAT. PLANTAR C'
POSTERIOR TIBIAL D
DORSAL E
ANTERIOR TIBIAL F
POPLITEAL G
LAT. H /MED. CIRCUMFLEX FEMORAL H'
PROFUNDA FEMORIS I
FEMORAL J
EXTERNAL ILIAC K
SUPERIOR L /INFERIOR GLUTEAL L'
OBTURATOR M
INTERNAL ILIAC N
RIGHT COMMON ILIAC O
INFERIOR VENA CAVA P

SUPERFICIAL VEINS:*
DIGITAL Q /METATARSAL Q'
DORSAL VENOUS ARCH R
LATERAL MARGINAL S
MEDIAL MARGINAL T
GREAT SAPHENOUS U
SMALL SAPHENOUS V

The flow of blood in the deep veins of the lower limb is generally an uphill course. In concert with gravity, prolonged horizontal positioning of the legs (and other conditions) can result in slowed flow (stasis) in the deep veins resulting in venous distention and inflammation (phlebitis). Formation of clots in these veins (thrombophlebitis, deep vein thrombosis) can result in clotted fragments being released into the circulation (emboli). These emboli continue up the venous pathway of ever-increasing size, easily pass into the right heart, and are pumped into the progressively smaller vessels of the lung and become stuck (pulmonary embolism).

While deep veins generally follow the arteries (venae comitantes), superficial veins do not. They travel with cutaneous nerves in the superficial fascia and many are easily visualized in the limbs. The blood in these long veins has to overcome gravity for a considerable distance, and their valves often come under weightbearing stress. Happily, there exist a number of communicating vessels (perforating veins, or perforators, not shown) between superficial and deep veins, permitting runoff into the deep veins, significantly offsetting the effect of incompetent valves. Incompetent valves lead to pooling of blood and swelling in the lower superficial veins, with potential inflammation. In the chronic condition, the *saphenous veins* and their tributaries can become permanently deformed and dysfunctional (varicosities).

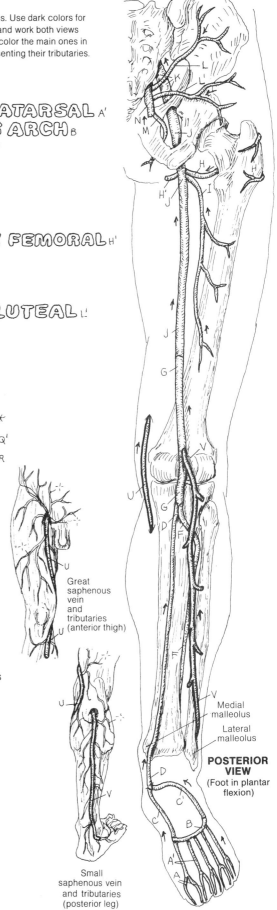

Coxal bone
L. common iliac v.
Inguinal ligament
Femur
Tibia
Fibula
Medial malleolus

ANTERIOR VIEW
(Dorsum of foot)

Lateral malleolus

Great saphenous vein and tributaries (anterior thigh)

Medial malleolus
Lateral malleolus

POSTERIOR VIEW
(Foot in plantar flexion)

Small saphenous vein and tributaries (posterior leg)

CN: Use blue for I and a dark color for J. (1)Color the veins draining the intestines, pancreas, gall bladder, and spleen. Note that there are both left and right gastro-epiploic (D, D[1]) and gastric (G, G[1]) veins. Color the darkly outlined directional arrows adjacent to blood vessels the color of the blood vessel. (2)After coloring the inferior vena cava, its tributaries, the tributaries of the superior vena cava, and directional arrows gray, color the three large arrows identifying anastomotic sites (include the esophageal veins passing posterior to the heart).

HEPATIC PORTAL SYSTEM: ∗
SUPERIOR RECTAL A
INFERIOR MESENTERIC B
PANCREATIC C
L. GASTRO-EPIPLOIC D
SPLENIC E
R. GASTRO-EPIPLOIC D[1]
SUPERIOR MESENTERIC F
R. GASTRIC G
L. GASTRIC G[1]
CYSTIC H
PORTAL I
HEPATIC V. J & TRIBUTARIES J[1]
INF. VENA CAVA ∗ / TRIBUTARY ∗[1]
SUP. VENA CAVA / TRIB. ∗[2]

ANASTOMOSES SITE ∗[3]

Organs of the gastrointestinal tract, the gall bladder, pancreas, and spleen are drained by tributaries of the *hepatic portal vein*. Within the liver, branches of this vein (like those of an artery) discharge blood into capillaries (sinusoids) surrounded by liver cells. These cells remove digested (molecular) lipids, carbohydrates, amino acids, vitamins, iron, etc. from the sinusoids and store them, alter their structure, and/or distribute them to the body tissues (and in the case of unnecessary molecules or degraded remains of toxic substances, the kidneys). The distribution process begins with the selective release of molecular substances from the liver cells into the small *tributaries* of the three *hepatic veins*. The hepatic veins join the inferior vena cava (IVC) immediately below the diaphragm.

The portal system of veins is so-called because it trans-*port*s nutrients and other molecules from the first capillary network in the intestines directly to the second capillary network (sinusoids) of the liver (without going through the heart first). Consequently, blood in the portal system passes through two capillary networks instead of one.

In certain diseases of the liver, the formation of scar tissue blocks flow through the sinusoids. As a result, blood begins to back up in the portal system. In the chronic condition, the veins enlarge significantly. The blood, seeking the path of least resistance, finds collateral routes to the heart, enlarging them (varices). These routes (noted by arrows here) occur by way of anastomoses between veins of the portal system and veins of the inferior caval, superior caval, azygos, and vertebral venous systems. Other anastomoses exist via a number of retroperitoneal veins and the umbilical vein (not shown). In the latter vessel, irregular varicose veins can appear on the surface of the abdominal wall (caput Medusae; see glossary).

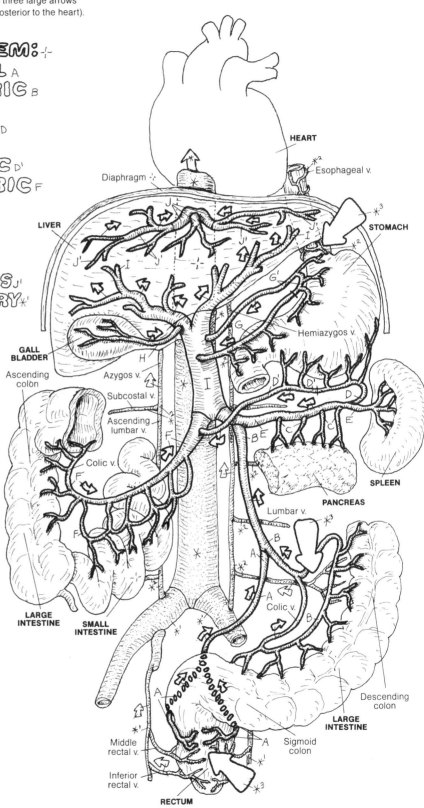

ANTERIOR VIEW
(Diagrammatic)

VI. CARDIOVASCULAR SYSTEM
REVIEW OF PRINCIPAL VEINS

CN: Use blue for the superior and inferior venae cavae (K and K¹). (1) Using preceding plates of the venous system for reference, color each vein and write its name in the appropriate lettered/numbered space with the same color; or write it in pencil (not pen—you may change your mind!) and circle the letter/number with the color used for the vessel. (2) Note that the figure is in the anatomical position, palms facing forward. (3) The superficial veins of the limbs are displayed on the figure's *right* side, the deep veins on the *left* side. On the *right*, the broken lines represent superficial veins on the posterior side. In both upper and lower limbs, the palmar/plantar veins are drained by deep veins. (4) Note in the thorax, the systemic veins are shown only on the figure's *right* side. The internal thoracic and anterior intercostal veins are not shown.

See glossary in the back of the book for answers.

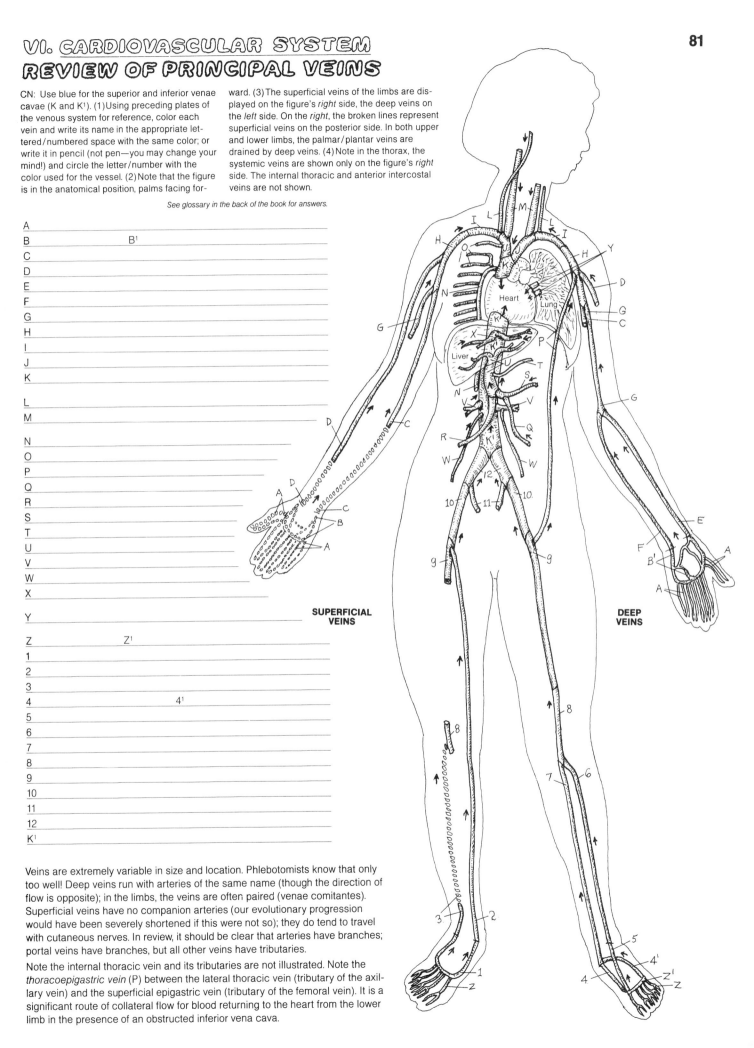

SUPERFICIAL VEINS

DEEP VEINS

A ___
B ___ B¹ ___
C ___
D ___
E ___
F ___
G ___
H ___
I ___
J ___
K ___
L ___
M ___
N ___
O ___
P ___
Q ___
R ___
S ___
T ___
U ___
V ___
W ___
X ___
Y ___
Z ___ Z¹ ___
1 ___
2 ___
3 ___
4 ___ 4¹ ___
5 ___
6 ___
7 ___
8 ___
9 ___
10 ___
11 ___
12 ___
K¹ ___

Veins are extremely variable in size and location. Phlebotomists know that only too well! Deep veins run with arteries of the same name (though the direction of flow is opposite); in the limbs, the veins are often paired (venae comitantes). Superficial veins have no companion arteries (our evolutionary progression would have been severely shortened if this were not so); they do tend to travel with cutaneous nerves. In review, it should be clear that arteries have branches; portal veins have branches, but all other veins have tributaries.

Note the internal thoracic vein and its tributaries are not illustrated. Note the *thoracoepigastric vein* (P) between the lateral thoracic vein (tributary of the axillary vein) and the superficial epigastric vein (tributary of the femoral vein). It is a significant route of collateral flow for blood returning to the heart from the lower limb in the presence of an obstructed inferior vena cava.

FETAL CIRCULATION

PLACENTA A
OXYGENATED BLOOD ➡ B
UMBILICAL VEIN B¹
DUCTUS VENOSUS C
DEOXYGENATED BLOOD ➡ D
DEOXY. & OXY. BLOOD ➡ E
FORAMEN OVALE F
DUCTUS ARTERIOSUS G
UMBILICAL ARTERY E¹

CN: Use red for B (oxygenated blood, represented by a dotted arrow) and B¹ (umbilical vein). Use blue for D (deoxygenated blood, represented by a light-lined arrow). Use purple for E (mixed deoxygenated and oxygenated blood, represented by a dark-lined arrow) and E¹ (umbilical artery). Use bright colors for C, F, and G. (1) Color the placenta and the large number 1, as well as the enlarged rectangular portion of the placenta magnified to show capillary exchange between fetal and maternal vessels. (2) Color the large numbers while coloring related structures and blood flow arrows. (3) Color the placenta and components of the umbilical cord in the uterus at lower left.

The fetus in the uterus does not breathe air; its lungs are deflated. This plate reveals how the fetus gets *oxygenated blood* to its system (in the absence of breathing air) and gets *deoxygenated blood* out of the body.

The *placenta* (numeral 1) is an organ in the uterus of a pregnant woman that provides gaseous and nutritional support for the fetus. The placenta communicates with the fetus by an umbilical cord (2). The vessel taking oxygenated blood from the placenta to the fetus is the *umbilical vein* (2) which runs to the underside of the liver (3) to join the portal vein. Here the oxygen-rich blood of the former is mixed with the deoxygenated blood of the latter. A vein existing only in the fetus (*ductus venosus*) diverts the blood directly to the hepatic vein bypassing the liver sinusoids. The mixed blood then enters the inferior vena cava to the right heart. The blood is directed to the left (systemic) side of the heart by two means: an opening in the interatrial wall (*foramen ovale*; 4); and a short vessel between the pulmonary trunk and the descending part of the aortic arch (*ductus arteriosus*; 5). Only a fraction of mixed blood gets pumped to the non-functioning (but living) lungs. The mixed blood leaves the heart via the aorta (6) to reach the body tissues. The oxygen-carrying capacity of fetal hemoglobin is particularly great in comparison with that of the adult; the fetal tissues get sufficient oxygenation from mixed blood to permit remarkably rapid growth.

Paired *umbilical arteries*, arising from the internal iliacs, return the deoxygenated blood from the fetus to the umbilical cord and placenta. After birth, due to altered hemodynamic patterns associated with breathing, the circulation in the fetal umbilical vessels and ducts of the newborn diminish significantly and the vessels soon thrombose. The umbilical vein atrophies to become the ligamentum teres in the falciform ligament (Pl. 106); the umbilical arteries become the medial umbilical ligaments (Pl. 75); the ductus venosus becomes the ligamentum venosum; revised circulation to the lungs induces closure of the foramen ovale; flow through the ductus arteriosus trickles down, the vessel closes, and becomes a ligamentous strand (ligamentum arteriosum; Pl. 66).

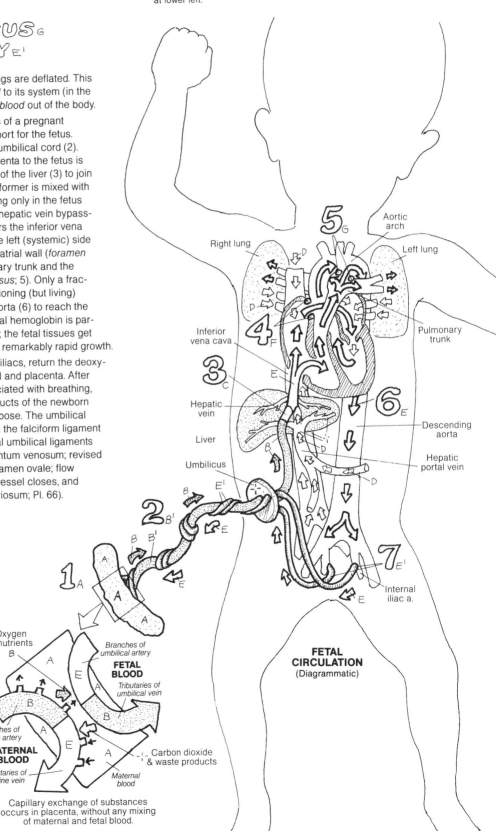

FETAL CIRCULATION
(Diagrammatic)

Right lung — Aortic arch — Left lung — Pulmonary trunk — Inferior vena cava — Hepatic vein — Liver — Umbilicus — Descending aorta — Hepatic portal vein — Internal iliac a.

Uterine wall — Umbilical cord

Oxygen & nutrients
B
Branches of umbilical artery
FETAL BLOOD
Tributaries of umbilical vein
Branches of uterine artery
MATERNAL BLOOD
Tributaries of uterine vein
Carbon dioxide & waste products
Maternal blood

Capillary exchange of substances occurs in placenta, without any mixing of maternal and fetal blood.

VII. LYMPHATIC SYSTEM
LYMPHATIC DRAINAGE & LYMPHOCYTE CIRCULATION

CN: Use blue for H, red for I, purple for L, and green for M. (1) Color over the light lines representing peripheral (superficial) lymph vessels (A). (2) Color each large step numeral in the diagram below with the related titles. In the bottom diagram, do not color the lymphocytes circulating in and between the blood and lymph capillaries.

SUPERFICIAL DRAINAGE: *
PERIPHERAL LYMPH VESSELS A
CERVICAL, B AXILLARY, B¹
INGUINAL NODES B²

DEEP DRAINAGE: *
LYMPHATIC TRUNK C
CYSTERNA CHYLI D
THORACIC DUCT E
RIGHT LYMPH DUCT F

The body is about 60% fluid (by volume), filling cells, vessels, and spaces. Fluid requires circulation. Some of the fluid of the blood, as well as some lymphocytes, leave the circulatory system and enter the tissue spaces. Some of this fluid, lipids, and lymphocytes (lymph), are recovered by thin-walled vessels (*lymphatic capillaries*) that form in the loose connective tissue spaces. Unlike the closed-loop blood capillary networks, these tiny vessels are closed at one end. They merge to form progressively larger lymphatic vessels that drain into large veins in the neck. These vessels constitute the lymphatic system. Certain lymphatic vessels enter and leave lymph-filtering stations called *lymph nodes*.

Region-draining *lymph trunks* converge into a dilated lymph sac (*cysterna chyli*) lying deep to the abdominal aorta on the first lumbar vertebra. The *thoracic duct* begins at the upper end of the sac, ascends the anterior surface of the vertebral column, and drains into the left subclavian vein at its junction with the internal jugular vein. The *right lymph duct* terminates similarly on the opposite side. It drains the dotted area.

LYMPHOCYTE CIRCULATION: *
GENERATIVE ORGAN G
VENOUS BLOOD H
ARTERIAL BLOOD I
LYMPHOID TISSUE J
PERIPHERAL TISSUE K
CAPILLARY NETWORK L
LYMPH VESSELS M
LYMPH NODE M¹

Lymphocytes are one of the principal cells of the immune system. The circulation scheme reveals the primary pathway for the dissemination of lymphocytes from their *generative organs* (*bone marrow*, *thymus*) into the *lymphoid tissues* and organs as well as organs and tissues in general (*peripheral tissues*). Such a circulation pattern provides for maximum exposure of lymphocytes to microorganisms and subsequent body defense operations (immune responses).

Formed and developed in the bone marrow and thymus (1), lymphocytes leave with the *venous blood* to enter the circulation. By way of *arterial blood* (2), lymphocytes enter the *capillary networks* of the lymphoid tissues (3) and other peripheral tissues (4). The lymphocytes may remain in or migrate from the lymphoid organs/peripheral tissues, entering blood capillaries or lymph vessels. From lymph capillaries, the lymphocytes flow with the lymph fluid into regional lymph nodes (5). Here they may become resident or they may depart the node and merge with other lymph vessels to join the lymph ducts (6) that connect with the blood circulatory system.

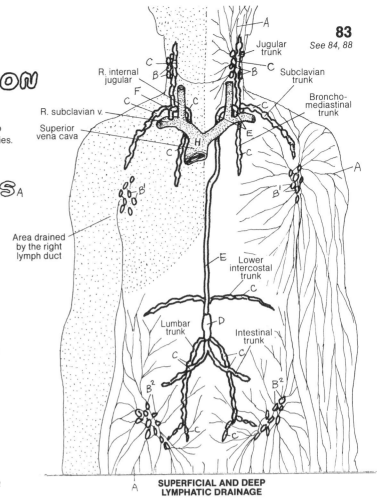

SUPERFICIAL AND DEEP LYMPHATIC DRAINAGE

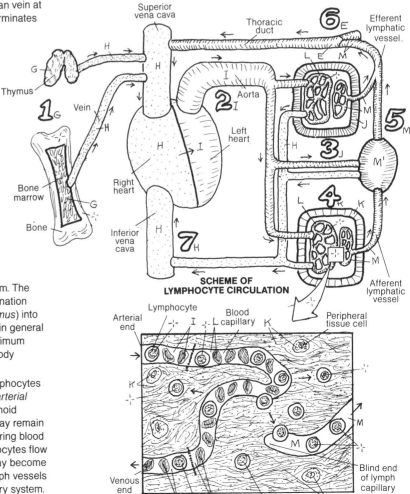

SCHEME OF LYMPHOCYTE CIRCULATION

CN: Use green for D, the same colors for bone marrow (A) and thymus (B) used on Plate 83, and very light colors for H-L. (1) The structures depicting mucosal associated lymphoid tissue (E) are generalizations; more accurate representations can be seen on Plate 89. (2) The three lymphocyte types have identifying letters drawn into their nuclei. Color over the entire cell in all cases.

PRIMARY ORGANS: *
BONE MARROW A
THYMUS B

The lymphoid system is the anatomical component of the immune system, and functions in defense against microorganisms entering the body as well as the destruction of cells or cell parts no longer recognizable as "self." Lymphoid tissues and organs are predominantly collections of lymphocytes and related cells (see below) often supported by a meshwork of reticular fibers and cells.

The red bone marrow and thymus are primary lymphoid organs. The *bone marrow* contains the precursors of all lymphocytes and disburses lymphocytes into the circulation. It consists largely of great varieties of blood cells in various stages of maturation, phagocytes, reticular cells and fibers, and fat cells. Some of the lymphocytes mature and undergo structural and biochemical revision (differentiation) in the bone marrow to become B lymphocytes. Some undifferentiated lymphocytes migrate via the blood to the thymus to become T lymphocytes before re-entering the circulation. Others become large lymphocytes, enter the circulation, and migrate to secondary lymphoid organs.

The *thymus* is located in the superior and anterior (inferior) mediastinum. It receives uncommitted lymphocytes from the bone marrow. The thymus is actively engaged in T lymphocyte proliferation and differentiation during embryonic and fetal life as well as the first decade of extrauterine life. The thymus begins to undergo degeneration (involution) after puberty.

SECONDARY ORGANS: *
SPLEEN C
LYMPH NODE D
MUCOSAL ASSOCIATED
LYMPHOID TISSUE (M.A.L.T.) E
TONSILS / ADENOIDS F
APPENDIX G

Secondary lymphoid organs are structures predominantly populated by lymphocytes that migrated from the primary lymphoid organs. The structural arrangement of these organs ranges from encapsulated, complex structures, like the *spleen* and *lymph nodes*, to a diffuse disposition of lymphocytes throughout the loose connective tissues (*mucosal associated lymphoid tissue*). These secondary organs represent satellite sites for lymphocytic activation when challenged by antigens. The *spleen* processes incoming blood. Its lymphocytes and phagocytes react rapidly to the presence of microorganisms and aged red blood corpuscles. *Lymph nodes* screen lymph from incoming (afferent) lymphatic vessels, much in the same manner as the spleen processes blood. Diffuse and nodular masses of lymphoid tissue in the mucosal lining of open cavities constitute *mucosal associated lymphoid tissue (M.A.L.T.)*. Such tissue includes *tonsils* and *adenoids* situated close to the epithelial layer at portals of entry to certain open (nasal, oral, and pharyngeal) cavities. They are active in "marking" incoming microorganisms for subsequent destruction. Unencapsulated masses of lymphocytes and related cells occur in varying concentrations throughout the mucosa, forming distinct masses (diffuse lymphoid tissue). The vermiform *appendix* incorporates multiple lymphoid follicles in its mucosa. The density of these cell collections varies with the degree of immune responsivity required.

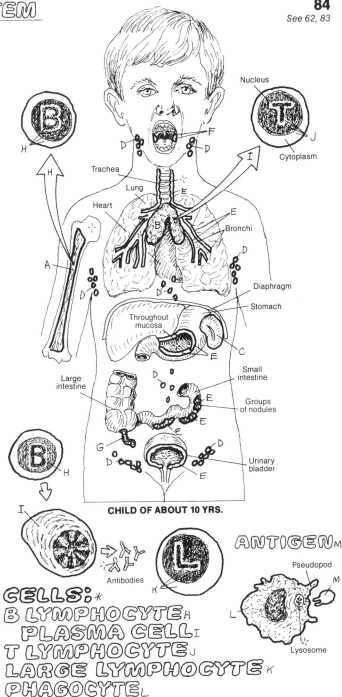

CHILD OF ABOUT 10 YRS.

CELLS: *
B LYMPHOCYTE H
PLASMA CELL I
T LYMPHOCYTE J
LARGE LYMPHOCYTE K
PHAGOCYTE L

Activated *B lymphocytes* (B, bone marrow–derived) differentiate along specific lines, one of which is the transformation into *plasma cells*. Both cells secrete protein molecules, called antibodies, into the tissue fluids. Antibodies interact with and enhance the destruction of elements that induced their activation and synthesis. Such elements are called *antigens*. The term antigen is restricted to those elements (molecules, cells, microorganisms) that induce activation of lymphocytes (immune response). *T lymphocytes* (T, thymus-derived) differentiate, upon stimulation by antigen, into one of a number of different activated cells, one of which stimulates and regulates specific and non-specific body defense operations (helper function; T_H cell).

Large, granular *lymphocytes* (natural killer or NK cells) destroy tumor cells or cells infected with virus. They can be activated by T lymphocytes to lyse target cells, as in graft rejection. Mononuclear *phagocytes* are cells that destroy by phagocytosis antigen as well as particles that do not induce an immune response. They present antigen to lymphocytes (antigen-presenting cells) for identification and destruction.

CN: Use pink for E and the same colors used on Plate 84 for cells D, F¹, G, and I¹. Radial lines surrounding a cell indicate activation. All elements shown have been magnified and schematized for coloring. (1) Begin with the title *microorganism* in the upper right corner.

NATURAL IMMUNITY: *
ANATOMIC BARRIER B
COMPLEMENT C
PHAGOCYTE D
INFLAMMATORY RESPONSE E

Immunity is an anatomic and physiologic state of security against disease. *Natural* immunity exists independent of any specific microorganismal interaction with a lymphocyte. Shortly before birth and following, one progressively *acquires* a specific immunity following each lymphocyte's encounter with antigen and resulting activation. Phagocytes participate in both natural and acquired immunity; lymphocytes participate in acquired immunity and enhance natural immunity.

Natural immunity operates indiscriminately against *microorganisms* and degenerated cells/cell parts. *Anatomic barriers* (1), such as skin or mucous membranes, physically resist microorganismal invasion. *Phagocytes* approach their prey from the blood (2) or connective tissues (3), engulf them (4, phagocytosis) and destroy them with lysosomal enzymes (5). *Complement* is the name given to several soluble proteins found in the body fluids which when activated attach to microorganisms, enhancing their phagocytosis. Tissue irritation, e.g., disruption by a splinter, induces an inflammatory response which involves both natural and acquired immunity.

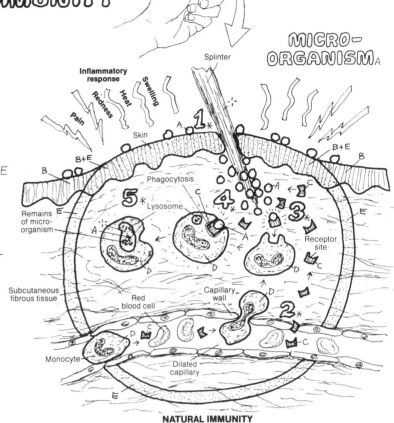

NATURAL IMMUNITY

ACQUIRED IMMUNITY: *

Acquired immunity involves diverse but specific lymphocyte responses to the presence of *antigen*. A specific lymphocytic reaction to antigens (immune response) is characterized by the activation and proliferation of lymphocytes followed by the destruction of antigens. Two kinds of acquired immunity are possible based on lymphocyte types: *humoral immunity* and *cellular immunity*. Inherent in both kinds of immunity are: specificity and diversity of response, retention of cellular memory of antigen, and the ability to recognize self from non-self among the body's proteins.

HUMORAL IMMUNITY: F
B LYMPHOCYTE F¹
PLASMA CELL G
ANTIBODY H

CELLULAR IMMUNITY: I
T LYMPHOCYTE I¹
HELPER CELL (TH) J
CYTOLYTIC CELL (TC) K

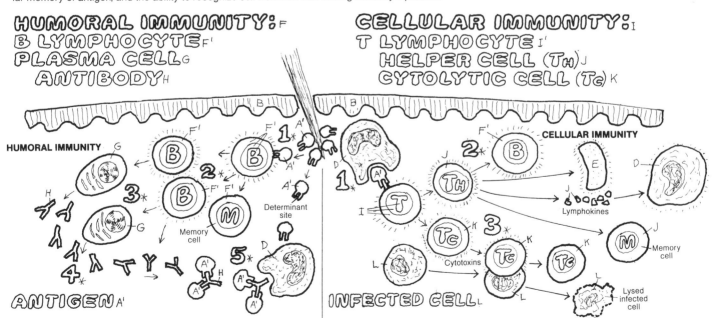

HUMORAL IMMUNITY

CELLULAR IMMUNITY

ANTIGEN A¹

INFECTED CELL L

Humoral (fluid-related) immunity is characterized by *B lymphocytes* being activated by antigen (1), proliferating, forming memory (M) cells, secreting antibody (2), and forming *plasma cells* (3) which secrete *antibody* (4). Antibodies are complex proteins formed in response to a specific antigen and attached to it at the antigenic determinant site (5), facilitating its phagocytosis.

Cellular immunity is characterized by *T lymphocytes* being activated by antigens attached to antigen presenting cells (phagocytes, 1). Activated

T cells differentiate into *helper T lymphocytes* (T_H) and *cytolytic T lymphocytes* (T_C, *CTL*). Helper T lymphocytes (2) enhance humoral immunity by activating B cells, augment the inflammatory response, activate phagocytes with stimulating factors (lymphokines), and form memory (M) cells. Cytolytic T lymphocytes (3) bind to and destroy *infected cells*, and form memory cells. Memory cells recognize specific structural characteristics of the antigens encountered ("memory") and facilitate rapid immune responses on subsequent exposure to those antigens.

CN: Use red for H, blue for I, green for J, and the same color as used on preceding plates for mature T lymphocytes (G). (1) Color the material on red marrow at the bottom of the plate. Then color the red marrow (K) portion of the newborn's arm bone. Color the thymus section and the diagrammatic description of lymphocyte maturation in the thymus. Note that the borders of the diagram represent the cortex and medulla of the thymus. Then color the schematic overview of thymic function. (2) Color the lowest drawing associated with red marrow.

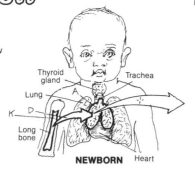

NEWBORN

Thyroid gland
Trachea
Lung
Long bone
Heart

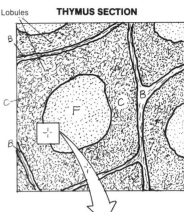

Lobules

THYMUS SECTION

THYMUS A
FIBROUS SEPTA B
CORTEX C
UNDIFFERENTIATED LYMPHOCYTE D
IMMATURE T LYMPHOCYTE E
MEDULLA F
MATURE T LYMPHOCYTE G
ARTERIAL VESSEL H
VENOUS VESSEL I
LYMPH VESSEL J

The *thymus* seeds the entire body with T lymphocytes, the protagonists of cellular immunity. It consists of two lobes of glandular tissue in the anterior and superior mediastinum. The thymus is functional and relatively large in the late fetus/newborn (15 gms), continues to grow and function during the pre-teen years, and declines in size and activity in the following years.

The functional thymus consists of microscopic lobules partitioned by *fibrous septa* containing blood vessels. Each lobule has an outer *cortex* dense with lymphocytes and a much less dense central *medulla*. The *epithelial cells* of the lobule form a structurally supporting "reticular" network. Distinctive concentric rings of keratinized epithelial cells (Hassal's corpuscles) are seen in the medulla; although associated with degenerative signs in aging, they may support lymphocyte differentiation. *Arterial vessels* bring *undifferentiated lymphocytes* into the cortex. The cells migrate into the medulla, showing signs of differentiating into T cells. In the inner cortex, the cells are largely *immature* (but committed) *T cells*. The medulla contains largely *mature T cells*. These cells leave the thymus by venules (*venous vessels*) to enter the systemic circulation. Some T lymphocytes enter *lymph vessels* destined for mediastinal lymph nodes and beyond. The thymus also produces a number of factors (hormones) stimulating lymphocyte differentiation.

Epithelial cell
Phagocyte

MATURATION OF T LYMPHOCYTES

Hassal's corpuscle

SCHEMATIC OVERVIEW

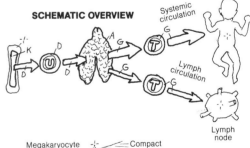

Systemic circulation
Lymph circulation
Lymph node

RED MARROW K
LYMPHOCYTE PRECURSOR L
GROWTH FACTOR M
LYMPHOCYTE D'
SINUSOID I'

Red marrow (recall Plate 17) is densely populated with a great variety of blood cells in various stages of development. The supporting framework of marrow is reticular fibers and cells. Fed by arterioles from the nutrient artery of the bone, the capillaries within the marrow are enlarged to the extent of being small sinuses (*sinusoids*). They reveal transient cytoplasmic "pores" for the immediate passage of cells into the circulation. Among the developing blood cells are *lymphocyte precursors*. These are stimulated to divide by certain *growth factors*. The progeny of these cells are mostly small and some large *lymphocytes*. Presumably, B lymphocyte maturation occurs in the bone marrow, whereas T lymphocyte maturation occurs in the thymus. The lymphocytes (large, undifferentiated pre-T, and B) enter the sinusoids and the venous outflow to be distributed body-wide.

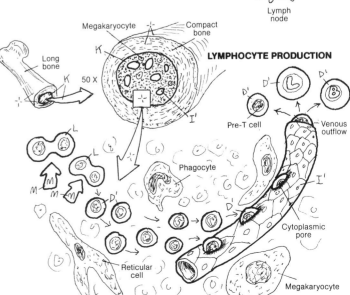

Megakaryocyte
Compact bone
Long bone
50 X

LYMPHOCYTE PRODUCTION

Pre-T cell
Venous outflow
Phagocyte
Cytoplasmic pore
Reticular cell
Megakaryocyte

SPLEEN

CN: Use red for E, purple for F, blue for G, the same colors as previously used for H, I, K and L, the color used on Plate 85 for antigen (M), and very light colors for C and D. (1) When coloring the schematic representation of spleen structures, note the underlying brackets which designate the structures fitting within the white pulp and red pulp regions.

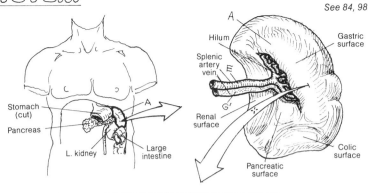

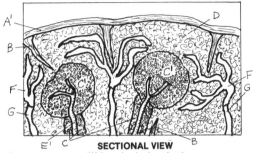

SPLEEN A
CAPSULE A'
TRABECULA B
WHITE PULP C
LYMPHOID FOLLICLE C'
RED PULP D

BLOOD VESSELS: *
ARTERY E
ARTERIOLE E'
VENOUS SINUSOID F
VENULE G
VEIN G'

CELLS: *
T LYMPHOCYTE H
B LYMPHOCYTE I
MITOTIC LYMPHOCYTE J
PHAGOCYTE K
PLASMA CELL L

SECTIONAL VIEW
(White and red pulp regions)

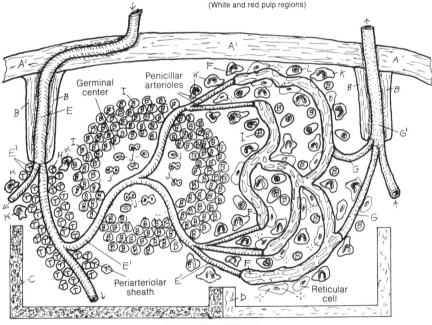

SPLEEN STRUCTURES
(Schematic)

The soft, blood-filled, dark purple *spleen* lies posteriorly in the upper left abdominal quadrant, just above the left kidney, at about the level of the 11th and 12th ribs. It is generally about the size of your closed fist. The *capsule* of the spleen projects inward extensions (*trabeculae*) that support the organ and incoming/outgoing vessels. The microscopic view of the spleen is complicated by the endless sea of lymphocytes and phagocytes, and in this organ, red blood cells.

Small, downstream branches of the *splenic artery* travel in the fibrous trabeculae; branching *arterioles* become enveloped in lymphocytes (periarteriolar sheath) and branch among *lymphoid follicles*. These follicles, the arterioles, and their cellular sheaths constitute the *white pulp*. These follicles enlarge with antigenic stimulation; large *mitotic lymphocytes* (in various stages of cell division) begin to appear in the central part of each follicle (germinal center) following stimulation, creating a zone less dense than the surrounding, cell-packed area. As the straight (penicillar) arterioles leave the white pulp, they become lined with phagocytes and dilate to form *venous sinusoids* (closed theory; illustrated). An alternate "open theory" holds that the vessels open into large spaces drained by the sinusoids. Cells (reticular cells, *plasma cells*, phagocytes, and some B lymphocytes) form long, irregu-

lar cellular strands (splenic cords) amidst interconnected sinusoids; these are collectively called *red pulp*. The sinusoids drain into *venules*, forming tributaries of trabecular veins; these form the tributaries of the *splenic vein*.

As blood flows into the arterioles, antigens and old red blood cells are greeted by mononuclear *phagocytes* and myriad *T lymphocytes*, setting off cellular immune responses in the periarteriolar sheaths. Snaking into the follicles of the white pulp, the vessels are surrounded by *B lymphocytes*; antibodies are formed and attach to antigens which are isolated from the circulation and phagocytosed (humoral immunity). The lumens of vessels in the red pulp are "strained" by the lining phagocytes; old red blood cells are sequestered and phagocytosed, and antigens are met by antibodies, and they are phagocytosed.

Clearly, antibody production and phagocytosis are major activities of the spleen. Although antibody and complement enhance phagocytosis of red blood cells here, the splenic phagocytes are not dependent on T cells and antibody to function. Systemic infection markedly increases the output of lymphocytes, causing palpable splenic enlargement (splenomegaly). Removal of the spleen (splenectomy) is not a benign event, as the body absent a spleen may have reduced immune capabilities.

LYMPH NODE

LYMPH NODE A
  CAPSULE A¹
    TRABECULA B
  RETICULAR NETWORK C
  CORTEX D
    LYMPH FOLLICLE E
      GERMINAL CENTER F
  MEDULLA G
    PARACORTEX H
LYMPH VESSELS: *
  AFFERENT LYMPH VESSEL I
  LYMPH SINUS J
  EFFERENT LYMPH VESSEL K
BLOOD VESSELS: *
  ARTERY L
  VEIN M / VENULE M¹

LYMPHOID CELLS: *
  PHAGOCYTE N
  T LYMPHOCYTE O
  B LYMPHOCYTE P
  MITOTIC LYMPHOCYTE Q
  PLASMA CELL R

LYMPH S
ANTIGEN T

CN: Use red for L, blue for M, green for S (if you have additional greens, use them on I–K), the same colors as previously used for N–R, and light colors for D–H. (1) As you color the regions of the lymph node, use the same color on the large numerals accompanying the insets identifying the dominant cell in that region. In the paracortex, note the small circles representing venules (M¹).

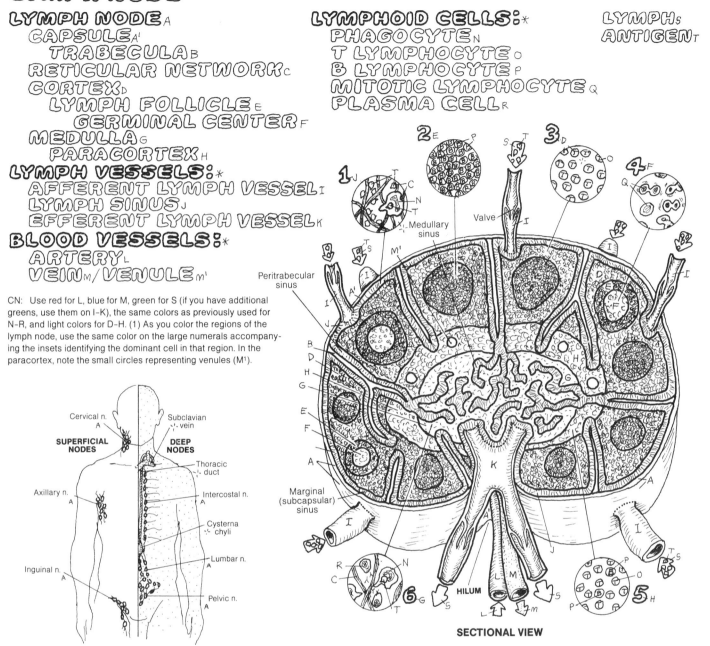

SECTIONAL VIEW

The *lymph node* has a fibrous *capsule* from which *trabeculae* invade the organ, dividing it incompletely into compartments. Fine reticular fibers and cells spread out from the trabeculae to form a thicket of interwoven branches throughout the node (*reticular network*). This intricate weave of fibers supports the dense populations of lymphocytes throughout the node. Lymph percolates through parts of the reticular network called *lymph sinuses* (only the marginal sinuses appear to be endothelial-lined). The reticular fibers in these sinuses (1) form a spatial framework from which *phagocytes* can readily engage *antigens* in the lymph flow.

The node interior is characterized by an outer *cortex* and an inner *medulla*. The cortex reveals a group of particularly dense masses of *B lymphocytes* (2, *lymphoid follicles*) existing among a more sparse array of largely *T lymphocytes* in the interfollicular areas (3). In the presence of significant amounts of antigen, the follicles develop *germinal centers*; here are seen *mitotic lymphocytes* in varying degrees of mitosis (4). The outer part of the medulla (*paracortex*) has more diffuse arrangements of phagocytes, T cells, and some B cells (5). The endothelial cells of the venules in the paracortex are specialized and provide lymphocyte homing receptors that influence the localization of T and B cells within the node. The medulla contains a concentrated

array of interconnecting sinuses, with phagocytes and *plasma cells* in significant numbers (6).

Lymph enters the nodes by *afferent vessels* with valves controlling unidirectional traffic. As the lymph meanders through the throngs of reticular fibers in the sinuses, phagocytes pick off the antigens and present them to the T cells in the interfollicular areas. Activated B cells in the follicles, facilitated by helper (T_H) cells, transform into plasma cells and memory cells. The plasma cells and B cells secrete antibody with receptors which bind a portion of the antigen (antigenic determinant). Binding of antibody to the antigen facilitates destruction of the antigen. Major stimulus promotes formation of the germinal centers. Further immune activity occurs in the paracortical and inner medullary areas. Lymph leaves the medullary sinuses and the node by way of the *efferent vessels*. Lymphocytes also enter the node by small *arteries*; these cells can migrate into the sinuses from the venules while the remaining blood leaves the node by *veins*.

In summary, the lymph node is the site of both humoral-mediated (B cell) and cell-mediated (T cell) immune responses to antigens in the lymph. Palpable enlargement of cervical lymph nodes during an upper respiratory infection, for example, gives testimony to the existence of such mechanisms operative in the face of microorganismal invasion.

MUCOSAL ASSOCIATED LYMPHOID TISSUE (M.A.L.T.)

CN: Use green for C, the same colors as previously used for E-J, and use light colors for A and B. (1) Begin with the representations of normal and inflamed tonsils, using pink and red colors, respectively. Include the circular insets identifying the dominant cells within the follicle and germinal center. Color the antigens (I) activating the follicle at center right, and related follicle parts, including insets. (2) Color the lower illustrations.

PRIMARY FOLLICLE A
GERMINAL CENTER B
EFFERENT LYMPH VESSEL C
LYMPHOID CELLS: *
MITOTIC LYMPHOCYTE D
PHAGOCYTE E
B LYMPHOCYTE F
T LYMPHOCYTE G
PLASMA CELL H

Unencapsulated lymphoid tissue, consisting of lymphocytes, plasma cells, and phagocytes (often termed "inflammatory cells"), not bound by fibrous capsules and ranging throughout the loose connective tissues of the mucosae in variable densities, is called mucosal associated lymphoid tissue (M.A.L.T.) or diffuse lymphoid tissue. The cells of this tissue are a mobile lot, and can increase in numbers rapidly in response to the presence of microorganisms. Often, in the setting of chronic antigen exposure, well organized follicles (also called nodules), similar to those seen in lymph nodes and the spleen, may form. Some examples of M.A.L.T. are shown here.

TONSIL K / INFLAMED TONSIL K'

Masses of *primary* lymphoid *follicles* surrounding infoldings of the pharyngeal mucosa are called *tonsils*. Tonsils have no definitive lymph sinuses; however, lymph capillaries can be seen draining into *efferent lymphatic vessels*. With antigenic stimulation, inflammation of the tonsil occurs (tonsillitis). A common event, an inflamed tonsil is swollen, red (often with streaks of vessels ranging across the mucosal surface), and painful. As the follicles are activated, germinal centers are formed, numbers of *B* and *T* lymphocytes increase, *phagocytes* and *plasma cells* appear, and *antibody* is produced. Considered a culturally accepted "rite of passage" in the past, tonsillectomies are accomplished now only for good cause (obstructed airways, chronic infections). These organs respond quickly to the presence of microorganisms.

PEYER'S PATCHES *

Aggregates of lymphoid follicles in the submucosa of the distal ileum are called Peyer's patches. Seen sporadically throughout the intestine, lymphatic follicles are more concentrated here. With antigenic stimulation, these follicles increase in much the same manner as tonsils.

VERMIFORM APPENDIX L

The *vermiform appendix* is a thin, tubular extension of the cecum (large intestine). It contains a number of lymphoid follicles which extend from the submucosa up to the epithelial lining of the mucosa. The mucosa of the appendix experiences fairly frequent insults (tomato and chile seeds, popcorn kernels, and ingested foreign matter) and inflammatory events are fairly common (appendicitis). The structure swells, reddens, and is often quite painful. Classical immune responses occur (formation of germinal centers, and so on). Due to the thin walls of the appendix, inflammations induced by acute infections can rupture through to the peritoneum (peritonitis). Surgical removal of the appendix (appendectomy) is common. There is no evidence that depressed immune activity occurs following appendectomy; neighboring intestinal lymphoid tissues can apparently fill any defense voids created by appendectomy.

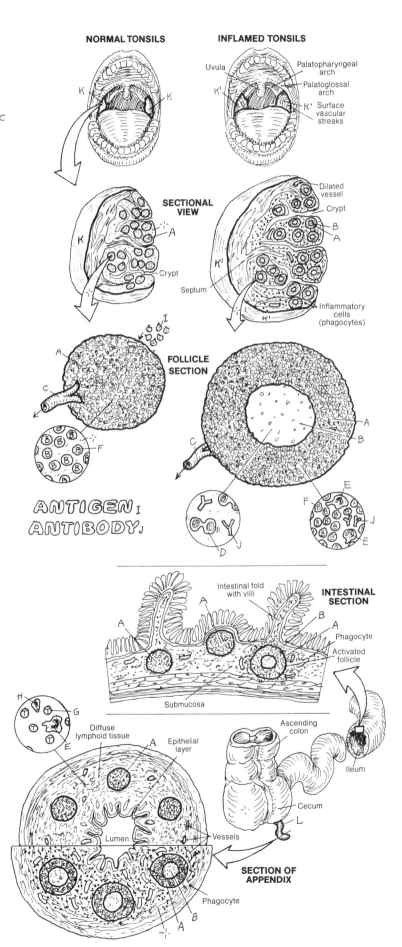

NORMAL TONSILS

INFLAMED TONSILS

Uvula

Palatopharyngeal arch

Palatoglossal arch

Surface vascular streaks

SECTIONAL VIEW

Crypt

Dilated vessel

Crypt

Septum

Inflammatory cells (phagocytes)

FOLLICLE SECTION

ANTIGEN I
ANTIBODY J

Intestinal fold with villi

INTESTINAL SECTION

Phagocyte

Activated follicle

Submucosa

Ascending colon

Ileum

Cecum

Diffuse lymphoid tissue

Epithelial layer

Lumen

Vessels

SECTION OF APPENDIX

Phagocyte

HIV-INDUCED IMMUNOSUPPRESSION

CN: You are advised not to color this plate until you have colored the previous plates on the Immune System. See the glossary for explanation of new terms. Use the same colors used on the previous plates (with different subscripts) for P, T, B, PL, Ab, Ag, and IR.
(1) Begin with the HIV infection and work your way to the activation stage. (2) Note that the broken lines indicate the destruction of T lymphocytes and the end or reduction of their influence on other cells or processes. The downward-pointing arrows also reflect that negative effect; the only upward-pointing arrow relates to an increase in antigen population.

HIV INFECTION: *
T-H LYMPHOCYTE T-H
HIV-1 VIRUS H
PROVIRUS H'
PHAGOCYTE P

EFFECTS OF AIDS ON: *
T LYMPHOCYTES T
MEMORY CELLS T'
PHAGOCYTOSIS P'
INFLAMMATORY RESPONSE IR
ANTIGENS Ag
NEUROTRANSMITTERS N
B LYMPHOCYTES B
MEMORY CELLS B'
PLASMA CELLS PL
ANTIBODIES Ab
LARGE LYMPHOCYTES L
MITOTIC REGULATION MR

gp 120 (antigenic part)

CD4 (receptor) — Cell DNA — Viral core — Reverse transcriptase — Viral RNA — Unintegrated viral DNA — Proviral (Integrated) DNA

INFECTION *

Cell membrane — Nucleus

LATENCY *

Viral RNA — Synthesis of viral protein — Assembly of infectious particles

ACTIVATION (AIDS) *

Blood vessel — P

Capillary — Brain cell — Dividing cancer cell — Neoplasm

At this time (early 1993), a catastrophic disease, called Acquired Immuno-Deficiency Syndrome (AIDS), is being experienced throughout the world, characterized by a marked reduction of functions by immune cells (immunosuppression). The disease is caused by an infection of phagocytes and lymphocytes with human immunodeficiency virus (HIV). The virus is transmitted from one person to another via body fluids, mainly blood and semen. Transmission is effected by male homosexual/bisexual activity (70%), intravenous drug use with shared, blood-contaminated syringes (17%), heterosexual intercourse with HIV-infected partners (4%), intravenous blood transfusions (2.5%), intravenous use of blood products for clotting disorders (1%), and transmission from infected mother to fetus by HIV transfer across placental membranes (?%).

Once in the body fluids, *HIV* surface molecules (glycoprotein or GP 120) attach to specific (CD4) receptors on the surface of the cell membranes of T_H lymphocytes and mononuclear *phagocytes*. Fusion of the virus to the cell and endocytosis of the virus (*infection*) usually follow. The HIV then converts its genetic material (RNA) into DNA by means of an enzyme called reverse transcriptase attached to the viral RNA. This new viral DNA segment is then incorporated into the DNA of the host cell. This integrated DNA is called proviral DNA. It is capable of producing viral RNA which provides the "blueprint" (instructions) for the synthesis of viral proteins in the host cell, and the subsequent construction of infective virus (viral particles, virions). At any point prior to or after formation of the viral proteins, the production of pro-viral materials may be suspended (*latency*), or at least progression of the manifestations of the disease seems to slow. Two to ten years may pass during which the normal activities of daily living can be carried on without the symptoms/signs of life-threatening immunosuppression. When infected cells disburse the viral particles to other T cells and phagocytes, or when those infected cells cease responding to antigenic stimuli, the disease becomes *activated* (AIDS).

Immune function is initially suppressed due to the destruction of *T lymphocytes* by the infective process (1) and the rapid depression of cellular

☆ OPPORTUNISTIC INFECTIONS *

immunity (2-12). T memory cells decline in numbers, and related antigen memory is impaired (2). T cell-enhanced *phagocytosis* is diminished (3) and *inflammatory responses* are limited (4), permitting an increase in *antigen* numbers and activity (5). Phagocytes of the brain and spinal cord (microglia) are particularly prone to HIV infection, resulting in defective *neurotransmitter functions*, memory loss and other neurologic deficits (encephalopathy) (6). *B lymphocytes* fail to proliferate in response to antigen (7) due to depleted T-related stimuli, sharply reducing their numbers and those of B *memory cells* (8), *plasma cells* (9) and *antibodies* (10). *Large lymphocytes* are reduced both in number and activity (11), disabling *mitotic regulation* (12) and permitting formation of neoplasms (cancer). The global effect of immunosuppression is microorganismal access to the unprotected body *(opportunistic infections)*. Many of these infections themselves are immunosuppressive. Early in the course of AIDS, it is not unusual to see the lymphoid cells respond rapidly to these infections with increased immune cell and antibody production; unfortunately, many of these activated cells are themselves infected. In summary, HIV infection leads to decreased numbers and functions of helper T cells and phagocytes which, in turn, adversely affect many aspects of acquired and natural immunity, resulting in infections, tumors, neurologic dysfunction, and wasting.

IX. RESPIRATORY SYSTEM OVERVIEW

CN: Use red for L and light colors throughout. (1) Begin with the structures of the respiratory system. (2) Color the cross section of the trachea (D), including the respiratory mucosa (I). (3) Color the enlargement of the mucosa in the lowest view.

NASAL CAVITY_A
PHARYNX_B
LARYNX_C
TRACHEA_D
PRIMARY BRONCHI_E
BRONCHIAL TREE_F
R. LUNG_G L. LUNG_{G'}
DIAPHRAGM_H

The *respiratory tract* conducts air to the respiratory units of the lungs where it can readily be absorbed by the blood, and it removes carbon dioxide–laden air from the air cells and exhausts it to the external atmosphere. It develops and refines sounds into potentially intelligible vocalization, and helps maintain acid-base balance of the blood by blowing off excess acid in the form of carbon dioxide. Nowhere in the body does the outside world, with all its creatures of microscopic dimension, have such easy access to the protected interior cavities of the body as it does at the air/blood interfaces of the lung. The respiratory tract has both air-conducting and respiratory (gas exchange) parts.

The air conduction tract includes an upper (*nasal cavity, pharynx, larynx*) and lower tract (*trachea, primary bronchi and bronchial tree*). The upper tract is lined with respiratory mucosa except in the lower pharynx where it has a stratified squamous epithelial surface. Except for the nose and pharynx, the skeleton of the respiratory tract is cartilaginous down to the smallest airways (bronchioles) where the cartilage is replaced by smooth muscle. Those parts associated with gaseous exchange are the smallest bronchioles and alveoli (respiratory units) and take up much of the *lung's* volume.

The muscular *diaphragm* provides much of the force necessary for inspiration and expiration of air. One quarter of that force is generated by the intercostal muscles moving the ribs.

RESPIRATORY MUCOSA:_I
PSEUDOSTRATIFIED COLUMNAR EPITHELIUM_J
LAMINA PROPRIA_K
BLOOD VESSEL_L
GLAND_M

The mucosa of the respiratory tract is largely *pseudostratified columnar* and (in the bronchioles) cuboidal *epithelia* with mucus-secreting *goblet* (unicellular gland) *cells* and *cilia*. Here excreted mucus traps foreign particulate matter, inhaled air is hydrated (mixed with water) putting oxygen in solution, and the air is heated from underlying vessels. These epithelial cells are supported by a loose fibrous, *glandular, vascular lamina propria*, replete with fibroblasts and cells of the lymphoid system. Deep to this connective tissue layer is the supporting tissue (bone in nasal cavity, muscle in the pharynx, hyaline cartilage in the trachea, larynx, and bronchi, smooth muscle in the bronchioles, and thin fibers supporting the air cells).

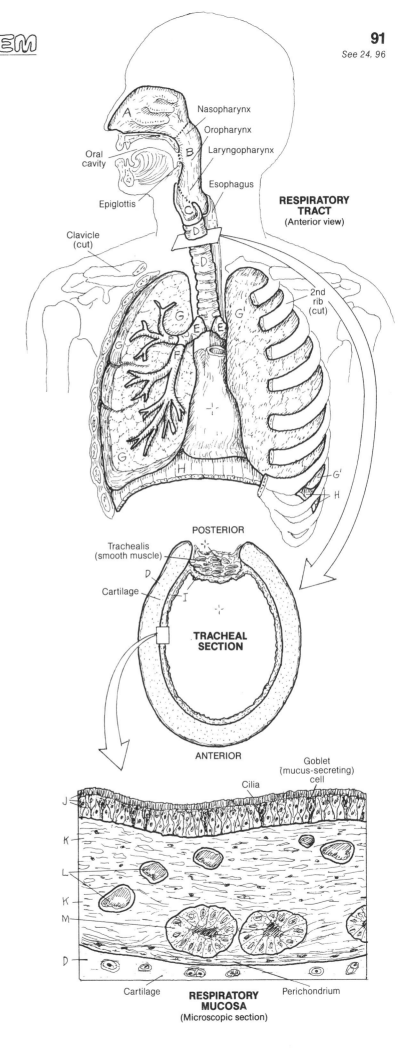

Nasopharynx
Oropharynx
Laryngopharynx
Esophagus
Oral cavity
Epiglottis
Clavicle (cut)
RESPIRATORY TRACT
(Anterior view)
2nd rib (cut)

POSTERIOR
Trachealis (smooth muscle)
Cartilage
TRACHEAL SECTION
ANTERIOR

Goblet (mucus-secreting) cell
Cilia
RESPIRATORY MUCOSA
(Microscopic section)
Cartilage
Perichondrium

CN: Use very light colors for H and I. (1) Begin with the upper illustration. (2) Color the nasal septum and its structure in the nasal cavities diagram. (3) Color the elements of the lateral wall of the nasal cavity and relations in the lowest illustration.

EXTERNAL NOSE: *
NASAL BONE A
CARTILAGE OF NASAL SEPTUM B
LATERAL NASAL CARTILAGE C
ALAR CARTILAGE D
FIBRO-FATTY TISSUE E

NASAL SEPTUM: *
CARTILAGE OF NASAL SEPTUM B
ALAR CARTILAGE D
PERPENDICULAR PLATE OF ETHMOID BONE F
VOMER BONE G

NASAL CAVITY & RELATIONS: *
NASAL BONE A
FRONTAL BONE H
SPHENOID BONE I
CRIBRIFORM PLATE OF ETHMOID F'
VESTIBULE OF NOSE D'
SUPERIOR CONCHA J
MIDDLE CONCHA K
INFERIOR CONCHA L
HARD PALATE M
SOFT PALATE N
LATERAL WALL O*

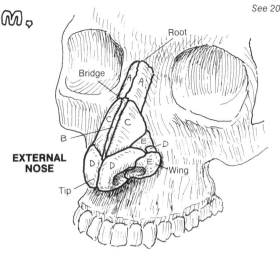

EXTERNAL NOSE

Root, Bridge, A, C, E, D, B, Tip, Wing

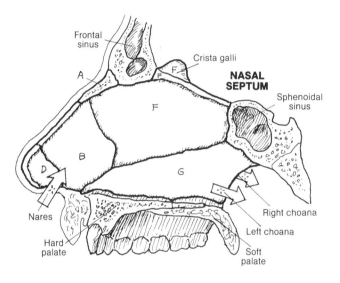

NASAL SEPTUM

Frontal sinus, Crista galli, A, F, Sphenoidal sinus, B, D, G, Nares, Hard palate, Right choana, Left choana, Soft palate

The nose is a largely cartilaginous affair external to the skull proper. Its orifices (nares, or nostrils) open into the nasal cavity of the skull which is a bony tunnel divided by a partly cartilaginous *nasal septum*. The nasal cavity opens into the muscular pharynx through two bony-walled posterior apertures called choanae. The nose, situated as it is in front of the face, often receives the brunt of a facial impact. In such event, it is not unusual for the *cartilage of the nasal septum* (septal cartilage) to break off from the *perpendicular plate of the ethmoid*. This "deviated septum" may obstruct air flow through the narrowed half of the cavity. The skin-lined *vestibule* of the nose has long hairs (vibrissae) that serve to discourage entrance of foreign bodies. The nasal cavity is carpeted with a mucosal lining characterized by ciliated epithelial cells that secrete mucus and whose cilia sweep small particulate matter down into the nasopharynx. The bony *conchae* (so called because of their resemblance, in frontal section, to the conch shell) increase the surface area of the nasal cavity, significantly boosting the local temperature and moisture content. The *inferior concha* on each side is attached to the ethnoid bone by an immovable joint (suture); the *superior and middle conchae* are part of the ethmoid bone. The spaces under the conchae (meatuses) are open to paranasal sinuses (air-filled cavities), the subject of the next plate. Note the roof of the nasal cavity (*cribriform plate*) transmits the olfactory nerve fibers; resting on or near this plate are the frontal lobes of the brain. Note that the floor of the nasal cavity is the *palate* which is also the roof of the oral cavity. The *soft palate* is a muscular extension of the bony palate, and plays a role in swallowing.

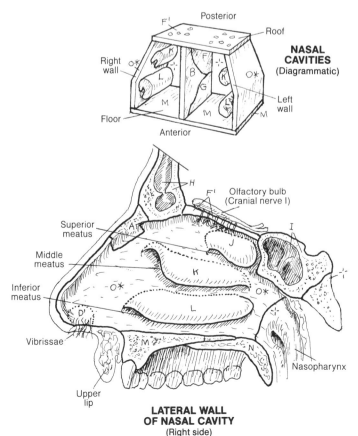

NASAL CAVITIES
(Diagrammatic)

Posterior, Roof, F', Right wall, O*, K, F, L, B, K, O*, Left wall, G, M, M, M, Floor, Anterior

LATERAL WALL OF NASAL CAVITY
(Right side)

Olfactory bulb (Cranial nerve I), H, F', Superior meatus, Middle meatus, Inferior meatus, A, J, K, O*, L, O*, I, Vibrissae, M, D', N, Nasopharynx, Upper lip

PARANASAL AIR SINUSES

CN: Use the same colors for the bones A and B, and conchae F, G, and H, that were used for those structures on Plate 92. (1) Color the sinus drainage sites in the lateral wall of the nasal cavity. Include the edges of the conchae which have been cut away to reveal the meatuses and related drainage sites. (2) Color the coronal section. Note that it is a composite view, showing openings into the nasal cavity that do not appear in any one single coronal plane. Even so, this view cannot show the relations of the sphenoid sinus and opening, nor the mastoid air cells and the auditory tube. (3) Color the lower drawings. Note that nasolacrimal duct and the duct of the frontal sinus are shown on one side only.

AIR SINUSES: *
FRONTAL A
SPHENOID B
ETHMOID C
MAXILLARY D
MASTOID E

NASAL CONCHAE: *
SUPERIOR F
MIDDLE G
INFERIOR H

OPENING OF AUDITORY TUBE I
NASOLACRIMAL DUCT J
NASAL SEPTUM K
NASAL CAVITY L *

The skull has a number of cavities in it. You are familiar with some of them (mouth, nose, external ear, orbits), but perhaps not so familiar with others. The frontal, sphenoid, maxillary, ethmoid, and temporal bones have variably sized cavities, all of which directly or indirectly communicate with the nasal cavity. These are the *paranasal air sinuses*, to be distinguished from the venous sinuses of the dura mater. These air sinuses serve to lighten the skull and they add timbre to the voice. They are lined with respiratory-type epithelium, which is continuous with the epithelium of the nasal cavity. The mucus secretions from these epithelial linings pass down canals and enter the nasal cavity just under the conchae (meatuses). Their specific drainage sites are indicated by the arrows. Should these passageways become blocked by inflammation and swelling, the pressure builds within the sinuses to a point where considerable pain can be experienced (sinusitis, sinus headache). Agents that constrict the blood vessels (decongestants) help to reduce the swelling and reestablish proper drainage. The *mastoid air cells*, in the mastoid process of the temporal bone, drain into the middle ear (tympanic) cavity, communicating by way of the auditory (pharyngotympanic) tube with the nasopharynx just posterior to the nasal cavity. The *nasolacrimal duct* receives secretions from the lacrimal gland which functions to keep the covering (conjunctiva) of the eye globe moist. These tears drain into slits at the medial aspect of the eyelids, which open into sacs that narrow into the nasolacrimal ducts. These ducts pass downward along the lateral walls of the nasal cavity and open into the meatus of the inferior concha on each side—and that explains how it is that one blows one's nose after one cries.

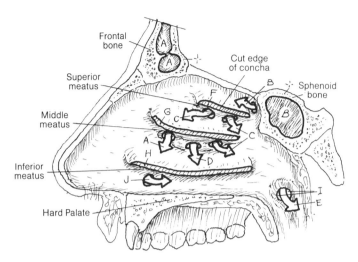

SINUS DRAINAGE SITES
(Right lateral wall of nasal cavity, nasal conchae removed)

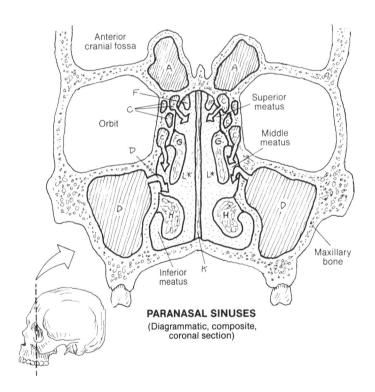

PARANASAL SINUSES
(Diagrammatic, composite, coronal section)

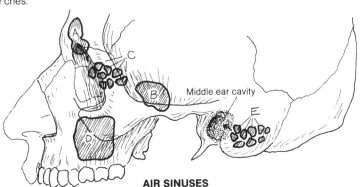

AIR SINUSES

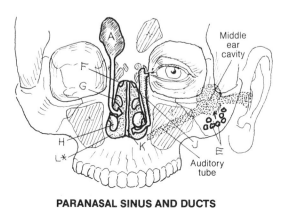

PARANASAL SINUS AND DUCTS

CN: Use dark or bright colors for N, O, and Q. (1) Begin with the overview diagram in the upper right corner. (2) Complete the large composite sagittal section. Take note of the surrounding structure as a frame of reference (not to be colored). (3) Color all six laryngeal views simultaneously. Note that the title for the laryngeal cavity (H¹) is at the lower right of the plate.

PHARYNX: A
NASOPHARYNX B
PHARYNGEAL TONSIL C
OROPHARYNX D
PALATINE TONSIL E
LARYNGOPHARYNX F

The *pharynx* is an incomplete tube of mostly skeletal (constrictor) muscle and fibrous tissue, appearing to hang from the edges of the choanae (posterior nasal apertures) at the base of the skull. Posteriorly, it is supported by fascia in front of the sphenoid bone and the upper six cervical vertebrae. It is the posterior and inferior continuation of the *nasal cavity*; it is open to the *oral cavity* anteriorly. Inferiorly, it continues as the *esophagus* behind and the *larynx* in front. Most of pharynx is lined with stratified squamous epithelium except the nasopharynx (respiratory lining). Coordinated muscular activity in the pharynx underlies the mechanism of swallowing (deglutition).

Masses of partially encapsulated lymphoid tissue incompletely encircle the nasal and oral openings into the pharynx (Waldeyer's ring), i.e., at the opening of the auditory tube (tubal tonsils), roof of the nasopharynx (adenoids), between the palatoglossal and palatopharyngeal pillars (palatine tonsils; see Plate 99), and the posterior tongue (lingual tonsils). See tonsil function in Plate 89.

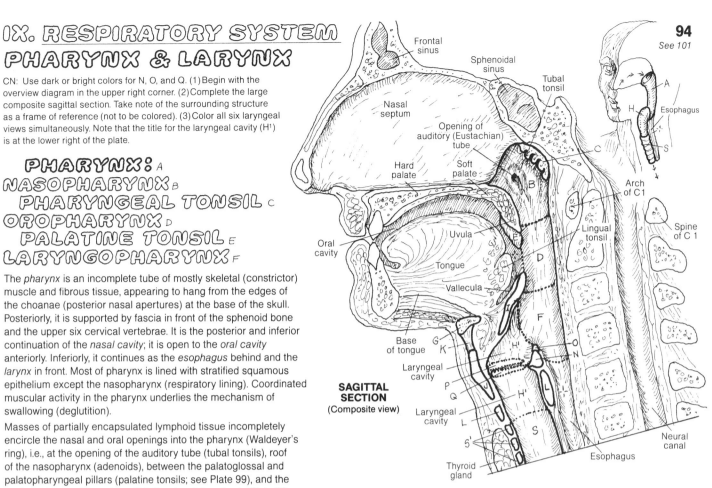

SAGITTAL SECTION (Composite view)

HYOID BONE G
LARYNX: H
EPIGLOTTIS I
THYROID CARTILAGE J
THYROHYOID MEMBRANE K
CRICOID CARTILAGE L
CRICOTHYROID LIGAMENT M
ARYTENOID CARTILAGE N
CORNICULATE CARTILAGE O
VESTIBULAR FOLD P
VOCAL FOLD Q
RIMA GLOTTIS R ✱
TRACHEA, CARTILAGE S¹

The *larynx* provides a mechanism for sound production, manipulation of sound waves, and protection from inadvertent aspiration (inhaling) of solid matter. The larynx is supported by a framework of hyaline cartilage connected by ligaments. Although associated with the larynx, the hyoid bone is not a laryngeal structure.

The *thyroid cartilage* is composed of two laminae which together are V-shaped when looking at them from above. The *arytenoid cartilages* articulate with the top of the cricoid, pivoting on it. The *vocal folds* are mucosa-lined ligaments stretching between thyroid and arytenoid cartilages. They are abducted/adducted by the movement of the arytenoid cartilages. In breathing they are abducted; in coughing, they are momentarily fully adducted (closing the *rima*) permitting intrathoracic pressure to build; opened rapidly by abduction of the folds, the rima experiences hurricane-force winds from the depths of the respiratory airway (explosive cough). During phonation, the vocal folds are generally adducted, varying somewhat with pitch and volume. The *vestibular folds* (false vocal folds) are fibrous and move only passively.

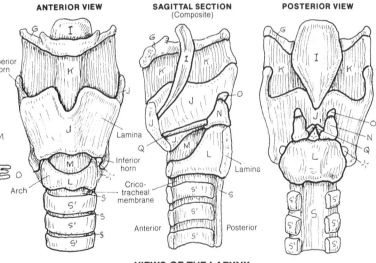

ANTERIOR VIEW

SAGITTAL SECTION (Composite)

POSTERIOR VIEW

VIEWS OF THE LARYNX

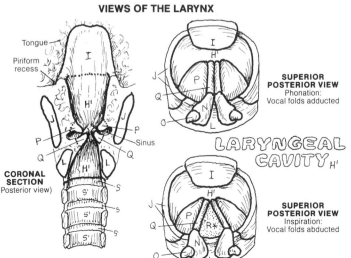

CORONAL SECTION (Posterior view)

SUPERIOR POSTERIOR VIEW
Phonation: Vocal folds adducted

LARYNGEAL CAVITY H¹

SUPERIOR POSTERIOR VIEW
Inspiration: Vocal folds abducted

See 24, 42, 65

IX. RESPIRATORY SYSTEM
LOBES & PLEURAE OF THE LUNGS

LOBES:
R. UPPER₄ R. MIDDLE₈ R. LOWER꜀
L. UPPER᷄ L. LOWER₍ₑ₎

PLEURAE:
VISCERAL PLEURA_F
PLEURAL SPACE
PARIETAL PLEURA_G

CN: Use bright colors for A-E, very light colors for F and G, and a reddish-brown color for H. In all of the illustrations the thickness of the pleurae (F and G) has been enlarged for coloring purposes. (1) Begin with the anterior view. Note that the ribs and intercostal muscles have been removed (see Plate 42). Sections of the pleurae have been stripped away and separated. The potential pleural space is between these layers; in the coronal and cross sections, this space is drawn as a dark line and not as a stucture to be colored. Similarly, the title is also left uncolored. A small section of the parietal pleura has been cut and pulled away to reveal the underlying visceral pleura and a portion of the costodiaphragmatic recess below the lung superficial to the diaphragm. (2) Color the coronal view, noting the left crus of the diaphragm, and the cardiac notch of the left lung. (3) Color the cross section of the lung lobes and pleurae (as seen from above), noting the vertebral level, and the roots of the lungs.

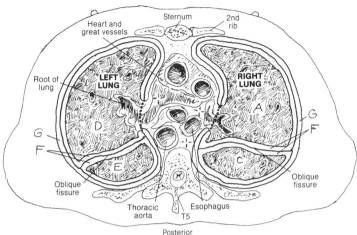

CROSS SECTION
(Through T5)

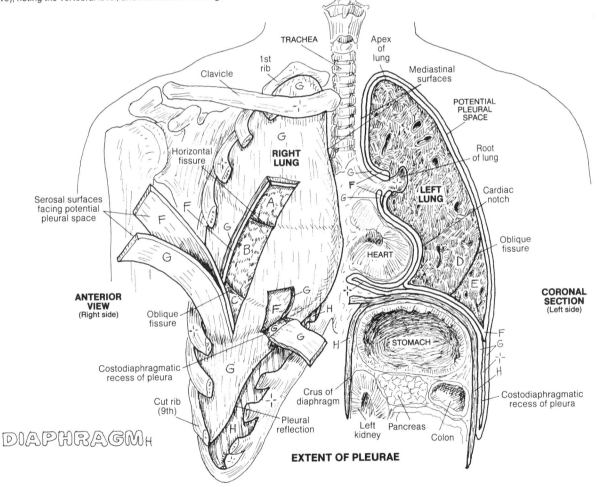

ANTERIOR VIEW
(Right side)

DIAPHRAGM_H

EXTENT OF PLEURAE

CORONAL SECTION
(Left side)

The lobes of the lungs are largely enveloped in *visceral pleura*, a thin serosal membrane which turns (reflects) off the lungs at their roots to become the *parietal pleura* which lines the inner surface of the chest wall, the lateral mediastinum, and much of the diaphragm. These serous membranes are in contact with each other, separated by a thin layer of serous (watery, glycoprotein) fluid. The interface of these membranes is potentially a cavity or space (*pleural space/cavity*). With certain diseases, the space is capable of expanding to accommodate increasing amounts of fluid (pleural effusion) at the expense of the lung, resulting in a reduction of total lung capacity. The serous fluid maintains surface tension between the pleural surfaces (resisting separation of visceral and parietal layers in contact with one another) and prevents frictional irritation between moving pleural membranes.

During quiet inhalation, the inferior and anterior margins of the visceral-pleura lined lungs do not quite reach the parietal pleura, leaving a narrow space or recess, i.e., the costomediastinal recess between the rib cage and the mediastinum (not shown), and the costodiaphragmatic recess between rib cage and diaphragm (see coronal section at lower right).

SEGMENTAL (TERTIARY) BRONCHI:
1 APICAL 2 POST. 3 ANT. 4 LAT.(R.L.) 4 SUP.(L.L.)
5 MED.(R.L.) 5 INF.(L.L.) 6 SUP. 7 MED. BASAL
8 ANT. BASAL 9 LAT. BASAL 10 POST. BASAL

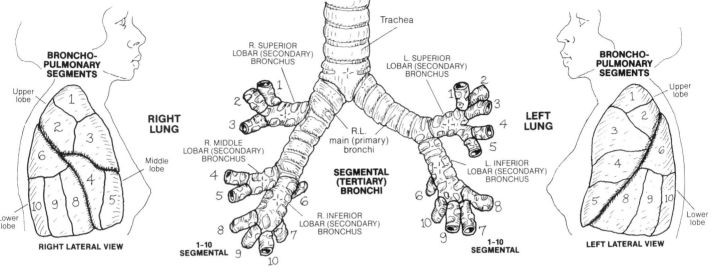

BRONCHO-PULMONARY SEGMENTS — Upper lobe, Middle lobe, Lower lobe — RIGHT LUNG — RIGHT LATERAL VIEW

Trachea — R. SUPERIOR LOBAR (SECONDARY) BRONCHUS — L. SUPERIOR LOBAR (SECONDARY) BRONCHUS — R.L. main (primary) bronchi — R. MIDDLE LOBAR (SECONDARY) BRONCHUS — R. INFERIOR LOBAR (SECONDARY) BRONCHUS — L. INFERIOR LOBAR (SECONDARY) BRONCHUS — SEGMENTAL (TERTIARY) BRONCHI — 1-10 SEGMENTAL — LEFT LUNG

BRONCHO-PULMONARY SEGMENTS — Upper lobe, Lower lobe — LEFT LATERAL VIEW

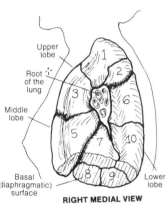

RIGHT MEDIAL VIEW — Upper lobe, Root of the lung, Middle lobe, Basal (diaphragmatic) surface, Lower lobe

CN: Save blue for E, purple for F, and red for G (in the respiratory unit below). (1) Use ten different colors for both lungs, and key those colors to the ten segmental bronchi of each lung. (2) Below, use the same color as above for the 7th segmental bronchus. Use one light color for the alveoli (D¹) and the alveolar sacs (D). Note in the gas exchange diagram, that red blood cells in the purple capillary (F) receive three different colors based on their stage of oxygenation.

The lower respiratory tract consists of the trachea and the bronchial tree, including the respiratory units which are engaged in gaseous exchange. The lungs are divided by connective tissue septa into triangular-shaped, surgically-resectable anatomical and functional units called *bronchopulmonary segments*, each served by a segmental bronchus, supplied by a segmental artery, and drained by segmental veins and lymphatics. Segments are of special significance to those interpreting lung sounds by stethoscope (auscultation) or listening to the sounds coming from the lungs when the chest wall is tapped (percussion). By such methods, sites of alveolar dysfunction/disease and levels of abnormal accumulations can often be determined.

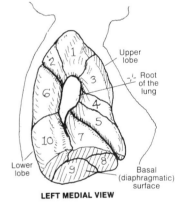

LEFT MEDIAL VIEW — Upper lobe, Root of the lung, Lower lobe, Basal (diaphragmatic) surface

BRONCHIOLE A
RESPIRATORY BRONCHIOLE B
ALVEOLAR DUCT C
ALVEOLAR SAC D & ALVEOLUS D¹
PULMONARY ARTERIOLE E
CAPILLARY NETWORK F
PULMONARY VENULE G

Within each bronchopulmonary segment, a segmental bronchus branches into several *bronchioles* (less than 1 mm in diameter, supported by smooth muscle instead of cartilage). These bronchioles give off smaller terminal bronchioles, characterized by ciliated cuboidal cells without glands. The terminal bronchioles represent the end of the air-conducting pathway. Each terminal bronchiole divides into two or more *respiratory bronchioles*, characterized by occasional alveolar sacs on their walls. Each respiratory bronchiole supplies a respiratory unit which is a discrete group of air cells (*alveoli*), arranged in *alveolar sacs*, fed by *alveolar ducts*. Extending from its source bronchiole, each respiratory bronchiole has more and more alveolar sacs, terminating as an alveolar duct opening into alveolar sacs. The walls of the air cells, composed of simple squamous epithelia supported by thin interwoven layers of elastic and reticular fibers, are surrounded by capillaries which arise from pulmonary arterioles and become the tributaries of pulmonary venules. The walls of these capillaries are fused to and structurally similar to those of the alveoli. Oxygen and carbon dioxide rapidly diffuse, on the basis of pressure gradients, through these walls.

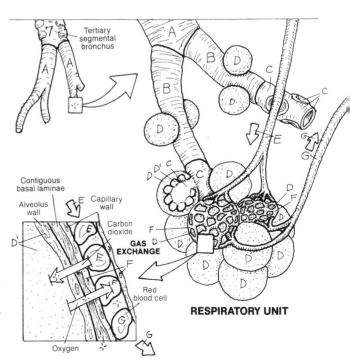

Tertiary segmental bronchus — Contiguous basal laminae — Alveolus wall — Capillary wall — Carbon dioxide — GAS EXCHANGE — Red blood cell — Oxygen — RESPIRATORY UNIT

CN: Use light colors throughout, except for a bright or dark color for E. (1) Begin with the illustration at far left (inspiration); note that the thoracic wall (A) is shown only in the far right diagram. Color the diaphragm, its location represented by broken lines. (2) Color the expiration illustration and the bucket handle analogy. (3) Finish with the illustration at far right.

THORACIC WALL: A
RIB & COSTAL CARTILAGE B
STERNUM C
THORACIC VERTEBRAE D

MUSCLES OF INSPIRATION: *
DIAPHRAGM E
EXTERNAL INTERCOSTAL F

MUSCLE OF EXPIRATION: *
INTERNAL INTERCOSTAL G

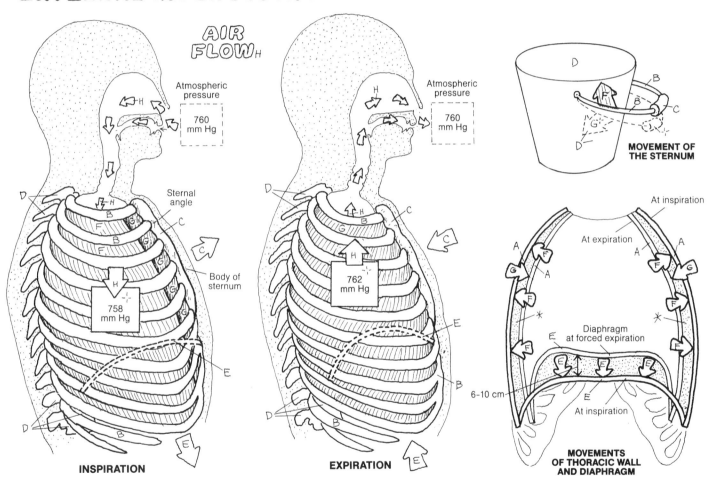

AIR FLOW H

Atmospheric pressure
760 mm Hg

Sternal angle

Body of sternum

758 mm Hg

INSPIRATION

Atmospheric pressure
760 mm Hg

762 mm Hg

EXPIRATION

MOVEMENT OF THE STERNUM

At inspiration

At expiration

Diaphragm at forced expiration

6-10 cm

At inspiration

MOVEMENTS OF THORACIC WALL AND DIAPHRAGM

The mechanism of respiration makes possible breathing which consists of inhalation (inspiration) and exhalation (expiration) phases. The physical principle underlying air movement in/out of the thorax is the inverse relationship of pressure and volume (as one goes up, the other goes down). Volume changes within the *thorax* alter the intrathoracic pressure 1-2 mm Hg above/below atmospheric pressure (outside the body) in quiet breathing, enough of a change to move about 500 ml of air with each breath. The thoracic *diaphragm* accomplishes about 75% of the inspiratory effort, the *external intercostals* 25%. Expiration is largely diaphragm and external intercostal relaxation/stretch, and lung elasticity, with some help from the *internal intercostals*. In inspiration, contraction of the diaphragm flattens the muscle and lowers the floor of the thorax, increasing the vertical dimension of the thoracic cavity. Contraction of the external intercostals

elevates the ribs, swinging the sternal body slightly outward at the sternal angle. This increases the transverse and anteroposterior dimensions of the thoracic cavity. These actions collectively increase the intrathoracic volume, momentarily lowering the pressure within. Given the relatively higher atmospheric pressure outside the head, *air* is induced to enter the respiratory tract to find lower pressure. The action of the bucket handle demonstrates the hinge action at the sternal angle and related rib elevation. In expiration, the relaxed diaphragm forms "domes" over the underlying liver and stomach, decreasing the vertical dimension of the thorax. Recoil/descent of the ribs decreases the transverse and anteroposterior dimensions. The thoracic volume is thus decreased, momentarily increasing the intrathoracic pressure above atmospheric. *Air* escapes to the outside, aided by the natural elastic recoil of the lungs.

X. DIGESTIVE SYSTEM
OVERVIEW OF THE SYSTEM

CN: When coloring the organs that overlap each other, use your lightest colors for D, E, T, V, and W. Each overlapping portion receives the color of both structures. (1) After coloring the alimentary canal, review the structures before completing the accessory organs. The central section of the transverse colon (J) has been removed to show deeper structures.

ALIMENTARY CANAL: *
ORAL CAVITY A
PHARYNX B
ESOPHAGUS C
STOMACH D
SMALL INTESTINE: *
DUODENUM E
JEJUNUM F
ILEUM G
LARGE INTESTINE: ÷
CECUM H
VERMIFORM APPENDIX H'
ASCENDING COLON I
TRANSVERSE COLON J
DESCENDING COLON K
SIGMOID COLON L
RECTUM M
ANAL CANAL N

ACCESSORY ORGANS: *
TEETH O
TONGUE P
SALIVARY GLANDS: *
SUBLINGUAL Q
SUBMANDIBULAR R
PAROTID S
LIVER T
GALL BLADDER U
BILE DUCTS V
PANCREAS W

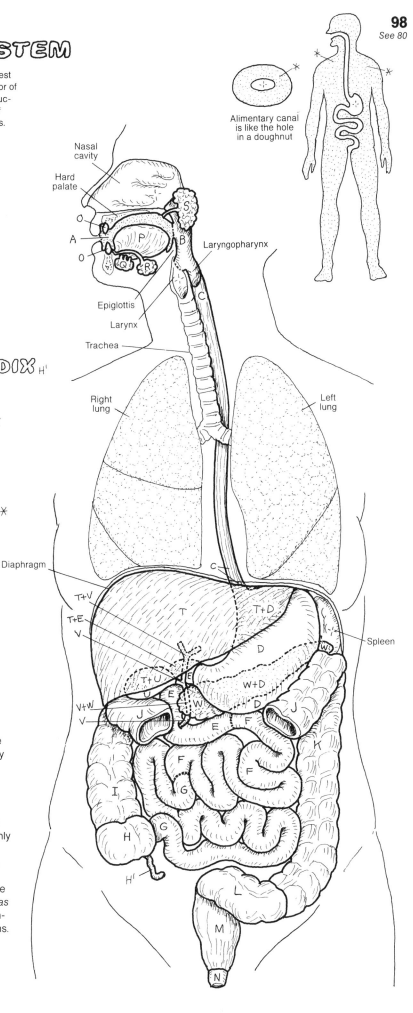

Alimentary canal is like the hole in a doughnut

The digestive system consists of an alimentary canal with accessory organs. The canal begins with the *oral cavity*. Here the *teeth* pulverize ingested food while it is softened and partly digested by *salivary gland* secretions. The *tongue* aids in mechanical manipulation of the food, and literally flips the food into the fibromuscular *pharynx* during swallowing.

The *esophagus* moves the bolus along to the *stomach* by peristaltic muscular contractions. Here the bolus is treated to mechanical and chemical digestion, then passed into the highly coiled *small intestine* for more enzymatic and mechanical digestive processes. Small molecular nutrients are extracted, absorbed by lining cells, and transferred to capillaries. Liver-produced bile, stored in the *gall bladder*, is discharged into the duodenum by *bile ducts*. Digestive enzymes from the *pancreas* enter the duodenum as well. The *large intestine* is mainly concerned with absorption of water, minerals, and certain vitamins. The non-nutritive residue of the ingested bolus is moved through the *rectum* and *anal canal* to the outside. Nutrients absorbed throughout the tract are transported to the *liver* by the hepatic portal system for processing and distribution to the body's cells.

ORAL CAVITY & TONGUE

CN: Use pink for K and very light colors for A, B, T, U, and V. Do not color the teeth. (1) When coloring the mouth, also color many of those structures that appear in the sagittal view. (2) It is not necessary to color all of the papillae of Q and R.

FRENULUM OF LIP_A
GINGIVA (GUM)_B
HARD PALATE_C
SOFT PALATE_D
UVULA_E
PALATOGLOSSAL ARCH_F
OROPHARYNX_G
PALATOPHARYN-GEAL ARCH_H
TONSILLAR FOSSA_I
PALATINE TONSIL_J
TONGUE_K
BUCCINATOR MUSCLE_L
BUCCAL FAT_M

ORAL CAVITY (Sagittal view)

Upper lip · Nasopharynx · Oral cavity · Laryngopharynx · Vocal cord · Trachea · Esophagus · Mandible

EPIGLOTTIS_N
LINGUAL TONSIL_O
PAPILLAE: *
CIRCUMVALLATE_P
FILIFORM_Q
FUNGIFORM_R
TASTE REGIONS: *
BITTER_S
SOUR_T
SALT_U
SWEET_V

The human mouth is concerned with vocalization as well as mastication and swallowing (deglutition). Its anterior half, including teeth, muscular *tongue* and related extrinsic muscles, *salivary glands*, *hard* (bony) *palate*, and *buccinator muscle* in the cheek wall, is concerned with wetting, macerating, and pulverizing ingested material. Thousands of mucous glands in the stratified squamous-lined mucosa of the mouth assist in these functions, as do the multiple, microscopically-towering papillae on the surface of the tongue, the latter forming an abrasive surface for mechanical digestion. The temporomandibular joints permit a fairly wide range of lower jaw motion and mouth opening (35–50 mm interincisal range in the adult). The posterior half of the mouth, including *soft* (muscular) *palate*, tongue, and *tonsils* between the muscular *arches*, is concerned with immune defense and propelling the mechanically treated food into the pharynx. Sense receptors (taste buds) buried among the papillae on the tongue surface are responsive to chemical stimuli dissolved in the saliva. These receptors are arranged in a pattern reflecting specific sensitivity to molecular variations, i.e., *bitter, sour, salt,* and *sweet* tastes.

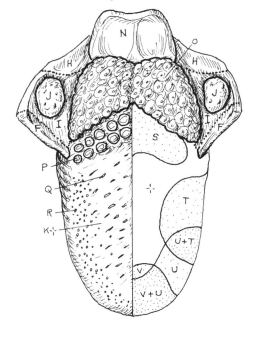

SALIVARY GLANDS: *
SUBLINGUAL_W
SUBMANDIBULAR_X
PAROTID_Y

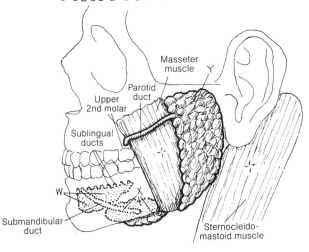

Masseter muscle · Parotid duct · Upper 2nd molar · Sublingual ducts · Submandibular duct · Sternocleidomastoid muscle

Salivary glands secrete a mixed water/mucus, enzyme-containing fluid into the mouth during periods of eating (or anticipated eating). Specialized muscle (myoepithelial) cells at the base of the glands stimulate secretion into the ducts following stimulation by autonomic nerves. The *sublingual glands* are the smallest of the three paired glands. Their ducts open on to the floor of the mouth, as do the ducts of the *submandibular gland*. The *parotid gland* is the largest, sending its duct across the masseter muscle, through the cheek, and into the oral cavity opposite the upper 2nd molar tooth.

ANATOMY OF A TOOTH

CN: Use yellow for F, red for G, blue for H, and light colors for A, B, and L.
(1) Begin with the anatomy of a tooth. Color gray the titles and arrows/bands arranged vertically. (2) Use only light colors on the teeth below. You may repeat colors used on the upper illustration. Note that the identifying letter and number labels are those used by the dental profession.

ENAMEL A
DENTIN B
PULP CAVITY C
   ROOT CANAL D
   PULP E
     NERVE F **ARTERY** G **VEIN** H
CEMENTUM I
PERIODONTAL MEMBRANE J
GINGIVA K
ALVEOLAR BONE L

The tooth is a hollow core of sensitive, mineralized *dentin* filled with a loose fibrous, vascular *pulp*, capped with insensitive mineralized *enamel* projecting above the *gingiva* (gum), buried within the bony socket (*alveolus*) of the mandible/maxilla and secured to that *periodontal*-lined bone by *cementum*. The bulk of the tooth is rooted in bone; the neck of the tooth is at the gum line. The crown is dentin covered with 1–1.5 mm thick enamel. Enamel is the hardest substance in the body, weighing in at 99% mineral content. It consists of circular rods arranged in a wave pattern stuffed with hydroxyapatite (bone) mineral crystals. During development, non-mineralized enamel (it mineralizes later) is secreted by cells which are worn off the enamel surface when the tooth erupts and becomes exposed.

Dentin is a bone-like material (70% mineral by weight) secreted in tubular form by cells at the dentin-pulp junction. With aging, the pulp diminishes in volume, replaced by dentin. The pulp cavity is filled with an embryonic connective tissue supporting *nerves*, *arteries*, and *veins* that supply the tooth. Pulp passes through a *root* (or *pulp*) *canal* to reach the apical or root foramen. The pulp is continuous with the periodontal membrane, a dense fibrous tissue similar to periosteum. Cementum is like bone, mineralized with a significant content of collagen fibers. It serves as an intermediate tissue between dentin and the periodontal membrane, with many of its fibers buried in alveolar bone. The gingiva, lined with keratinized stratified squamous epithelium, is part of the mucous membrane of the mouth, and surrounds the neck of each tooth. The gingival epithelial cells are attached to the tooth surface. The gingiva is firmly anchored to the periosteum of the underlying alveolar bone.

ADULT/CHILD DENTITION: *

CENTRAL INCISOR 8, 9, 24, 25, E, F, O, P
LATERAL INCISOR 7, 10, 23, 26, D, G, N, Q
CANINE 6, 11, 22, 27, C, H, M, R
1ST PREMOLAR 5, 12, 21, 28
2ND PREMOLAR 4, 13, 20, 29
1ST MOLAR 3, 14, 19, 30, B, I, L, S
2ND MOLAR 2, 15, 18, 31, A, J, K, T
3RD MOLAR (WISDOM) 1, 16, 17, 32

Two sets of teeth develop within a lifetime. The first set are deciduous (milk teeth), There are 10 in each jaw. The incisors are usually the first to erupt at about 6 months; the rest follow within 36 months after birth. Pressure from the permanent teeth induce osteoclastic resorption of the milk teeth roots, and subsequently the remaining crowns dislodge and fall out without pain or bleeding. In the permanent set, the *first molar* or *central incisor* erupts first (at about 6 years). The second molar erupts around 11 years of age; the *third molar* generally emerges about 18 years ("wisdom tooth").

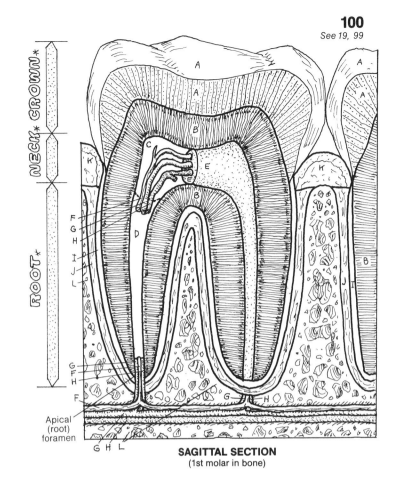

Crown — A
Neck
Root — B

SAGITTAL SECTION
(1st molar in bone)

Apical (root) foramen

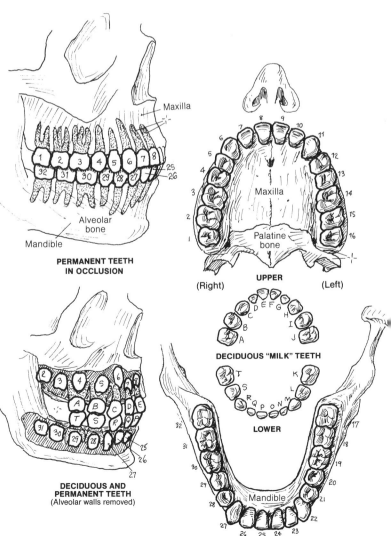

Maxilla

Alveolar bone

Mandible

PERMANENT TEETH IN OCCLUSION

Maxilla

Palatine bone

UPPER
(Right)   (Left)

DECIDUOUS "MILK" TEETH

LOWER

DECIDUOUS AND PERMANENT TEETH
(Alveolar walls removed)

Mandible

PHARYNX & ESOPHAGUS

CN: Use pink for K. (1) Color the three lower illustrations simultaneously. In the posterior view of the interior of the pharynx, the posterior pharyngeal wall is divided and retracted so you can note the relationship of internal pharyngeal structure to the constrictor muscles (A, B, C) and the subdivisions of the pharynx (D, G, I). Color gray the boluses of food in both upper and middle lower views, and the titles at upper right. (2) Follow the text when coloring the deglutition diagrams.

MUSCULAR WALL OF PHARYNX: *
SUPERIOR CONSTRICTOR A
MIDDLE CONSTRICTOR B
INFERIOR CONSTRICTOR C
INTERIOR OF PHARYNX: *
NASOPHARYNX D
SOFT PALATE E
UVULA F
OROPHARYNX G
PALATINE TONSIL H
LARYNGOPHARYNX I
ESOPHAGUS J
RELATED STRUCTURES: *
TONGUE K
HYOID BONE L
THYROID CARTILAGE M
CRICOID CART. N & MUS. N'
TRACHEA O
NASAL CAVITY P
EPIGLOTTIS Q
LARYNX R

ORAL CAVITY & BOLUS OF FOOD T *

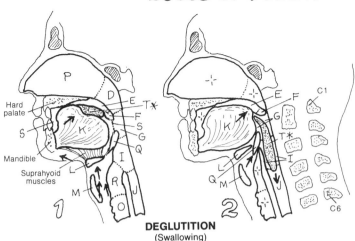

DEGLUTITION
(Swallowing)

The pharynx is a complex fibromuscular, mucosa-lined sac, open to the *nasal cavity* above, the *oral cavity* in front, and the *larynx* and *esophagus* below. Like three stacked pots, the *constrictor* muscles of the pharyngeal wall overlap one another posteriorly and posterolaterally. Several small muscles (not shown) reinforce the constrictor muscles structurally and functionally. The pharyngeal muscles are primarily concerned with deglutition. Swallowing begins with pushing the food bolus from the oral cavity into the *oropharynx* (1). This is done with the *tongue* assisted by the suprahyoid muscles pulling the *hyoid bone* and *larynx* upward. The soft palate (levator palati) then elevates and the *superior constrictors* contract, closing off the nasopharynx. Incarcerated in the oral pharynx, unable to return to the mouth or enter the nasal cavity, the bolus shoots into the laryngopharynx (2) with the aid of the *middle* and *inferior constrictors*, past the closed larynx (pinched off by the aryepiglottic folds) and into the esophagus.

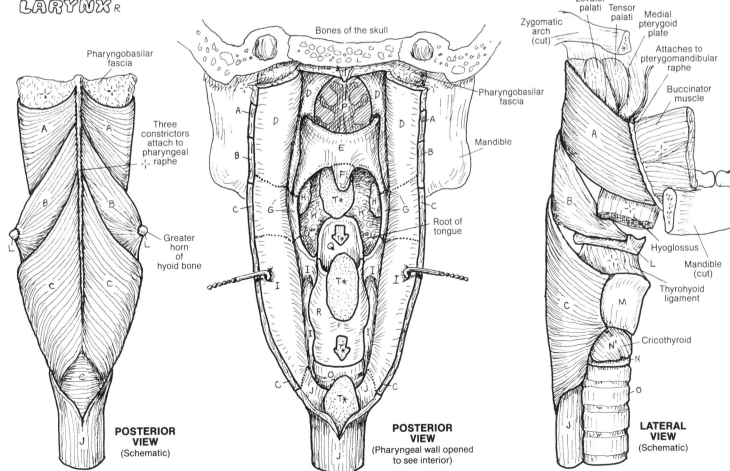

POSTERIOR VIEW
(Schematic)

POSTERIOR VIEW
(Pharyngeal wall opened to see interior)

LATERAL VIEW
(Schematic)

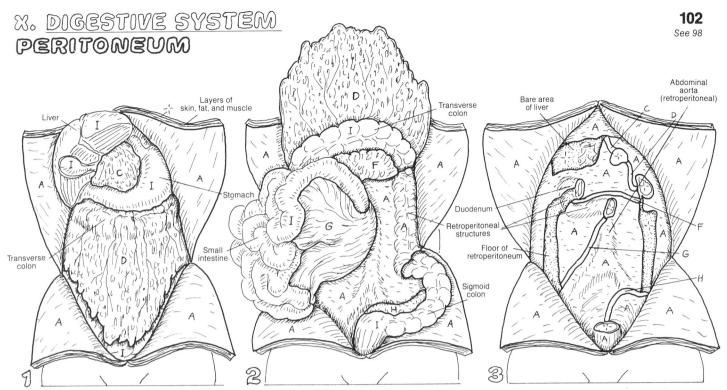

1

With the anterior abdominal wall opened through its deepest (*parietal peritoneal*) layer, the liver, stomach, and the fatty *greater omentum* are generally all that can be seen with the contents undisturbed. Lifting the liver exposes the *lesser omentum*, a double-peritoneal layer between stomach and liver. It is the anterior wall of the *omental bursa* (E). The greater omentum connects the transverse colon to the stomach.

2

With the greater omentum lifted, the double-layered, *transverse mesocolon* between transverse colon and the parietal peritoneum can be seen. Retracting the intestines to one side reveals the *common mesentery* between most of the small intestine and the parietal peritoneum on the posterior body wall. The sigmoid colon has a mesentery (*sigmoid mesocolon*) as well. Abdominal structures posterior to these mesenteries/omenta are retroperitoneal.

3

The parietal peritoneum of the posterior body wall is seen when all structures except retroperitoneal ones (aorta, inferior vena cava, kidneys, ureters, pancreas, duodenum, ascending/descending colon) are removed. Many nerves and vessels travel in this retroperitoneal space. As organs emerge from the peritoneum, they develop a mesentery to suspend them. The cut layers of several of them can be seen (C, D, F, G, and H).

CN: Use a very light color for A and I. (1) Color the upper three diagrams in numerical order. Note that the digestive organs are covered with visceral peritoneum (I). (2) Color the sagittal view. Use a darker gray or black for the omental bursa (E). The space of the peritoneal cavity (B) has been greatly exaggerated for clarity of peritoneal membranes.

PERITONEAL STRUCTURES:
PARIETAL PERITONEUM A
PERITONEAL CAVITY B*
LESSER OMENTUM C
OMENTAL BURSA E•
GREATER OMENTUM D
TRANSVERSE MESOCOLON F
COMMON MESENTERY G
SIGMOID MESOCOLON H
VISCERAL PERITONEUM I

Peritoneum is a serosal membrane of the abdominal cavity. The disposition of the peritoneum is similar to that of the serosal layers around heart (pericardium) and lungs (pleura): peritoneum attached to the body wall is *parietal*; peritoneum attached to the outer visceral wall is *visceral*. Structures deep to the posterior parietal peritoneum are retroperitoneal. Peritoneal layers suspending organs are called *mesenteries*; those suspending an organ from another organ are called *omenta* or *ligaments*. When coloring the sagittal view, the continuity of these peritoneal membranes can be appreciated. The cavity of the peritoneum is empty; it can fill with fluid in disease and trauma. The view at right shows intestines separated apart from one another; in life, they are as close together as strands of coiled wet rope. Vessels/nerves to the intestines and stomach travel in the mesenteries/omenta; they do not penetrate peritoneal layers. The source vessels are retroperitoneal. The *omental bursa* is a peritoneal-lined sac created by rotation of the stomach during fetal life. It is open on the right at the epiploic foramen between the lesser omentum and the parietal peritoneum. Here the omental bursa (lesser sac) communicates with the collapsed, empty peritoneal cavity (greater sac).

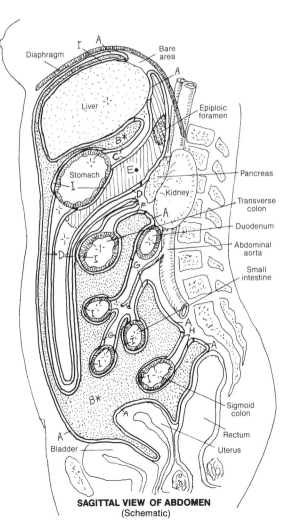

SAGITTAL VIEW OF ABDOMEN
(Schematic)

X. DIGESTIVE SYSTEM
STOMACH

CN: Use light colors for E–J, and O. (1) Color the regions of the stomach. (2) Color simultaneously the large view of the stomach and the section of the stomach wall. The layers of the wall in the large view have been enlarged to facilitate coloring. Note the oblique muscle (G) does not reach the duodenum. (3) Color the lower diagrams.

REGIONS: *
CARDIA A
FUNDUS B
BODY C
PYLORUS D
STOMACH WALL: *
MUCOSAL SURFACE (RUGAE) E
SUBMUCOSA F
MUSCULARIS EXTERNA: -:-
OBLIQUE M. G
CIRCULAR M. H
LONGITUDINAL M. I
SEROSA J

The stomach is the first part of the gastrointestinal tract. It acidifies ingested food to enhance protein digestion and kill microorganisms, secretes proteolytic enzymes (pepsin), mechanically manipulates digesting food, and induces secretion of bile and pancreatic enzymes. The *cardia* is the area of the gastroesophageal junction; though there is circular muscle here, it relaxes during swallowing. The *pylorus* is thickened with circular muscle near the duodenal junction (pyloric sphincter). Its regulatory function has been questioned.

While coloring note carefully the organization of the stomach wall; the mucosa is considered below. The fibrous, vascular *submucosa* provides some support for the larger vessels and nerves traveling in this layer. When the stomach is not full, the mucosa and submucosa are often thrown into a series of irregular folds (rugae). The different orientations of the muscularis externa layer provide complex and effective peristaltic movements during digestion.

The mucosa of the stomach, lined with simple columnar epithelial cells with microvilli, contains a subepithelial, vascular, loose, fibrous tissue layer (*lamina propria*) supporting the gastric glands. It contains numerous fibroblasts and particularly dense masses of lymphocytes (lymphoid follicles; see Plate 84). The *fundus* and *body* reveal tubular-shaped gastric glands that exhibit deep *gastric pits*, a neck (largely *mucous cells*) and a base containing mucous cells, *parietal cells* secreting hydrochloric acid, *chief cells* secreting the protein-lysing enzyme pepsin, and *enteroendocrine cells* secreting gastrointestinal regulatory hormones. The parietal cells also secrete intrinsic factor, a glycoprotein which binds with vitamin B_{12} and permits the latter's absorption in the ileum of the small intestine. Malabsorption of intrinsic factor leads to vitamin B_{12} deficiency which induces abnormal erythrocyte (RBC) development and subsequent pernicious anemia. The *pylorus* contains largely mucous glands as well as enteroendocrine cells secreting gastrin, a polypeptide. It stimulates secretion of pepsin and acid in the stomach, and augments gastric muscle contractions (increased motility). The smooth muscle fibers of the thin *muscularis mucosae* participate in the mechanical digestive process.

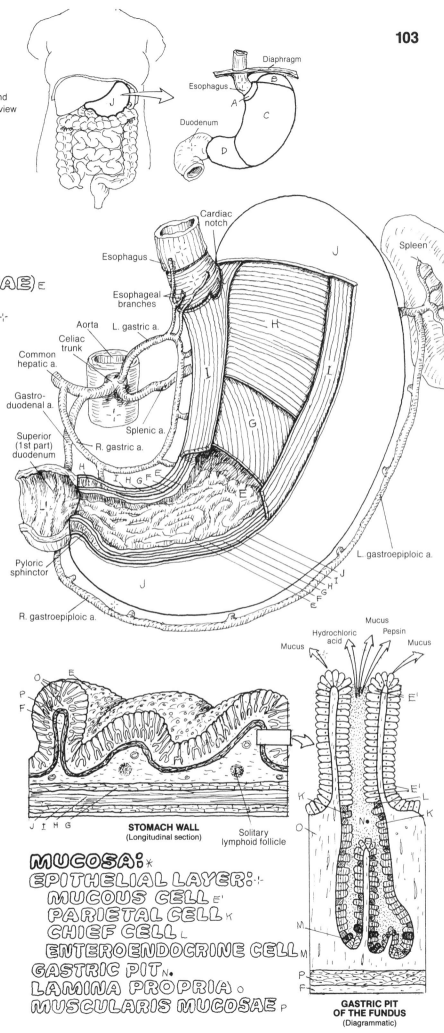

STOMACH WALL
(Longitudinal section)

MUCOSA: *
EPITHELIAL LAYER: -:-
MUCOUS CELL E'
PARIETAL CELL K
CHIEF CELL L
ENTEROENDOCRINE CELL M
GASTRIC PIT N.
LAMINA PROPRIA O
MUSCULARIS MUCOSAE P

GASTRIC PIT OF THE FUNDUS
(Diagrammatic)

104

SMALL INTESTINE

DUODENUM: A
SUPERIOR (1ST) PART B
DESCENDING (2ND) PART C
HORIZONTAL (3RD) PART D
ASCENDING (4TH) PART E
JEJUNUM F
ILEUM G

INTESTINAL WALL: *
MUCOSA: ⊹
PLICA CIRCULARE (FOLD) H
VILLI H'
LAMINA PROPRIA I
MUSCULARIS MUCOSAE J
LYMPHOID FOLLICLE K
SUBMUCOSA L
MUSCULARIS EXTERNA: ⊹
CIRCULAR M. M
LONGITUDINAL M. N
SEROSA D'

CN: Use green for K, red for R, purple for S, blue for T, yellow for U, and a very light color for H. (1) Begin with the three divisions of the small intestine. (2) Color the parts of the duodenum and the section of duodenal wall. The lamina propria (I) is identified and colored only in the enlarged view of the villi below.

The small intestine is a highly convoluted, thin-walled tube that undertakes much of the chemical and mechanical digestive process and almost the whole of the absorptive process of the entire gastrointestinal tract. The first part of the *duodenum* is suspended by the lesser omentum. The second and third parts are retroperitoneal. The fourth part emerges anteriorly to become embraced by the common mesentery, pulled upward/suspended by a band of smooth muscle at the duodenojejunal junction. The *jejunum* is highly coiled, suspended by the common mesentery between the peritoneal layers through which travel its blood and nerve supply and draining veins. The thinner but longer *ileum* is also suspended by the common mesentery. It opens into the cecum of the large intestine.

The mucosal surface of the jejunum (and to a lesser extent the duodenum and ileum) is characterized by circular folds (*plicae circulares*). Myriads of conical, finger-shaped *villi* (leaf-shaped in the duodenum) project from the surface of the jejunum; these diminish in number in the ileum. Among the epithelia lining the villi are hormone-secreting *enteroendocrine cells* (see glossary). The loose, vascular, fibrous *lamina propria* support the villi and its contents, including lymphatic capillaries (*lacteals*) and muscle fibers from the *muscularis mucosae*. *Lymphoid follicles* exist in both lamina propria and *submucosa*; they increase in number in the ileum where they form aggregates (Peyer's patches). At the base of the villi are tubular *intestinal glands*, the *ducts* of which open into the intervillous spaces. The submucosa of the small intestine is fibrous and *vascular*, and contains ganglia of autonomic motor neurons, the *axons* of which supply the muscularis externa. In the duodenum only, mucus-secreting glands occupy the submucosa.

ABSORPTIVE CELL H²
MUCOUS CELL O
ENTEROENDOCRINE CELL P
GLANDULAR DUCT Q.
ARTERY R
CAPILLARY S
VEIN T
NERVE U
LACTEAL K'

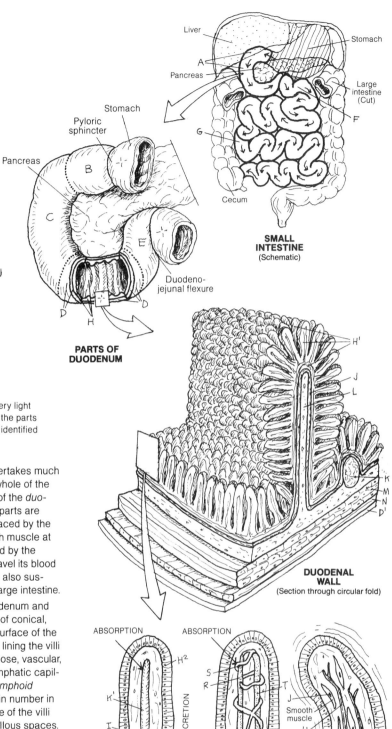

Liver
Stomach
A
Pancreas
Large intestine (Cut)
G
F
Cecum

SMALL INTESTINE
(Schematic)

Stomach
Pyloric sphincter
Pancreas
B
C
E
D
D
H
Duodeno-jejunal flexure

PARTS OF DUODENUM

H'
J
L
K
M
N
N'
D'

DUODENAL WALL
(Section through circular fold)

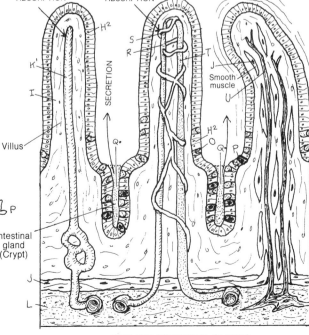

ABSORPTION
ABSORPTION
H²
S
R
K'
T
J
Smooth muscle
I
SECRETION
H²
U
Villus
Q.
O
Q.
P
Intestinal gland (Crypt)
J
L

VILLI AND INTESTINAL GLANDS
(Structures common to all villi are shown separately)

CN: By using the same colors for the parts of the intestinal wall you used on the preceding plate, you can demonstrate the similarity between the structures of the two intestines. The epithelium/mucous glands (N) should receive the same color as the villi (H¹) of Plate 104, and the serosa in both plates should receive the same color. Use a very light color for B. (1) Begin with the section above.

LARGE INTESTINE:✻
CECUM A
ILEOCECAL VALVE B
VERMIFORM APPENDIX C
ASCENDING COLON D
TRANSVERSE COLON E
DESCENDING COLON F
SIGMOID (PELVIC) COLON G
RECTUM H
ANAL CANAL I
INTERNAL SPHINCTER ANI J
EXTERNAL SPHINCTER ANI K

TAENIA COLI L
APPENDICES EPIPLOICA M

INTESTINAL WALL:✻
MUCOSA:⊹
EPITHELIUM/MUCUS GLANDS N
LAMINA PROPRIA O
MUSCULARIS MUCOSAE P
SUBMUCOSA Q
MUSCULARIS EXTERNA:⊹
CIRCULAR M. R
LONGITUDINAL M. L'
SEROSA D'

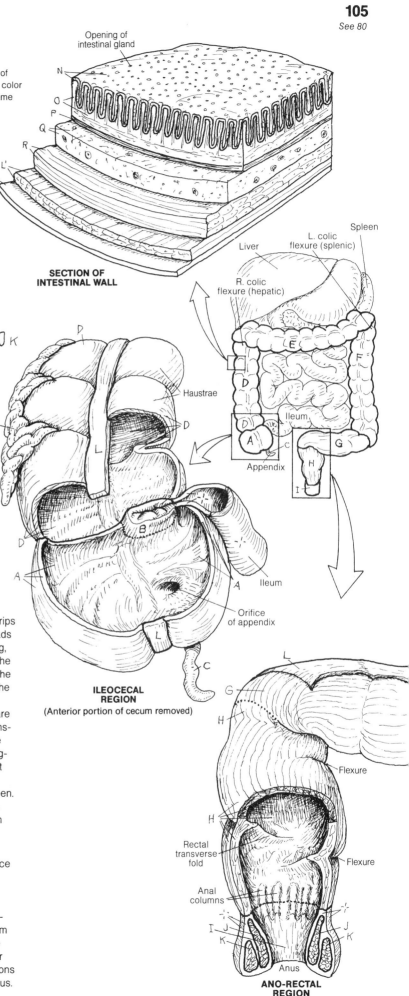

SECTION OF INTESTINAL WALL

Opening of intestinal gland

ILEOCECAL REGION
(Anterior portion of cecum removed)

Haustrae

Ileum

Orifice of appendix

Appendix

Spleen
L. colic flexure (splenic)
R. colic flexure (hepatic)
Liver
Ileum

Flexure

Rectal transverse fold

Flexure

Anal columns

Anus

ANO-RECTAL REGION

The large intestine is characterized by large sacculations (*haustrae*), strips of *longitudinal muscle* in the muscularis externa (*taenia coli*), and fat pads (*appendices epiploica*) attached to the serosal surface of the ascending, transverse, and descending colon (only). The large intestine begins at the *ileocecal valve* with the *cecum*, usually suspended by a mesentery, in the right lower abdominal quadrant. The function of the valve is not clear. The *vermiform appendix* varies in length (2–20 cm); it may lie anterior, posterior, or inferior to the cecum. The *ascending* and *descending colons* are retroperitoneal; the *transverse colon* is suspended by a mesentery (transverse mesocolon). Note the colic flexures and their relationships. At the pelvic inlet (not shown), the colon turns medially, gains a mesentery (sigmoid mesocolon), and is named the *sigmoid colon*. Variable in its extent and shape, it becomes the *rectum* at the level of the S3 vertebra. Here the haustrae, the appendices epiploica, and the taenia are no longer seen. About 12 cm long, the rectum has a dilated lower part (ampulla). Feces entering the rectum stimulate the desire for defecation; thus, the rectum is not a long-term storage site. As the rectum narrows inferiorly, it becomes the *anal canal* surrounded by *sphincter muscles*.

The intestinal wall of the large intestine is characteristic: *mucosal* surface without villi or plicae, underlying vascular *submucosa*, and two-layered *muscularis externa* lined with peritoneal *serosa*. The epithelial lining is simple columnar except in the anal canal where it becomes stratified squamous. The glands are tubular and mucus-secreting. Lymphoid follicles are seen in the lamina propria. At the anorectal junction, about 2 cm above the anus, a remarkably large number of veins can be seen in the lamina propria (not shown). Varicose dilatations of these veins (rectal or hemorrhoidal plexus) are called hemorrhoids. The large intestine functions in absorption of water, vitamins, and minerals, and the secretion of mucus.

LIVER

CN: Use blue for I, red for J, and yellow for K. Use very light colors for A, B, and L. (1) Color the two upper views simultaneously. (2) Color the group of lobules, and then the enlargement. Begin with the branches of the portal vein (I¹) at the bottom of the section. (3) Begin the overview of blood and bile with the arterial flow.

LOBES: *
RIGHT LOBE A
LEFT LOBE B
QUADRATE LOBE C
CAUDATE LOBE D
LIGAMENTS: *
CORONARY L. E
TRIANGULAR L. F
LESSER OMENTUM G
FALCIFORM L. H
PORTA: *
PORTAL VEIN I
HEPATIC ARTERY J
BILE DUCT K
LIVER LOBULE: L
TRIAD: *'
BRANCH OF PORTAL V. I'
BRANCH OF HEPATIC A. J'
BILE DUCT K'
SINUSOID I²
HEPATIC CELL L'
CENTRAL VEIN I³
TRIBUTARY OF HEPATIC V. I⁴

ANTERIOR VIEW

Diaphragm (Lifted)

Superior border

Inferior border

Gall bladder

Ligamentum teres (Round lig.)

Ligamentum venosum

POSTERIOR/INFERIOR VIEW (Visceral surface)

Esophageal impression

Gastric impression

Hepatic veins

Bare area (Nonperitoneal)

Inferior vena cava

Common bile duct

Ligamentum teres (Round lig.)

Gall bladder

Duodenal impression

Renal impression

Colic impression

LOBULES

LIVER LOBULE (Section removed)

Ductule

Triad

Inferior vena cava

Heart

Hepatic veins

Aorta

Diaphragm

R. hepatic duct

L. hepatic a.

Gall bladder

Common bile duct

Cystic duct

Portal system

Small intestine

ARTERIAL BLOOD J
VENOUS BLOOD I
BILE FLOW K

BLOOD AND BILE FLOW (Schematic)

The liver is the largest gland in the body. Wedge-shaped (rounded upper border, thin, sharp inferior border) when seen from the side, the liver occupies the whole of the upper right quadrant of the abdominal cavity. Weighing about 1.5 kg in health, it can weigh over 10 kg when diseased (chronic cirrhosis). It is relatively huge in young children, hence the protuberance of the upper abdomen in such persons. The liver is enveloped in visceral peritoneum except for a part of the right posterior surface which is flush against the fascia covering the diaphragm (bare area). The visceral peritoneum around the bare area turns or reflects upward (*coronary ligaments*) on to the diaphragm to become parietal peritoneum. The edges of the coronary ligaments are called the *triangular ligaments*. The two anterior leaves of the coronary ligaments join to become the *falciform ligament*; the two posterior leaves become the *lesser omentum* which encircles the *porta* of the liver. Here the hepatic *portal vein* and the *hepatic artery* approach the visceral surface of the liver and branch, and the common bile duct receives the common hepatic duct and cystic duct. The two-layered lesser omentum descends to support the pyloric end of the stomach and the first part of the duodenum. The falciform ligament is continuous with the parietal peritoneum of the anterior abdominal wall.

Connective tissue septa divide the liver cells/tissue (parenchyma) into irregular polyhedral *lobules*. Within each lobule, the cells form radially arranged cords; on two surfaces of these cords are sinusoids that converge onto a more or less *central vein*. At the corners of these lobules are the hepatic artery and portal vein branches, and bile ducts (called a *triad*). The portal vein branches feed into the sinusoids; the hepatic artery branches supply the cells; the *bile ducts* drain the bile ductules formed from tiny canaliculi (not shown) surrounding the cells. The liver cells discharge their products into the sinusoids (except bile), and absorb from these same sinusoids various nutrients and non-nutrients as well. Liver cells store and release proteins, carbohydrates, lipids, iron, and certain vitamins (A, D, E, K); they manufacture urea from amino acids, and bile from pigments and salts; they detoxify many harmful ingested substances. Bile is released from the cells into tributaries of bile ducts. The central veins are tributaries of larger veins which merge to form three *hepatic veins* at the posterior, superior aspect of the liver. These veins join the inferior vena cava just inferior to the diaphragm.

BILIARY SYSTEM & PANCREAS

CN: Use the same colors as were used on the preceding plate for the hepatic cells and bile ducts, and a very light color for H. (1) Color simultaneously the diagram of bile formation/transport and the large central illustration. (2) Color the diagram describing bile storage.

LIVER (HEPATIC) CELL A
BILE B
R. & L. HEPATIC DUCT C'
COMMON HEPATIC DUCT D
GALL BLADDER E
CYSTIC DUCT F
COMMON BILE DUCT G
PANCREAS H
PANCREATIC DUCT I
DUODENAL AMPULLA J

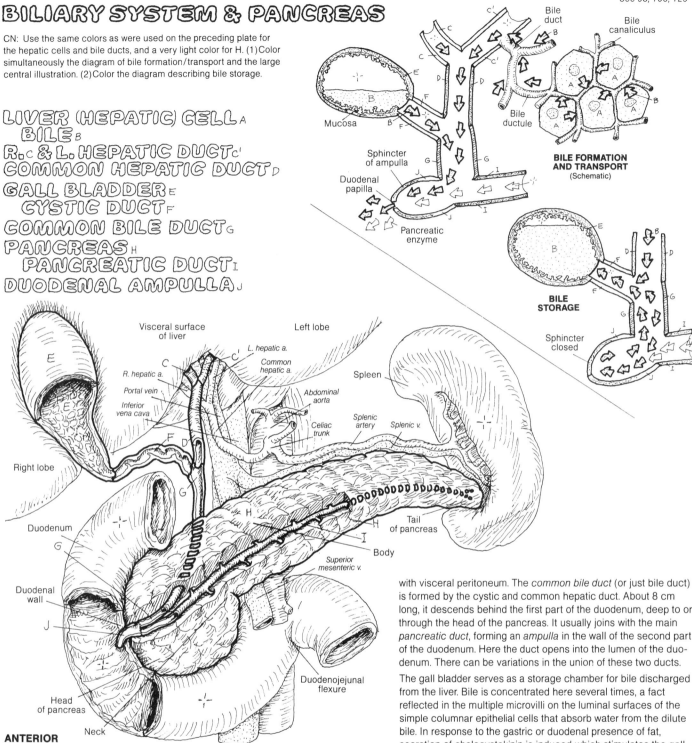

BILE FORMATION AND TRANSPORT
(Schematic)

BILE STORAGE

Sphincter closed

ANTERIOR VIEW
(Stomach removed)

The biliary system consists of an arrangement of *ducts* transporting *bile* from the *liver cells* that manufacture it to the *gall bladder* and to the second part of the duodenum.

It is worth repeating: bile is formed in the liver (not the gall bladder). It is a fluid consisting largely of water (97%), and bile salts and pigments (from the breakdown products of hemoglobin in the spleen). Once formed, bile is discharged from liver cells into surrounding *bile canaliculi*. These small canals merge to form bile ductules that join the bile ducts which travel in company with the branches of the portal vein and hepatic artery. The bile is brought out of the liver by the *right and left hepatic ducts* which merge at the porta to form the *common hepatic duct* which descends between the layers of the lesser omentum and receives the 4 cm-long *cystic duct* from the *gall bladder*. The gall bladder is pressed against the visceral surface of the right lobe of the liver, covered

with visceral peritoneum. The *common bile duct* (or just bile duct) is formed by the cystic and common hepatic duct. About 8 cm long, it descends behind the first part of the duodenum, deep to or through the head of the pancreas. It usually joins with the main *pancreatic duct*, forming an *ampulla* in the wall of the second part of the duodenum. Here the duct opens into the lumen of the duodenum. There can be variations in the union of these two ducts.

The gall bladder serves as a storage chamber for bile discharged from the liver. Bile is concentrated here several times, a fact reflected in the multiple microvilli on the luminal surfaces of the simple columnar epithelial cells that absorb water from the dilute bile. In response to the gastric or duodenal presence of fat, secretion of cholecystokinin is induced which stimulates the gall bladder to discharge its contents into the cystic duct. Peristaltic contractions of the duct musculature squirt bile into the duodenal lumen through the ampullary sphincter. Bile saponifies and emulsifies fats, making them water soluble and amenable to digestion by enzymes (lipases).

The *pancreas* is a gland in the retroperitoneum, consisting of a head, neck, body, and tail. Most of the pancreas consists of exocrine glands that secrete enzymes into the pancreatic duct tributaries and on into the duodenum at a rate of about 2000 ml per 24 hour day. These enzymes are responsible for a major part of chemical digestion in the small intestine (lipases for fat, peptidases for protein, amylases for carbohydrates, and others). Pancreatic secretion is regulated by hormones (primarily cholecystokinin and secretin) from entero-endocrine cells and by the vagus nerves (acetylcholine). The endocrine portion of the pancreas is covered in Plate 129.

URINARY TRACT: *
KIDNEY_A
URETER_B
URINARY BLADDER_C
URETHRA_D
  PROSTATIC U. (MALE)_D'
  MEMBRANOUS U.(MALE)_D²
  SPONGY U. (MALE)_D³

KIDNEY RELATIONS: *
SUPRARENAL
  GLAND_E
LIVER_F
DUODENUM_G
TRANSVERSE
  COLON_H
SPLEEN_I
STOMACH_J
PANCREAS_K
JEJUNUM_L

ANTERIOR VIEW
(Kidney surface areas)

Renal pelvis

Opening of ureter

Likely location of stones

Trigone of bladder

Prostate gland

Urogenital diaphragm

Penis

Scrotum

External urethral orifice

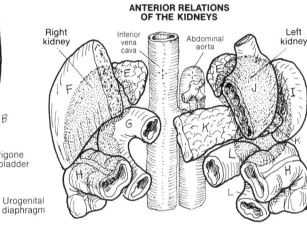

CN: Use very light colors for C and E-L.
(1) Color the three views of the urinary tract simultaneously. Note that the kidneys at the top of the plate are to be colored according to areas which are in contact with other organs. Also note that the ureters penetrate the posterior wall of the urinary bladder, and that these openings receive a color. (2) Color the anterior relations of the kidneys which are shown as shaded silhouettes. Color gray the arrows marking sites of potential obstruction by "stones."

ANTERIOR RELATIONS OF THE KIDNEYS

Right kidney
Inferior vena cava
Abdominal aorta
Left kidney

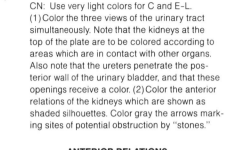

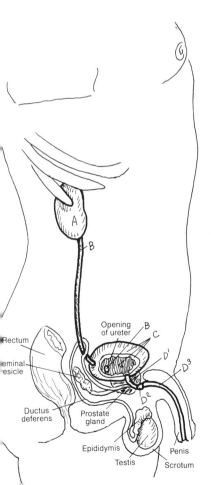

Rectum
Seminal vesicle
Ductus deferens
Prostate gland
Opening of ureter
Epididymis
Testis
Penis
Scrotum

The urinary tract consists of paired *kidneys* and *ureters* in the retroperitoneum, a single *urinary bladder*, and a *urethra*. The urinary tract represents a pathway for the elimination of metabolic by-products and toxic and other non-essential molecules all dissolved in a small volume of water (urine). The *kidneys* are not simply instruments of excretion; they function in the conservation of water and maintenance of acid-base balance in the blood. The process is a dynamic one, and what is excreted as waste in one second may be retained as precious in the next.

The *ureters* are fibromuscular tubes, lined by transitional epithelium. Three areas of the ureters are relatively narrow and are prone to being obstructed by mineralized concretions ("stones") from the kidney (see arrows).

The fibromuscular *urinary bladder* lies in the true pelvis, its superior surface covered with peritoneum. The mucosa is lined with transitional epithelium. The bladder can contain as little as 50 ml of urine and can hold as much as 700–1000 ml without injury; as it distends, it rises into the abdominal cavity and bulges posteriorly. The mucosal area between the two ureteral orifices and the urethral orifice is called the trigone.

The fibromuscular, glandular *urethra*, lined with transitional epithelium except near the skin, is larger in males (20 cm) than females (4 cm). Hence, urethritis is more common in men, cystitis is more common in women.

The urethra is described in three parts in the male (*prostatic, membranous and spongy*). The membranous urethra is vulnerable to rupture in the urogenital diaphragm with trauma to the low anterior pelvis.

12th rib

Ovary
Uterine tube
Uterus
Opening of ureter
Rectum
Vagina

KIDNEYS & RELATED RETROPERITONEAL STRUCTURES

CN: Use red for B, blue for L, and a very light color for X (use a color, not gray). (1) Color the various structures in the abdominal cavity. Part of the peritoneum (X), whose title is among the upper diagrams, is shown covering much of the right side. (2) At the upper right, note the relationship of the retroperitoneum to the parietal peritoneum.

KIDNEY A
   URETER A'
   URINARY BLADDER A²
AORTA B
   & BRANCHES: B
   CELIAC A. & BRS. C
   SUPRARENAL A. D
   SUP. MESENTERIC A. E
   RENAL A. F
   TESTICULAR A. G
   INF. MESENTERIC A. H
   COMMON ILIAC A. I
   INTERNAL ILIAC A. J
   EXTERNAL ILIAC A. K

INFERIOR VENA CAVA L
   & TRIBUTARIES: L
   INTERNAL ILIAC V. M
   EXTERNAL ILIAC V. N
   COMMON ILIAC V. O
   TESTICULAR V. P
   RENAL V. Q
   SUPRARENAL V. R
   HEPATIC VS. S

ORGANS & DUCTS: *
   ESOPHAGUS T
   SUPRARENAL GLAND U
   RECTUM V
   DUCTUS (VAS) DEFERENS W

The paired *kidneys* and *ureters* lie posterior to the *parietal peritoneum* of the abdominal cavity; they are, therefore, in the *retroperitoneum*. During fetal development, some abdominal structures arise in the retroperitoneum (e.g., kidneys), and some become retroperitoneal as a result of movement of visceral organs (e.g., ascending/descending colon, pancreas). The abdominal *aorta and its immediate branches* and the *inferior vena cava and its immediate tributaries* are all retroperitoneal. Arteries and veins travel between layers of peritoneum to reach the organs they supply/drain. Lymph nodes, lumbar trunks. and the cysternal chyli (not shown) are all retroperitoneal. The ureters descend in the retroperitoneum and under the parietal peritoneum to reach the posterior and inferior aspect of the bladder. Pelvic viscera and vessels lie deep to the parietal peritoneum.

The kidneys are encapsulated in perirenal fat, secured by an outer, stronger layer (renal fascia). Each kidney and its fascia are packed in pararenal fat. These compartments do not communicate between left and right. Such a support system permits kidney movement during respiration but secures them against impact forces.

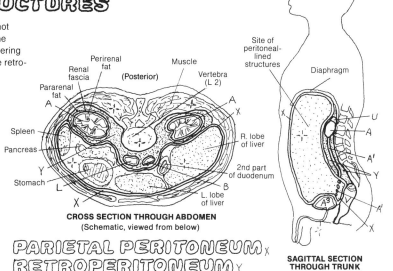

CROSS SECTION THROUGH ABDOMEN
(Schematic, viewed from below)

PARIETAL PERITONEUM X
RETROPERITONEUM Y

SAGITTAL SECTION THROUGH TRUNK
(Schematic)

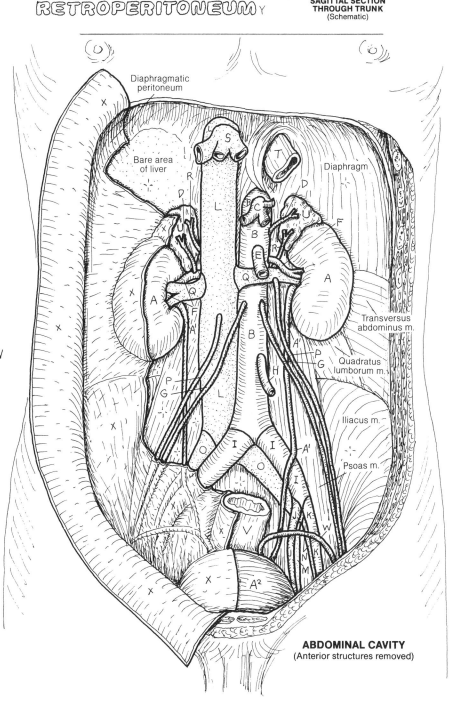

ABDOMINAL CAVITY
(Anterior structures removed)

XI. URINARY SYSTEM
KIDNEY STRUCTURE

CN: Use red for H, blue for I, yellow for K, and very light colors for B, E, F, G, and J. (1) Begin with the large illustration and note that the thickness of the renal capsule (A) has been greatly exaggerated for coloring purposes. Color the cut edges of blood vessels in the cortex (B). Also color the titles and arrows reflecting blood and urine flow. (2) Complete the overview diagram at the top of the plate.

PERIPHERAL PART: *
RENAL CAPSULE A
CORTEX B
MEDULLA (PYRAMID) C
PAPILLA D

INNER (CENTRAL) PART: *
MINOR CALYX E
MAJOR CALYX F

HILUS: *
RENAL PELVIS G
RENAL ARTERY H
OXYGENATED BLOOD H'
RENAL VEIN I
DEOXYGENATED BLOOD I'
RENAL SINUS J

URETER K
URINE K'

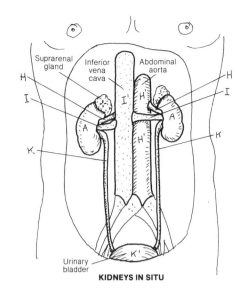

Suprarenal gland
Inferior vena cava
Abdominal aorta
Urinary bladder

KIDNEYS IN SITU

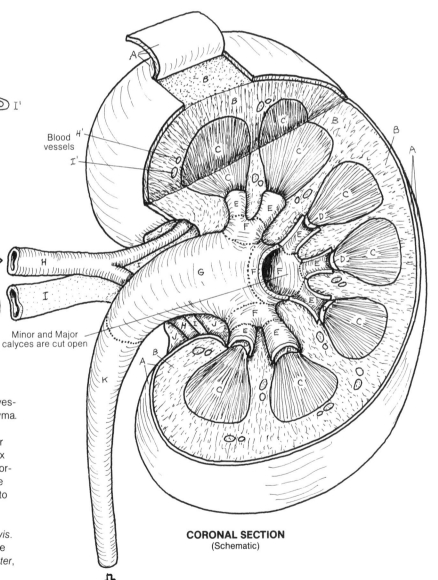

Blood vessels H' I'

1300 mL/min H' (Into both kidneys)

1299 mL/min I' (Out of both kidneys)

Minor and Major calyces are cut open

CORONAL SECTION
(Schematic)

0.7 mL/min K'

The kidney consists of filtering capsules, tubules and blood vessels tightly pressed together into what is called the parenchyma. The parenchyma of the kidney consists of an outer *cortex* covered on its surface by a thin fibrous *capsule*, and an inner *medulla* consisting of pyramids of straight tubules. The cortex reaches down between the pyramids (renal columns). The cortex consists of convoluted tubules and filtering capsules. The apex of each medullary pyramid forms a *papilla* which fits into the small cup-shaped funnel called the *minor calyx*. These funnels, numbering 8–18, open into three much larger *major calyces* all of which open into the cavity called the *renal pelvis*. In the concavity of the kidney (the hilus), in an area called the *renal sinus*, the renal pelvis narrows to form the proximal *ureter*, sharing the area with the renal artery and vein.

Renal blood flow (the amount of blood flowing through the kidneys) is about 1300 mL per minute (both kidneys). About 125–130 mL of plasma is filtered into the renal tubular systems each minute. Less than 1% of that filtered plasma (about 0.7 mL) is actually excreted as *urine*. Clearly, the kidney is in the water conservation business!

XI. URINARY SYSTEM
URINIFEROUS TUBULE

CN: Use red for E, yellow for K, and the same colors used on the preceding plate A–D.
(1) Complete the drawing above. (2) In the enlarged wedge-shaped section, one uriniferous tubule is shown; actually, thousands are packed in each such section. Color all directional arrows the color of the adjacent vessel. (3) In the diagram below, color the capsule space gray but not the arrows representing filtrate.

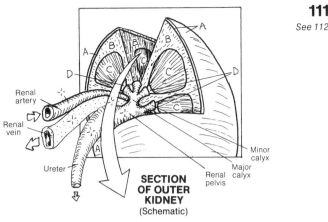

Renal artery
Renal vein
Ureter
Minor calyx
Major calyx
Renal pelvis

SECTION OF OUTER KIDNEY
(Schematic)

KIDNEY SECTION: *
CAPSULE A
CORTEX B
MEDULLA C
PAPILLA D

NEPHRON: -/-
RENAL CORPUSCLE: *
GLOMERULUS E
GLOMERULAR (BOWMAN'S) CAPSULE: F
PARIETAL LAYER F
VISCERAL LAYER (PODOCYTES) F¹
CAPSULAR SPACE F² *
PROXIMAL TUBULE: G
CONVOLUTED PART G
STRAIGHT PART G¹
LOOP OF HENLE H
DISTAL TUBULE: I
STRAIGHT PART I
CONVOLUTED PART I¹
COLLECTING TUBULE J
URINE K

BLOOD VESSELS: *
INTERLOBULAR ARTERY E¹
AFFERENT ARTERIOLE E²
EFFERENT ARTERIOLE E³

The functional unit of the kidney is the nephron (one million per kidney). Each nephron and a collecting tubule constitute a uriniferous (renal) tubule; each nephron consists of a renal corpuscle and tubules leading to a collecting tubule. The renal corpuscles are in the *cortex*; the tubules are in both the cortex and the *medulla*. Each corpuscle consists of a *glomerular capsule* invaginated by a cluster of specialized capillary-like vessels (*glomerulus*). The vessel leading into each glomerulus is an *afferent arteriole*, a downstream, 5th-order branch of the renal artery. Its entrance into the capsule is the vascular pole. The *efferent arteriole* departs the vascular pole, its blood destined for the tubular capillary plexus (next plate).

The capsule is shaped like a soft, rubber, partly flat hollow ball pushed in on one side so that it has an outer and an inner layer to it. The inner layer (of cells) is called the *visceral layer*; the outer layer the *parietal layer*; the interior is the *capsular space* which opens into the proximal tubule (urinary pole). The visceral layer is intimately and complexly interwoven with the glomerular vessels. Each cell in the layer has the shape of a centipede, with a "body" containing the nucleus, and multiple "legs" (cell membrane-lined cytoplasmic extensions called primary processes). These processes incompletely encircle the glomerular vessels, leaving slits (interdigitations) among the processes. The "legs" have "feet" (called foot or secondary processes) which attach to the porous vessel wall in such a way as to leave filtration spaces among them. These highly modified, simple squamous epithelial cells of the visceral layer are called *podocytes*. Plasma escapes the glomerular vessel through the pores, then rushes through the filtration slits to enter the capsular space. This non-cellular plasma filtrate enters the proximal tubule.

We continue with the structure and function of the parts of the uriniferous tubule, in conjunction with the vascular system, in the next plate.

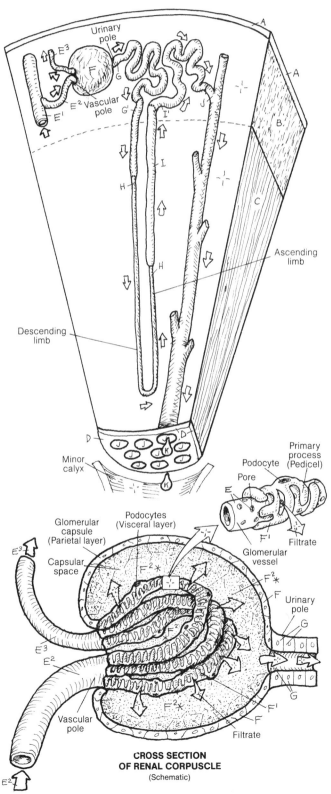

Urinary pole
Vascular pole
Descending limb
Ascending limb
Minor calyx

Primary process (Pedicel)
Podocyte
Pore
Filtrate

Glomerular capsule (Parietal layer)
Podocytes (Visceral layer)
Capsular space
Glomerular vessel
Urinary pole
Vascular pole
Filtrate

CROSS SECTION OF RENAL CORPUSCLE
(Schematic)

CN: Use red for A and blue for I, and a very light color for F. (1)Color the major blood vessels (A–D), noting that the corresponding veins receive only one color (blue). (2)Color the nephrons in the larger illustration gray. Note that the glomerular capsules have been opened to reveal the glomeruli (F). All arrows receive the color of the adjacent vessels.

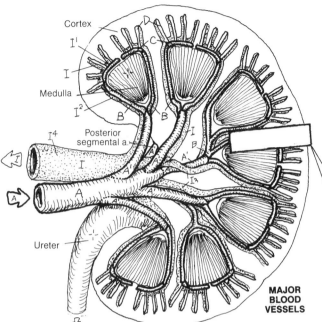

MAJOR BLOOD VESSELS

RENAL A. A
 SEGMENTAL A. A'
INTERLOBAR A. B
ARCUATE A. C
INTERLOBULAR A. D
AFFERENT ARTERIOLE E
GLOMERULUS F
EFFERENT ARTERIOLE G
PERITUBULAR CAPIL. PLEXUS H
VASA RECTA H'

INTERLOBULAR V. I
ARCUATE V. I'
INTERLOBAR V. I²
SEGMENTAL V. I³
RENAL V. I⁴

The nephron consists of the renal corpuscle and tubules less collecting tubules. The renal corpuscle was considered in Plate 111. Refer to that plate as necessary for the following paragraph. The *proximal tubule* close to the capsule of origin is highly *convoluted*. The pyramidal-shaped cuboidal cells of this tubule absorb 85% of the vitamins, amino acids, small proteins, glucose, sodium chloride, and water that came in with the plasma filtrate. The straight part of the tubule descends to become the *loop of Henle,* the cells of which absorb large amounts of water. The cells of the *distal tubule* are relatively impermeable to water but not to minerals. Here the filtrate tends to become diluted. The reabsorption of sodium (Na^+) is facilitated by aldosterone from the adrenal cortex and by dietary restriction of salt. The cells of the *collecting tubule* are permeable to water (taking it out of the tubule) in the presence of antidiuretic hormone (ADH, vasopressin). It is significantly less permeable in the absence of ADH, and even less so in the presence of diuretic medication.

The renal vascular pattern explains how the tubular cells can recover large amounts of fluid/solute from the filtrate and secrete it into the peritubular capillary plexus, to the extent that only 1% of the plasma filtered by the glomerular capsule reaches the calyces at any one moment. Follow the route of blood from the renal artery to and through the glomeruli to the *efferent arterioles.* The efferent arteriole in the upper and middle cortex branches into a *peritubular capillary plexus* that is intertwined around the convoluted tubules. The plexus is drained by *interlobular veins* which conduct the blood toward the *renal vein.*

In nephrons close to the medulla (juxtamedullary nephrons), the efferent arteriole (shown without its glomerulus of origin) may give off straight vessels (*vasa recta*) that "descend" into the medulla adjacent to the straight tubules. These vessels contribute to a separate peritubular capillary plexus in the medulla. Medullary vessels leaving the peritubular capillary plexuses form or contribute to "ascending" vasa recta which generally terminate by joining the *arcuate veins.* The relationship of the vasa recta to the loops of Henle is a critical factor in the success of water reabsorption. The circulatory pattern among the uriniferous tubules is a vital feature in the preservation of body water and the maintenance of chemical neutrality throughout the body.

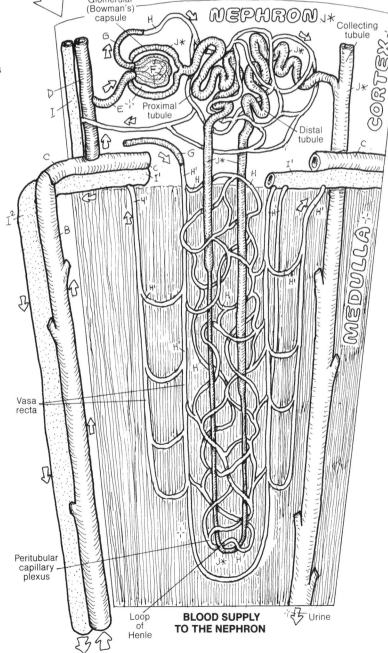

Glomerular (Bowman's) capsule

NEPHRON J*

Collecting tubule

CORTEX

Proximal tubule

Distal tubule

MEDULLA

Vasa recta

Peritubular capillary plexus

Loop of Henle

BLOOD SUPPLY TO THE NEPHRON

Urine

CN: Use red for L, blue for M, and very light colors for A, J, and K. (1) Color the upper views simultaneously. In the sagittal view, only the urethra is shown in the median plane. (2) The coverings of the spermatic cord in the illustration below actually consist of several layers (recall Plate 43). Color the parts of K and L seen deep to the pampiniform plexus (M).

SCROTUM_A
TESTIS_B
EPIDIDYMIS_C
DUCTUS DEFERENS_D
SEMINAL VESICLE_E
EJACULATORY DUCT_F
URETHRA_G
BULBOURETHRAL GLAND_H
PROSTATE GLAND_I
PENIS_J

The male reproductive system consists of the primary organs, the *testes* (testicles), suspended within a sac of skin and thin fibromuscular tissue (the *scrotum*); a series of ducts; and a number of glands. Development of the male germinating cells (sperm) in the testes requires a slightly lower-than-body temperature (about 35° C or 95° F); this is achieved by their separation from the warmer body cavities. The temperature within the scrotum can be adjusted slightly by the contraction/relaxation of smooth muscle (dartos muscle) in the scrotal wall, tightening or loosening the tension of the wall about the testes. Mature sperm are stored in the *epididymis*; with stimulus, sperm cells are induced to move into and through the *ductus* (vas) *deferens* by rhythmic contractions of the smooth muscle in the ductal wall. Within the ductus deferens, the sperm pass through the abdominal wall (via the inguinal canal) and pelvic cavity to enter the prostatic *urethra* via the pencil-point shaped *ejaculatory duct*. Here the nutrient-rich secretions of the *prostate gland* and *seminal vesicles* are added to the population of sperm in the prostatic urethra, forming semen. Prior to the release of the semen (ejaculation), the *bulbourethral glands* add secretions to the urethra. The *penis* and scrotum constitute the external genital organs.

Enlargement of the prostate is common (prostatic hypertrophy/hyperplasia) in men 50 years and older. The glands and connective tissues surrounding the urethra are subject to thickening and blocking urine flow (benign prostatic hypertrophy). Neoplastic growth (prostatic carcinoma) is less common (5–15% of men with prostatic hypertrophy) and occurs in the more peripheral tissues of the prostate.

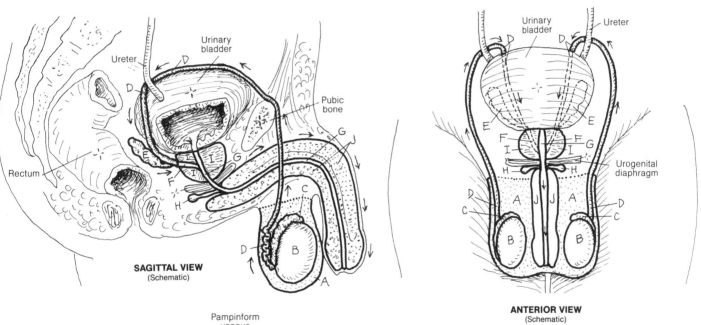

SAGITTAL VIEW
(Schematic)

ANTERIOR VIEW
(Schematic)

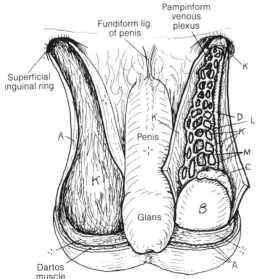

**ANTERIOR VIEW
OF SCROTUM**
(With cord dissected)

SPERMATIC CORD: *
  COVERINGS_K
  CONSTITUENTS: ⊹
    DUCTUS DEFERENS_D
    TESTICULAR ARTERY_L
    TESTICULAR VEIN_M

The *testicular artery* and *vein*, and some nerves and lymphatics, join the *ductus deferens* just before entering the deep ring (intra-abdominal orifice) of the inguinal canal. The collection of these form the constituents of the spermatic cord. Passing through the inguinal canal, they become invested by a representative layer from each of the abdominal wall layers (less rectus); these are the coverings of the spermatic cord and testes (here represented as one layer; see Plate 43). In a vasectomy procedure, the ductus deferens is identified within the cord and it alone is divided. A number of techniques (ligatures, cauterization, folding and burial, and so on) exist to prevent the natural tendency of the transected duct sections to recanalize.

XII. REPRODUCTIVE SYSTEM
TESTIS

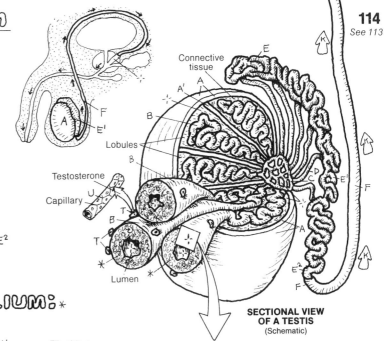

CN: Use the colors employed for the testis, epididymis, and ductus deferens on the previous plate with those same structures here (A, E, and F). Use red for U, and light colors for G, H, I, S, and T. (1) Note that the spermatogenic epithelium is colored gray in the cross section through the tubules above and that the tubular lumen is not to be colored.

TUNICA ALBUGINEA a
  SEPTUM a'
SEMINIFEROUS TUBULE b
RETE TESTIS c
EFFERENT DUCT d
EPIDIDYMIS HEAD e BODY e' TAIL e²
DUCTUS DEFERENS f

SPERMATOGENIC EPITHELIUM: *
  SPERMATOGONIUM g
  PRIMARY SPERMATOCYTE h
  SECONDARY SPERMATOCYTE i
  SPERMATID j
  SPERMATOZOON k
    HEAD:
      ACROSOME l
      NUCLEUS m
    TAIL:
      NECK n
      MIDDLE PIECE o
        MITOCHONDRION p
      PRINCIPAL PIECE q
      END PIECE r
  SERTOLI (SUPPORTING) CELL s
  BASEMENT MEMBRANE b'
  INTERSTITIAL CELL (OF LEYDIG) t
  BLOOD VESSEL u

SECTIONAL VIEW OF A TESTIS (Schematic)

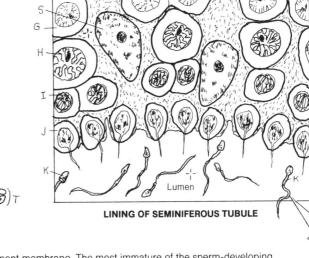

LINING OF SEMINIFEROUS TUBULE

SPERMATOZOON (Sperm cell)

The *testes* (testicles) arise on the posterior abdominal wall during fetal development; as the developing body lengthens, they appear to "descend" into outpocketings of the anterior abdominal wall (scrotum). The testes have two principal functions: development of male germ cells (sperm or *spermatozoa*) and the secretion of testosterone, the male sex hormone.

Each testis has a dense, fibrous, outer capsule (*tunica albuginea*) from which *septa* are directed centrally to compartmentalize the testis into lobules. One to four highly coiled *seminiferous tubules* exist in each lobule. These tubules converge toward the posterior side of the testis, straighten (tubuli recti), and join a network of epithelial-lined spaces (*rete testis*). *Efferent ducts* leave the rete to form the head of the *epididymis*. The convoluted epididymal duct (*head, body, tail*) is lined with pseudostratified columnar epithelium, one type of which contains long, immobile cilia (stereocilia). At the lower portion of the epididymis, the tubule turns upward to form the *ductus deferens*. The epididymis provides nutritional support for immature sperm cells and stores sperm. The wall of the ductus deferens, lined with pseudostratified columnar epithelium with stereocilia, contains significant smooth muscle, the rhythmic contractions of which drive sperm toward the prostate gland during emission.

Each seminiferous tubule consists of a lumen with walls of compact, organized masses of cells (*spermatogenic epithelia* and *supporting/(Sertoli) cells*) encapsulated by a thin fibrous, basement membrane. The most immature of the sperm-developing cells are the *spermatogonia*. These divide and the daughter cells are pushed out toward the lumen of the tubule. These cells differentiate into *primary spermatocytes,* the largest of the developing germ cells. These divide to become *secondary spermatocytes,* at which time the chromosome number is reduced from 46 to 23 (meiosis). Each pair of newly formed secondary spermatocytes rapidly divides again to form four *spermatids.* These small cells mature by developing tails, condensing their nuclei and cytoplasm, and developing acrosomal caps (with enzymes to break down the wall of the ovum and permit penetration). The mature sperm cell (*spermatozoan*) consists of a *head* of 23 chromosomes (nucleus) including the acrosome, a *middle piece* containing mitochondria to power cell movement, and the rest of the *tail* (fibers containing microtubules; the end piece is essentially a single flagellum) whose flagellations provide the cell's motive force. Mature sperm are driven out of the tubules through the tubular network into the highly convoluted *epididymis.*

The *interstitial cells* dispersed in the vascular loose connective tissue around the tubules include fibroblasts as well as the secretory cells (*of Leydig*), which are known to produce and secrete testosterone into adjacent capillaries. This male sex hormone stimulates the development of ducts and glands of the reproductive tract at puberty (generally between 11 and 14 years of age) as well as secondary sex characteristics.

CN: Use blue for I, red for J, yellow for K, and very light colors for D, E, and G. (1) Color the two upper views simultaneously, noting that the superficial and deep fascia (G, H) have been omitted from the coronal view. (2) Color the structural view and the cross section.

URETHRA: *
PROSTATIC U. A
MEMBRANOUS U. B
SPONGY U. C

PENIS: *
CORPUS CAVERNOSUM D
CRUS OF PENIS D'
CORPUS SPONGIOSUM E
BULB OF PENIS E'
GLANS PENIS E²
PREPUCE (FORESKIN) F

RELATED STRUCTURES: *
SUPERFICIAL FASCIA G
DEEP FASCIA H
VEIN I ARTERY J NERVE K
SUSPENSORY LIG. L
LEVATOR ANI
(PELVIC DIAPHRAGM) M
UROGENITAL DIAPHRAGM N
BULBOURETHRAL GLAND O

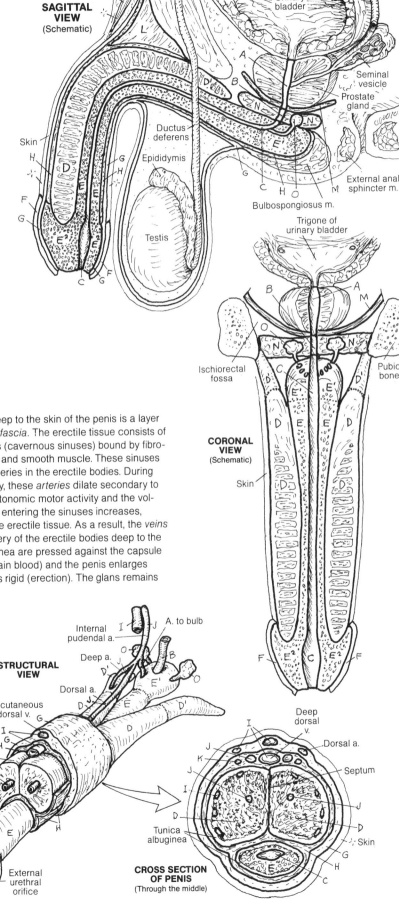

The *urethra* in the male has an extensive 20 cm (or so) course from the neck of the bladder to the external urethral orifice at the end of the penis. The *prostatic urethra* receives urine from the urinary bladder, sperm from the ejaculatory ducts, seminal fluid from the seminal vesicles, and secretions from the prostate via several ducts. Reflex contraction of the bladder neck muscles prevent voiding of urine during the expulsion of semen. The urethra continues through the pelvic diaphragm and into the thin, fibromuscular urogenital diaphragm as the *membranous urethra*. The *spongy urethra* passes through the penis. Numerous mucus glands exist in the urethral mucosa.

The *penis* consists of three bodies of erectile tissue, ensheathed in two layers of fasciae. The *corpora cavernosa* (the two lateral bodies) arise from the ascending rami of the pubic bones; the central *corpus spongiosum* arises as a *bulb* suspended from the inferior fascia of the *urogenital diaphragm* (perineal membrane). Each body consists of erectile tissue with a fibrous capsule (tunica albuginea); the corpus spongiosum contains the urethra as well. The three bodies are bound together in a dense stocking of *deep perineal fascia* and hang as a unit suspended by the deep *suspensory* and more superficial fundiform

ligaments. Deep to the skin of the penis is a layer of *superficial fascia*. The erectile tissue consists of lakes of veins (cavernous sinuses) bound by fibroelastic tissue and smooth muscle. These sinuses are fed by arteries in the erectile bodies. During sexual activity, these *arteries* dilate secondary to increased autonomic motor activity and the volume of blood entering the sinuses increases, expanding the erectile tissue. As a result, the *veins* at the periphery of the erectile bodies deep to the tunica albuginea are pressed against the capsule (unable to drain blood) and the penis enlarges and becomes rigid (erection). The glans remains non-rigid.

FEMALE REPRODUCTIVE SYSTEM

CN: (1) Color the two (upper) views of the internal reproductive structures simultaneously. In the sagittal view, color the double line representing the peritoneum in gray. (2) In the lower drawings, color the vestibule (N) gray after coloring the other structures located in that area (L–P). (3) In the dissected view of the external structures, take note of the surrounding musculature, none of which is colored.

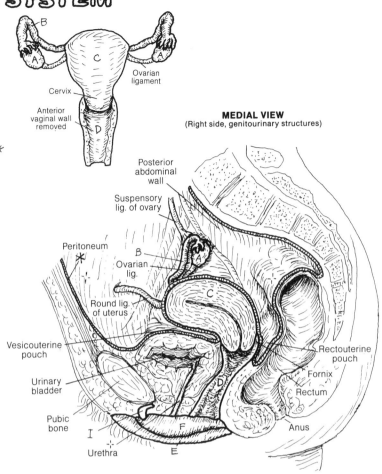

MEDIAL VIEW
(Right side, genitourinary structures)

INTERNAL STRUCTURES: *
OVARY ᴀ
UTERINE (FALLOPIAN) TUBE ʙ
UTERUS ᴄ
VAGINA ᴅ

The primary organ of the female reproductive system is the *ovary* which produces the female germ cells (ova) and secretes the hormones estrogen and progesterone. Each ovary, like the testis, arises on the posterior abdominal wall (adjacent to the kidneys) during early fetal development. It also descends along that wall, like the testis, but is interrupted early in its journey by a ligament and is retained in the pelvis. The *uterus* serves as a site for implantation and nourishment of the embryo/fetus. The *uterine tubes* provide a conduit for the freshly fertilized or unfertilized ovum enroute to the uterus. The *vagina*, a fibromuscular sheath, receives the semen from the penis and transmits it to the uterus and acts as a birth canal from the uterus to the outside for the newborn.

EXTERNAL STRUCTURES: *

LABIUM MAJUS ᴇ   CLITORIS ɪ   BULB OF THE VESTIBULE ʟ
LABIUM MINUS ꜰ   GLANS ɪ'   VESTIBULAR GLAND/DUCT ᴍ
FRENULUM ɢ   BODY ᴊ   VESTIBULE: ɴ*
PREPUCE ʜ   CRUS ᴋ   URETHRAL ORIFICE ᴏ
VAGINAL ORIFICE ᴅ' /HYMEN ᴘ

The female external genitals collectively constitute the vulva. They are located within the perineum. The *labia majora* are fat-filled folds of skin largely obscuring the cavity/space between them (vestibule) that contains the *urethral* and *vaginal orifices*. Medial to the labia majora are thin folds of skin (*labia minora*) which approach the *clitoris* anteriorly and split around it, forming the *frenulum* and *prepuce* of the clitoris. Like the penis, the clitoris has a *crus* (pl. crura) arising from each ischiopubic ramus; the two crura join in the midline to form the *body* or corpus. The body is capped by a skin-covered, vascular, sensitive *glans*. These clitoral components contain erectile tissue (less in the glans) enclosed in fascial coverings; their erection or rigidity is accomplished by the same mechanism operative in the penis. The clitoris, unlike the penis, does not incorporate the urethra. The *vestibular bulbs* are homologous to the bulb of the penis, but separated into two erectile bodies. They are covered by the bulbospongiosus muscle, and protrude into the vagina during sexual stimulation. The *vaginal orifice* is completely or incompletely covered or surrounded by a rim of mucosa called the *hymen*. Remnants of it (as shown) are often retained in the sexually active female.

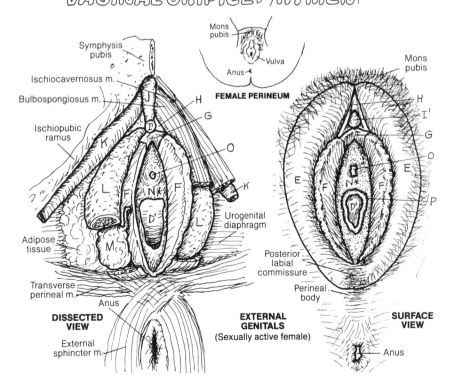

FEMALE PERINEUM

DISSECTED VIEW

EXTERNAL GENITALS
(Sexually active female)

SURFACE VIEW

CN: Use the colors from the preceding plate for the ovary (A) and uterine tube (M). Use red for K and R, yellow for L, blue for S, and very light colors for C-J, M, O and P. (1) Color the development of the female germ cell in both upper and lower views of the sectioned ovary. The oocyte (C) is colored through ovulation. In the large illustration, color the background stroma (B) gray; do not color the blood vessels in the stroma.

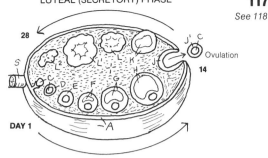

LUTEAL (SECRETORY) PHASE

28 · DAY 1 · 14

Ovulation

FOLLICULAR (PROLIFERATIVE) PHASE

OVARIAN CYCLE

OVARIAN STRUCTURES:*
EPITHELIUM / TUNICA ALBUGINEA A
CONNECTIVE TISSUE STROMA B*
OOGENIC EPITHELIUM:-:-
OOCYTE / OVUM c
PRIMORDIAL FOLLICLE D
PRIMARY FOL. E
SECONDARY FOL. F
MATURING FOL. G
MATURE (GRAAFIAN) FOL. H
ATRETIC FOL. I
RUPTURED FOL. J
DISCHARGED OVUM J'
CORPUS HEMORRHAGICUM K
YOUNG CORPUS LUTEUM L
MATURE CORPUS LUTEUM L'
CORPUS ALBICANS L²-:-

RELATED STRUCTURES:*
UTERINE TUBE M / FIMBRIAE M'
BROAD LIGAMENT N
MESOSALPINX O
MESOVARIUM P
SUSP. LIG. OF OVARY Q
OVARIAN A. R / V. S
UTERINE A. R' / V. S'
OVARIAN LIG. T

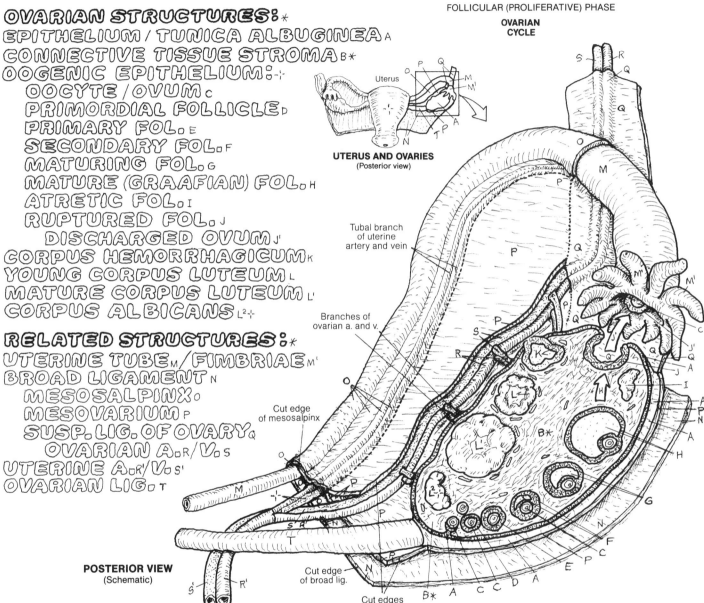

Uterus

UTERUS AND OVARIES
(Posterior view)

Tubal branch of uterine artery and vein

Branches of ovarian a. and v.

Cut edge of mesosalpinx

POSTERIOR VIEW
(Schematic)

Cut edge of broad lig.

Cut edges of mesovarium

Development of female germ cells and the secretion of the hormones estrogen and progesterone are the functions of the *ovary*. Confined by the thin but dense fibrous *tunica albuginea*, lined with epithelium, many ovarian follicles in various stages of development can be seen in the *connective tissue stroma*. A follicle consists of an immature epithelial germ cell (*oocyte*) surrounded by one or more layers of non-germinating cells. These germ cells were seeded in the ovary early in development—over 400,000 of them. Of these, only 500 or so will mature, the rest stopping short in their development and degenerating with their follicular cells (*atretic follicles*). Development of an ovum starts with the *primordial follicle*—an oocyte with one layer of follicular cells. The oocyte increases in size and maturity as the follicle cells increase in number. In *secondary follicles*, a small lake (antrum) filled with follicular fluid appears. This antrum continues to increase/expand at the expense of the follicle cells, which are pushed away from the oocyte

except for a layer of cells (*mature follicle*). Those cells in the outermost part of the follicle secrete estrogen during the proliferative phase of the reproductive cycle. On about the 14th day of that cycle, the ovum (surrounded by a glycoprotein coat, the zona pellucida, and some follicular cells) bursts from the follicle into the waiting fingers (*fimbriae*) of the *uterine tube*. The *ruptured follicle* involutes, and some bleeding and clotting goes on (*corpus hemorrhagicum*) as the follicle cells transition, characterized by accumulating large amounts of lipid. This newly formed structure (*corpus luteum*) secretes estrogen and progesterone during the secretory phase of the cycle, and in the event of pregnancy, will support the developing embryo/fetus for up to three months with these secretions. Should pregnancy not ensue, the corpus luteum will involute and degenerate as the corpus *albicans*. Follicles or corpora albicans/lutea collectively relating to two or three different but sequential cycles can be seen in the ovary at one time.

XII. REPRODUCTIVE SYSTEM
UTERUS, UTERINE TUBES, & VAGINA

CN: Use red for N, blue for O, and light colors for D, E, and Q. (1) Begin with the left half of the large illustration. Only parts of the ovarian and uterine veins are shown. Nerves and lymph vessels that may accompany arteries and veins are not shown. (2) Color the two views of the anteflexed and the retroflexed uterus.

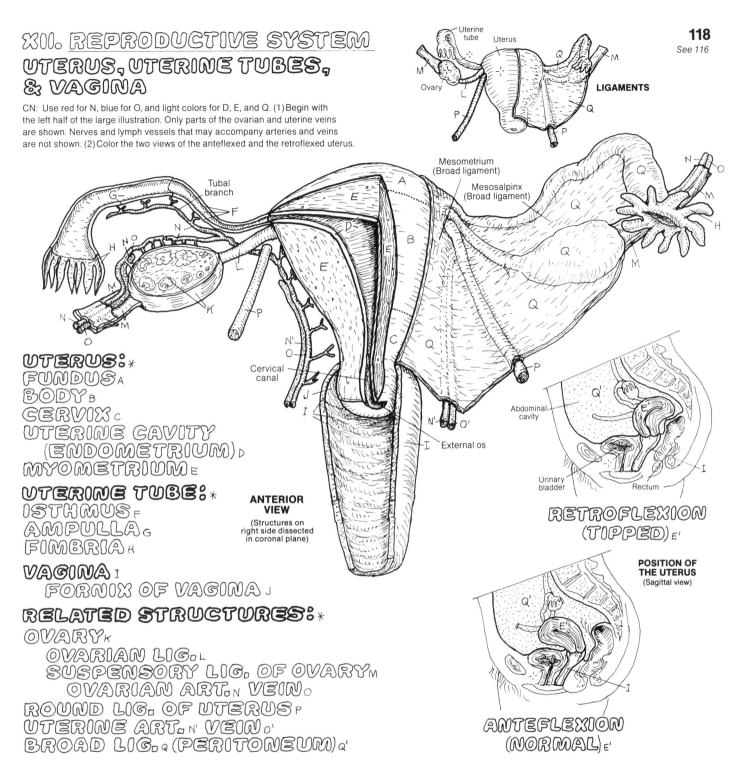

LIGAMENTS

Uterine tube
Uterus
Ovary

Tubal branch

Mesometrium (Broad ligament)
Mesosalpinx (Broad ligament)

Cervical canal

External os

ANTERIOR VIEW
(Structures on right side dissected in coronal plane)

Abdominal cavity

Urinary bladder
Rectum

RETROFLEXION (TIPPED) E'

POSITION OF THE UTERUS
(Sagittal view)

ANTEFLEXION (NORMAL) E'

UTERUS: *
FUNDUS A
BODY B
CERVIX C
UTERINE CAVITY (ENDOMETRIUM) D
MYOMETRIUM E

UTERINE TUBE: *
ISTHMUS F
AMPULLA G
FIMBRIA H

VAGINA I
FORNIX OF VAGINA J

RELATED STRUCTURES: *
OVARY K
OVARIAN LIG. L
SUSPENSORY LIG. OF OVARY M
OVARIAN ART. N VEIN O
ROUND LIG. OF UTERUS P
UTERINE ART. N' VEIN O'
BROAD LIG. Q (PERITONEUM) Q'

The *ovaries, uterus, uterine tubes,* and *vagina* make up the internal organs of reproduction in the female. The ovaries are suspended on the posterior layer of the *broad ligament* by a peritoneal extension (mesovarium), and supported by the *suspensory ligament of the ovary* (a lateral extension of the broad ligament and mesovarium), the *ovarian ligament*, and the *round ligament* (from the lateral wall of the uterus to the medial wall of the ovary). In this view, the ovaries have been brought to the horizontal to better clarify their relationship to the uterine tubes. The uterine tubes, suspended in a part of the broad ligament (mesosalpinx), are lateral extensions of the uterus, lined with ciliated columnar epithelium supported by connective tissue and smooth muscle. The rhythmic contractions of this muscle aid the ovum in its trek from the *fimbriae* to the *uterine cavity* and the lining cells support it nutritionally. The tube shows three rather distinct parts: the distal *fimbriae* (finger-like projections) which "catch" the discharged ovum and whisk it into the tubular lumen; the *ampulla* or widest part of the tube; and the *isthmus* whose lumen narrows as it enters the uterine wall/cavity.

The uterus is a pear-shaped structure whose neck (*cervix*) fits into the upper part of the vagina and whose *body/fundus* is bent (*anteflexed*) and tilted (*anteverted*) anteriorly over the bladder. Backward bending/tilting

(*retroflexion/retroversion*) of the uterus is not uncommon, particularly in women who have given birth. The retroflexed uterus predisposes to mild slipping into the vagina (prolapse) when the uterus is more in the axis of the cervix/vagina. Such an event is generally resisted by the pelvic and urogenital diaphragms, the perineal body, and numerous fibrous ligaments (broad ligament, and condensations of the pelvic fasciae, not shown) mooring the uterus and its tubes to the pelvic wall and sacrum. The wall of the uterus is largely smooth muscle (*myometrium*) lined with a glandular surface layer of variable thickness (*endometrium*) that is extremely sensitive to the hormones estrogen/progesterone.

The vagina is an elastic, fibromuscular tube with a mucosal lining of stratified squamous epithelium. The anterior and posterior mucosal surfaces are normally in contact. The anterior vaginal wall incorporates the short (4 cm) urethra. Remarkably, the mucosa of the vagina lacks glands; secretory activity during sexual stimulation is derived from a transudate of plasma from the local capillaries and from glands of the cervix. The vaginal lining reveals few sensory receptors. Where the cervix fits into the vagina, a circular moat or trough is formed around it (*fornix, fornices*). The fibroelastic posterior fornix is capable of significant expansion during intercourse.

FEMALE REPRODUCTIVE CYCLE

CN: Use yellow for B, red for G, and a very light color for A. (1) Color the time bar of the menstrual cycle at the bottom of the main diagram. Color the arrows C and D in the drawing on hormonal influences above. Then color the hormonal curves C and D in the main diagram, followed by the different follicular stages of the ovarian cycle (A, B), noting how these hormones influence the follicular changes. (2) Color the arrows E and F and the endometrium in the diagram above. Color the curves E and F in the main diagram, followed by the uterine structures in the menstrual cycle, noting how these hormones influence endometrial growth and menstruation. Color only the epithelial surface, glands, and vessels of the endometrium. (3) The days indicated are approximate. The hormonal curves reflect relative plasma hormone levels and are not absolute values.

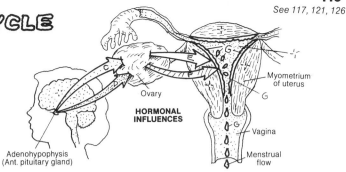

HORMONAL INFLUENCES

Adenohypophysis (Ant. pituitary gland)
Ovary
Myometrium of uterus
G
Vagina
Menstrual flow

OVARIAN CYCLE: *
PRIMORDIAL FOL. A
PRIMARY FOL. A'
SECONDARY FOL. A²
MATURE FOL. A³
OVULATION A⁴
CORPUS LUTEUM B, B'
CORPUS ALBICANS B²

HORMONAL CYCLE: *
FSH C
LH D
ESTROGEN E
PROGESTERONE F

MENSTRUAL CYCLE: *
PHASES:
MENSTRUATION G
PROLIFERATIVE H
SECRETORY I
ENDOMETRIUM:
EPITHELIUM J
GLAND I'
SPIRAL ARTERY G'

OVARIAN CYCLE

Ovum
OVULATION

HORMONAL CYCLE

THICKNESS OF UTERINE ENDOMETRIUM *

Menstrual blood
Secretions
Glandular pit
MENSTRUAL CYCLE

DAY 1 G | G | **4** | H | **14** | I | **21** | I | **28** G

The 28-day human female reproductive cycle, initiated and maintained by hormones, involves significant alterations in ovarian (follicular) and uterine (endometrial) structure. The cycle is characterized by a period of endometrial breakdown and discharge *(menstruation)* which begins at about 12 years of age (menarche) and ends at about 45 years of age (menopause). The progressive changes that occur in the ovary and uterus during each cycle serve to develop and release the female germ cell for possible fertilization by the male germ cell and to prepare the *endometrium* for implantation of the fertilized ovum.

The menstrual period constitutes the first five days of the cycle. Note the loss of endometrial tissue and attendant bleeding during this time. Endometrial regrowth begins on about the 5th day of the menstrual cycle, and is precipitated by hormones from the ovarian follicles. The ovarian cycle is regulated by hormones from the adenohypophysis (anterior pituitary gland), specifically *follicle stimulating hormone (FSH)* and *luteinizing hormone (LH)*. During the last few days of the previous cycle and the first several days of the next, these hormones stimulate follicular development.

As the selected follicle develops, it begins to produce *estrogen* on about the 7th day. Estrogen enters the circulation and influences endometrial growth *(proliferative phase)*. On about the 14th day of the menstrual cycle, the combined "spikes" of increased concentrations of FSH, LH, and estrogen induce *ovulation*: bursting of the mature follicle and release of the immature ovum into the fimbriae of the uterine tube. After ovulation, the burst follicle undergoes significant reconstruction *(corpus luteum)*

influenced by LH. On about the 21st day, this body secretes *progesterone* as well as estrogen, the combination of which have remarkable influence on endometrial glandular development *(secretory phase)*. This phase is characterized by the development of numerous secretory cells in the *epithelium*, a connective tissue stroma edematous with secretions from developing *glands*, and *spiral arteries* taking a tortuous course about the many glands: a condition conducive to nutritional support for an implanted fertilized ovum. If fertilization occurs (on about day 16), the corpus luteum becomes the principal source of hormones supporting development of the embryo and will continue as such for the next 90 days or so or until the placenta is capable of producing its own hormones.

In the absence of fertilization, on about day 26 the corpus luteum begins to involute (forming a *corpus albicans*), and estrogen/progesterone levels drop. Lacking hormonal stimulation, the endometrium experiences reduced glandular secretion in the presence of continued fluid absorption by the local veins, and the tissues collapse. The spiral arteries are flexed by these events, rupture, and hemorrhage with considerable hydraulic force; epithelial lining, glands, and fibrous tissues are disrupted and the structural integrity of the endometrium is largely destroyed. The vessels rapidly constrict, and bleeding is generally limited. The broken tissue (menstruum: mostly glandular tissue and secretions), blood, and one or more unfertilized ova gravitate toward the vagina. After 3–5 days of menstruation, only about 1 mm (in height) of endometrium is left for regeneration. The ovary will attend to that.

CN: Use yellow for E, pinks, tans, or browns, for J and K, and light colors for A, D, E, and G. (1) Color the two illustrations of the breast and underlying breast structures, simultaneously. (2) Color the arrows indicating the direction of lymph flow, and the lymph nodes of the chest. If you wish, you may color over the network of lymph vessels. (3) Color the diagrams of breast development. (4) Color the enlargement of glands and ducts in the lower right corner.

RIB_A CLAVICLE_{A'}
INTERCOSTAL MUSCLE_B
PECTORALIS MAJOR M._C
DEEP FASCIA_D
SUPERFICIAL FASCIA (FAT)_E
SUSPENSORY LIGAMENT_F
GLANDULAR LOBE_G
LACTIFEROUS DUCT_H
LACTIFEROUS SINUS_I
NIPPLE_J
AREOLA_K

LYMPHATIC DRAINAGE_L

The breast (in both males and females) is an area of fatty (adipose and loose areolar) fibrous tissue, and associated nerves, blood and lymphatic vessels, in the subcutaneous fascia overlying the pectoralis major muscle on the anterior chest wall. The fatty tissue is supported by extensions of the deep fascia overlying the muscle (*suspensory ligaments*) and functions most prominently in the young, well-developed, post-pubescent (after puberty) female breast. Packed within the adipose tissue is a collection of branching ducts (*lactiferous ducts*). In the male and in the non-pregnant (non-lactating) female, these ducts are undeveloped. There are few or no glands (alveoli) associated with these ducts in these populations. At puberty, the increased secretion of estrogen from the ovaries (and perhaps the adrenal glands) influences an enlargement of the *nipple* and *areola* in the female, and a generally marked increase in local fat proliferation. As a result, the breast enlarges to some degree (highly variable).

In the early stages of pregnancy, the lactiferous duct system undergoes profound proliferation, and small, inactive tubular and alveolar (*tubuloalveolar*) glands form, opening into alveolar ducts. A *lobule* consists of a number of these ducts and glands; a *lobe* (of which there are 15-20) consists of a number of lobules and an interconnecting *interlobular duct*. The interlobular ducts converge to form as many as 20 lactiferous ducts. These ducts dilate to form *lactiferous sinuses* just short of the nipple, and then narrow again within the nipple. These sinuses probably function as milk reservoirs during lactation. The nipple consists of pigmented skin with some smooth muscle fibers set in fibrous tissue. Erection of the nipple may enhance flow of milk through the ducts. The circular areola, also pigmented more highly than the surrounding skin, contains sebaceous glands that may act as a skin lubricant during periods of nursing. In the latter stages of pregnancy, the alveolar glands undergo maturation and begin to form milk. Milk production peaks after delivery of the newborn, and is the result of the action of several hormones influencing the gland cells. The letdown and excretion of milk results from a neuroendocrine reflex mechanism that is initiated by the baby sucking the nipple.

The *lymphatic* vessels are an important part of the breast: they drain the fat portion of the milk produced during lactation, and they transfer infected material or neoplastic (cancer) cells from the breast to more distant parts. The potential lymphatic avenues for metastasis or spread of infection are shown above.

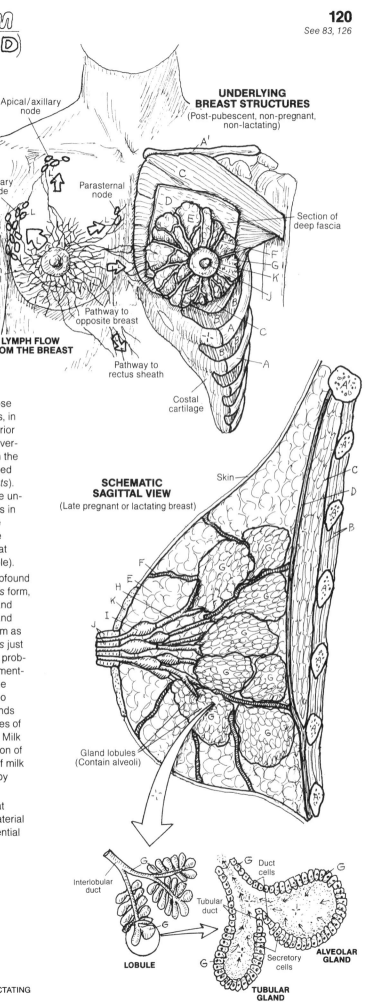

UNDERLYING BREAST STRUCTURES
(Post-pubescent, non-pregnant, non-lactating)

Apical/axillary node
Axillary node
Parasternal node
Section of deep fascia
Lymph vessel
Pathway to opposite breast
LYMPH FLOW FROM THE BREAST
Pathway to rectus sheath
Costal cartilage

SCHEMATIC SAGITTAL VIEW
(Late pregnant or lactating breast)
Skin

Gland lobules (Contain alveoli)

Interlobular duct
LOBULE
Tubular duct
TUBULAR GLAND
Duct cells
Secretory cells
ALVEOLAR GLAND

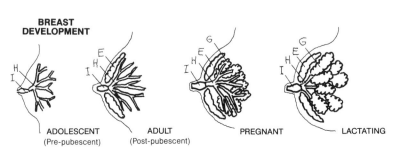

BREAST DEVELOPMENT

ADOLESCENT (Pre-pubescent)

ADULT (Post-pubescent)

PREGNANT

LACTATING

EMBRYONIC DEVELOPMENT (1)

CN: Use light colors throughout. (1) Follow the events from fertilization to implantation. The number of days cited in this and the following two plates are days after (post-) fertilization. Fertilization occurs about 14 days after the last day of menstruation; physicians date fetal age by time since last menstrual period (LMP). Thus, fetal age according to LMP is 14 days earlier than true (post-fertilization) age.

ZONA PELLUCIDA A
FERTILIZATION (1ST STAGE): *
FEMALE PRONUCLEUS B
HEAD OF SPERM C
MALE PRONUCLEUS C'
BLASTOMERE
(CLEAVAGE) STAGE: *
2-CELL D
4-CELL E
MORULA F
BLASTOCYST: *
TROPHOBLAST G
INNER CELL MASS H
BLASTOCOELE I

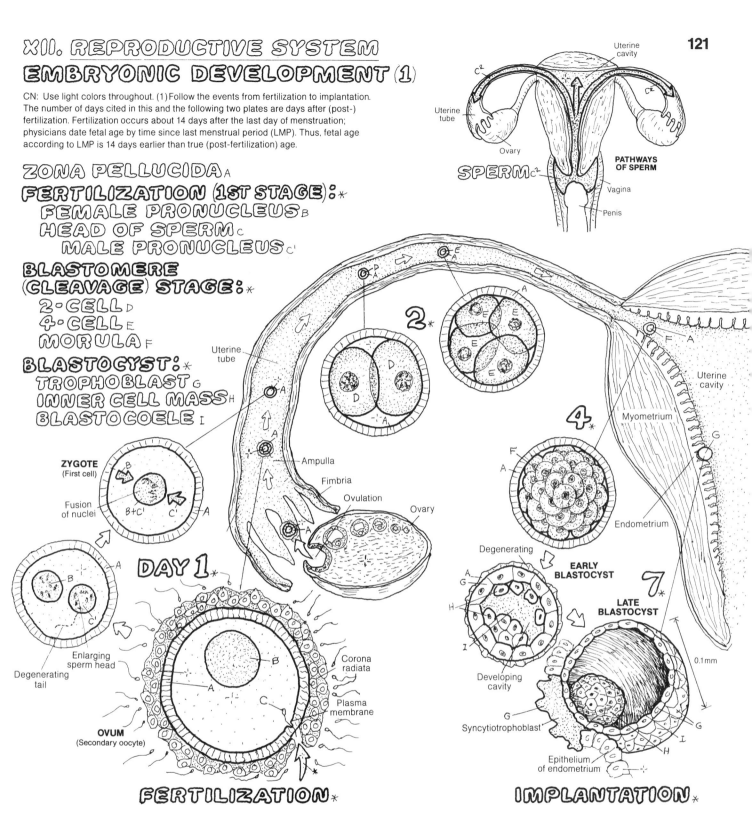

Following ovulation, the *ovum* enters the uterine tube and proceeds toward the uterus. It reaches the ampulla of the tube in about 30 minutes. If sperm-laden semen has been deposited in the fornices of the vagina in the preceding several minutes to several (24) hours, several hundred of the original 50 million or more sperms will successfully reach the ampulla. Over a period of several hours, the sperms become activated and with the aid of sperm-produced enzymes, one of the sperms will penetrate the corona radiata (retained follicular cells) and *zona pellucida* of the ovum, fuse with the plasma membrane (leaving its cell membrane attached to the ovum's plasma membrane), and enter the ovum. This event is called fertilization. As the tail breaks down and disappears, the *head of the sperm* enlarges, and forms the *male pronucleus*. The nucleus of the ovum is the *female pronucleus*. The two pronuclei approach each other, fuse nuclear membranes, and form a single nucleus. The male and female

chromosomes join up in the metaphase stage of the first mitotic division of the fertilized ovum. The zygote is the name given to the first cell of the new individual.

The zygote undergoes division (cleavage stage) to form two *blastomeres*. Over the next two days or so post-fertilization, within the restraints of the zona pellucida, the cells divide to form a *four cell blastomere* and again to form eight cells, and so on, until a ball of cells (*morula*) is formed. After about five days, the cells within the morula disperse enough to accommodate progressively enlarging fluid-filled cavities. Some cells are pushed aside to form a peripheral rim of cells (*trophoblast*) enclosing a large single cavity (*blastocoele*); some cells form an *inner cell mass* within the blastocoele. This multicellular structure is called the blastocyst. The blastocyst enters the uterus and implants in the endometrium on about the 7th day post-fertilization.

CN: Use the same color as on the previous plate for trophoblast (C) and note that the syncytiotrophoblast (D) is now given a separate color. Use yellow for F. Complete each drawing before proceeding to the next.

2 LAYER EMBRYONIC DISC: *
EPIBLAST A
HYPOBLAST B
TROPHOBLAST C
SYNCYTIOTROPHOBLAST D
AMNION E /AMNIOTIC CAVITY E'
YOLK SAC: PRIMARY F SECONDARY F'
EXOCOELOMIC MEMBRANE G
EXTRAEMBRYONIC MEMBRANE H
CONNECTING STALK H'
EXTRAEMBRYONIC COELOM I

3 LAYER EMBRYONIC DISC: *
ECTODERM A'
MESODERM J
ENDODERM K

On day 11 post-fertilization, the inner cell mass gives rise to a flat embryonic disc, consisting of a layer of columnar cells (*epiblast*), and an adjacent layer of cuboidal cells (*hypoblast*). The epiblast will develop almost entirely into the embryo. The *amniotic cavity* develops among the *trophoblast* cells adjacent to the epiblast; the roof of the cavity is called the *amnion*. The embryo and subsequent fetus will develop within this cavity. The trophoblast also gives rise to the *primary yolk sac*; the cells lining this sac are continuous with those of the hypoblast. Though it has no yolk, the sac probably has a nutritional function for the embryonic disc. Cells of the trophoblast form an *extraembryonic mesoderm* tissue (*membrane*) that largely fills the cavity once known as the blastocyst.

By day 14, the primary yolk sac diminishes in size, replaced by a *secondary yolk sac*. Cavities within the extra-embryonic membrane form a single cavity (it looks paired, but the connection between yolk sacs does not create two cavities). This cavity (I) surrounds the amnion/amniotic cavity and the yolk sac except where the amnion retains a *connecting stalk* to the trophoblast layer.

By day 16, the epiblast undergoes significant changes. The primary yolk sac is gone. Cells emerge from the epiblast and migrate into the area between the epiblast and hypoblast and into the hypoblast itself. The cells between are embryonic *mesoderm* cells; the cells migrating into the hypoblast layer form embryonic *endoderm*. The remaining epiblast cells become embryonic *ectoderm*. The earlier two-layered embryonic disc has formed into a three-layered embryonic disc. These three layers are called germ layers and give rise to the cells and tissues of the body. From ectoderm forms the skin and related glands, nervous system, the hypophysis, lens of the eye, and the inner ear. From mesoderm forms bones, muscle and the connective tissues, lymphoid organs, blood, the urogenital system, and serous membranes. From endoderm forms the epithelial part of the gastrointestinal system and respiratory system as well as the epithelia of the pharynx and thyroid.

By day 24 post-fertilization, the once flat embryonic disc has rounded to form within the amniotic cavity an embryo with a definitive head end and tail end, secured to the chorion (C, D, H) by the connecting stalk. As the lateral folds of embryonic mesoderm encircle the ventral (anterior) part of the embryo to form the antero-lateral abdominal walls, the yolk sac is pinched off and formation of the primitive gut begins. By the end of three weeks post-fertilization, the gastrointestinal tract, brain, and heart have begun their development.

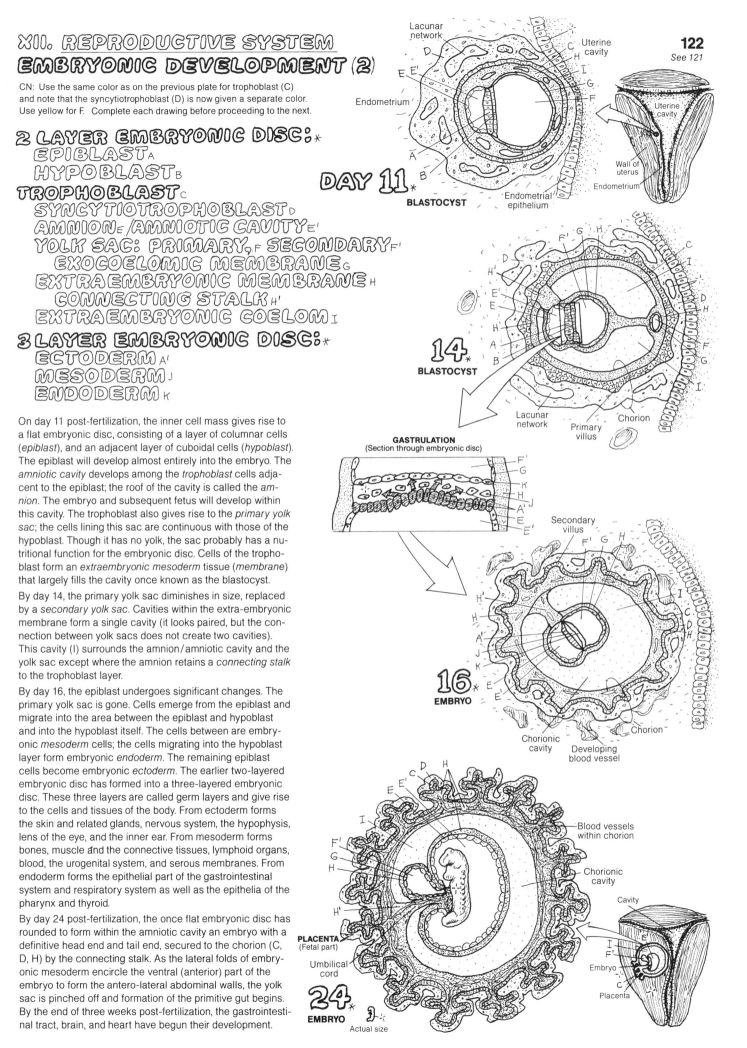

DAY 11 *
BLASTOCYST

Lacunar network
Uterine cavity
Endometrium
Endometrial epithelium
Uterine cavity
Wall of uterus
Endometrium

14 *
BLASTOCYST

Lacunar network
Primary villus
Chorion

GASTRULATION
(Section through embryonic disc)

16 *
EMBRYO

Secondary villus
Chorionic cavity
Developing blood vessel
Chorion

24 *
EMBRYO

Actual size
PLACENTA
(Fetal part)
Umbilical cord
Blood vessels within chorion
Chorionic cavity
Cavity
Embryo
Placenta

CN: Use the same colors for structures B and C that were given to them on the preceding plate. Use the color given to "connecting stalk" for the umbilical cord (A), and use the color given to "trophoblast" for chorion (D). (1) Color the embryonic coverings. The uterine cavity is colored gray, though it is actually lined with the decidua capsularis (E). Note that the *amniotic cavity* (C¹), *chorionic cavity* (D¹), and the embryo/fetus, are left uncolored. (2) The umbilical cord is composed of different blood vessels but receives one color (A). The band representing the uterine wall (below) is colored with both G and H.

EMBRYO:
UMBILICAL CORD A
YOLK SAC B
AMNION C & CAVITY C'
CHORION D / CAVITY D' / VILLI D²

UTERUS:
ENDOMETRIAL DECIDUA: *
D. CAPSULARIS E
D. BASALIS F
D. PARIETALIS G
MYOMETRIUM H
CERVIX I

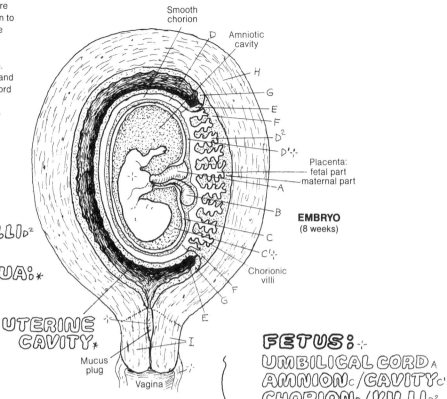

Smooth chorion · Amniotic cavity · Placenta: fetal part / maternal part · **EMBRYO (8 weeks)** · Chorionic villi · UTERINE CAVITY * · Mucus plug · Vagina

The developing embryo (called the fetus after eight weeks of development) lies within and is supported, nurtured, and protected by membranes and sacs. These coverings have both maternal and fetal origins.

Those of fetal origin include the *amnion* and *chorion/chorionic villi*, the *umbilical cord*, the *yolk sac*. The chorion forms a sac around the early embryo; the cavity of the sac is the *chorionic cavity* (recall Plate 122). As the embryo grows, the chorionic sac is obliterated and the amnion and chorion fuse (amniochorionic membrane). The chorion exhibits villi circumferentially early on (e.g., 24 day embryo); in time, most of the villi are absorbed except for those in the developing placenta (eight week embryo), creating a smooth chorion around the amnion and a bushy one (the villi and an underlying chorionic plate) in the future placenta. This is the situation with the fetal membranes at term (40 weeks) as well.

The coverings of maternal origin (the *decidua*), are thickened, fairly distinct layers of the uterine mucosa (endometrium) in which the blastocyst implanted. Looking at the eight week embryo (above): the *decidua basalis* is integrated with the fetal villi to form the placenta. The *decidua capsularis* surrounds the embryo and its membranes. The *decidua parietalis* lines the uterine cavity, superficial to the *myometrium*.

The parietalis is continuous with the capsularis as shown. This is the situation with the maternal membranes at eight weeks.

When the fetus grows to the point of pushing the decidua capsularis against the parietalis, the uterine cavity is obliterated. The capsularis soon degenerates, leaving only the parietalis (lower illustration). This layer will be retained after birth as the basal endometrium. The decidua basalis and chorionic villi (placenta) will be discharged after birth.

The fetus develops within the fluid-filled *amniotic cavity*. The plasma-like fluid gives freedom to the embryo to develop its form without mechanical pressure. It also acts as a water cushion absorbing shock forces. Just prior to birth, the amniochorionic membrane surrounding the fetus bursts, sending a half liter or more fluid into the *vagina* and to the outside (breaking the "bag of waters"). Parturition (childbirth) generally occurs about 280 days (40 weeks) after fertilization.

FETUS:
UMBILICAL CORD A
AMNION C / CAVITY C'
CHORION D / VILLI D²

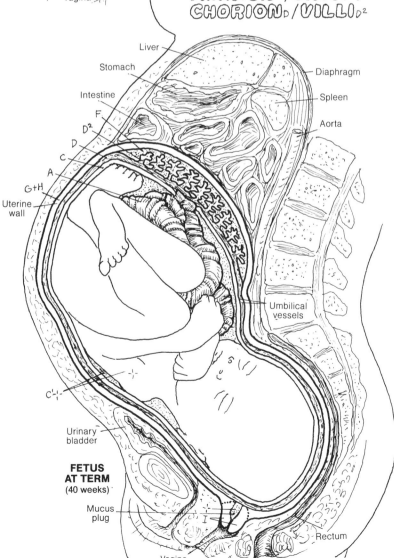

Liver · Stomach · Intestine · Diaphragm · Spleen · Aorta · Uterine wall · Umbilical vessels · **FETUS AT TERM (40 weeks)** · Urinary bladder · Mucus plug · Rectum · Vagina

XIII. ENDOCRINE SYSTEM
INTRODUCTION

CN: Use a very light color for C, and a darker one for D (actually located on posterior surface of thyroid). (1) After coloring endocrine glands and tissues, color the scheme at lower left.

ENDOCRINE GLANDS:*
HYPOPHYSIS (PITUITARY) A
PINEAL B
THYROID C
PARATHYROID (4) D
THYMUS E
ADRENAL (SUPRARENAL) (2) F
PANCREAS G
OVARY (2) H
TESTIS (2) I

ENDOCRINE TISSUES:*
HYPOTHALAMUS J
HEART K
KIDNEY (2) L
GASTROINTESTINAL TRACT M
PLACENTA N

Endocrine glands and tissues are discrete masses of secretory cells and their supporting tissue, and neighboring capillaries into which the cells secrete their hormones. These glands are without ducts. Some endocrine secretory cells exist in non-endocrine organs; they may be diffusely distributed (e.g., enteroendocrine cells of the *gastrointestinal tract*, interstitial cells of the *testis*), or they may occur as microscopic "islands" (e.g., *pancreatic islets*), or they may be a group of secretory neurons in the brain (*hypothalamus*). Others form single organs, the principal function of which is to secrete hormones (e.g., *thyroid*, *hypophysis*, and so on). Hormones are chemical agents usually effective among cells (target organs) some distance from their source. In concert with the nervous system, endocrine organs/tissues integrate and harmonize the activities of varied and sometimes seemingly unrelated organs and their activities by negative and positive feedback control mechanisms, resulting in growth, reproduction and related activity, and metabolic stability (a condition of the body's internal environment known as homeostasis).

Less well-known endocrine activity is seen in the *heart* (secretes atrial natriuretic peptide or ANP; increases sodium excretion and inhibits smooth muscle contraction), the *kidney* (secretes renin which is involved in the formation of substances that cause vasoconstriction and raise blood pressure; and erythropoietin which stimulates red blood corpuscle development), the *gastrointestinal tract* (produces numerous polypeptide hormones that regulate gastrointestinal motility and enzyme secretion), and the *placenta* (secretes human chorionic gonadotrophin or HCG in support of embryonic growth during the first 90 days post-fertilization; also secretes estrogen, progesterone, and a lactogenic and growth-stimulating hormone).

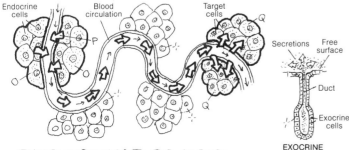

ENDOCRINE FUNCTION

Endocrine cells · Blood circulation · Target cells

Secretions · Free surface · Duct · Exocrine cells

EXOCRINE GLAND

ENDOCRINE GLAND O
HORMONAL SECRETION P
TARGET ORGAN Q

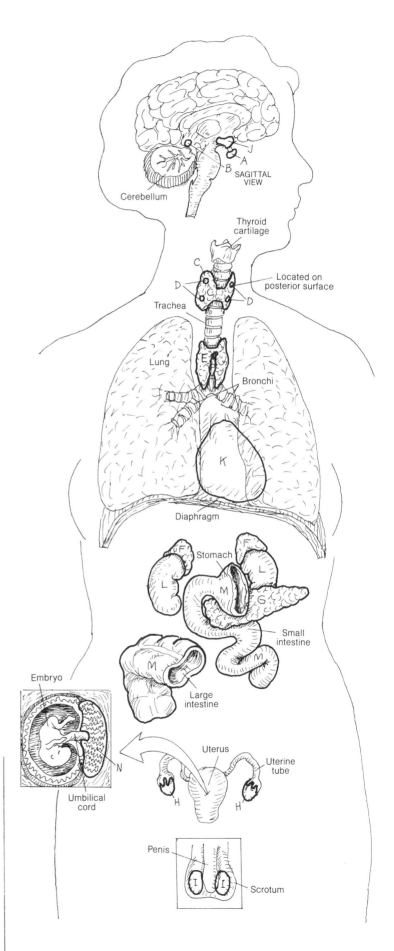

Cerebellum

SAGITTAL VIEW

Thyroid cartilage

Located on posterior surface

Trachea

Lung

Bronchi

Diaphragm

Stomach

Small intestine

Large intestine

Embryo

Umbilical cord

Uterus

Uterine tube

Penis

Scrotum

XIII. ENDOCRINE SYSTEM
HYPOPHYSIS & HYPOTHALAMUS

CN: Use red for H, blue for K, purple for I, and a very light color for J.
(1) Begin with the enlarged view of the hypophysis and hypothalamus.

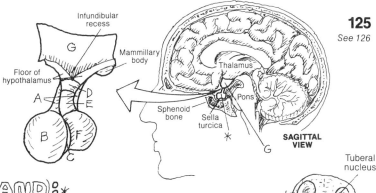

HYPOPHYSIS (PITUITARY GLAND): *
ADENOHYPOPHYSIS:
PARS TUBERALIS A
PARS DISTALIS (ANT. LOBE) B
PARS INTERMEDIUS C
NEUROHYPOPHYSIS:
MEDIAN EMINENCE D
INFUNDIBULAR STEM E
PARS NERVOSA (POST. LOBE) F
HYPOTHALAMUS G

The hypophysis (pituitary gland) is suspended from the *hypothalamus* of the brain by the infundibulum and fits into a bony recess of the sphenoid bone called the sella turcica. The hypophysis is about the size of four peas. The gland is derived from an upward extension of the developing roof of the mouth (adeno-hypophysis) and a downward migration of the floor of the hypothalamus of the brain (neurohypophysis). From above to below, the neurohypophysis includes the *median eminence,* the *infundibular stem* and the *pars nervosa* (posterior lobe). It is contiguous with the hypothalamus. The adenohypophysis includes the *pars tuberalis* which embraces the infundibular stem, the *pars distalis* (anterior lobe), and the *pars intermedius* which is rudimentary and appears to secrete no significant levels of hormones.

ADENOHYPOPHYSIS:
HYPOTHAL. SECR. NEURON/HORMONE G'
SUP. HYPOPHYSEAL ARTERY H
HYPOPHYSEAL PORTAL SYSTEM: I
CAPILLARY I'/PORTAL V. I²/SINUSOID I³
SECRETORY CELL J/HORMONES J'
INF. HYPOPHYSEAL VEIN K

The pars distalis of the adenohypophysis contains cells that secrete one of several hormones (see next plate). These cells are stimulated by *hormones* from *secretory neurons* in the hypothalamus. These hormones reach the pars distalis by way of a vascular portal system: *capillaries* in the median eminence form long and short *portal veins* that enter the pars distalis and form *sinusoids* around the *secretory cells.* Secretions from these cells enter the sinusoidal vessels which are drained by *inferior hypophyseal veins.*

NEUROHYPOPHYSIS:
HYPOTHAL. SECRETORY NEURONS OF:
SUPRAOPTIC NUCL./HORMONE G²
PARAVENTRIC. NUCL./HORMONE G³
CAPILLARY PLEXUS L
HYPOPHYSEAL VEIN K'

The pars nervosa of the neurohypophysis has no secretory cells of its own. Axons of secretory neurons in the *supraoptic* and *paraventricular nuclei* of the hypothalamus extend down through the infundibulum to *capillary networks* in the posterior lobe. There these axon terminals release oxytocin and antidiuretic *hormones* into the circulation (see next plate).

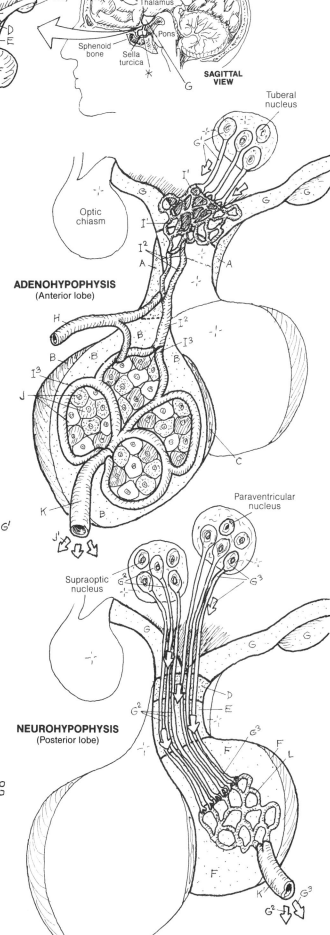

CN: Use a very light color for J, and the color from the previous plate for hypothalamic hormones (A) and secretions (A¹). (1) Begin with the arrows and circles representing those hormones and secretions: including the penetration of the cells of the anterior lobe. (2) Color the hypophyseal hormones. (3) Color the arrows representing the target organ hormones performing their feedback function.

HYPOPHYSEAL HORMONES:*
FOLLICLE STIMULATING H. (FSH) B
LUTEINIZING H. (LH) C
THYROID STIMULATING H. (TSH) D
ADRENOCORTICOTROPIC H. (ACTH) E
GROWTH H. (GH) F
PROLACTIN G
OXYTOCIN H
ANTIDIURETIC H. (ADH) I

TARGET ORGAN HORMONES: J
ESTROGEN K
PROGESTERONE L
TESTOSTERONE M
THYROXIN N
ADRENAL CORTICAL H. O

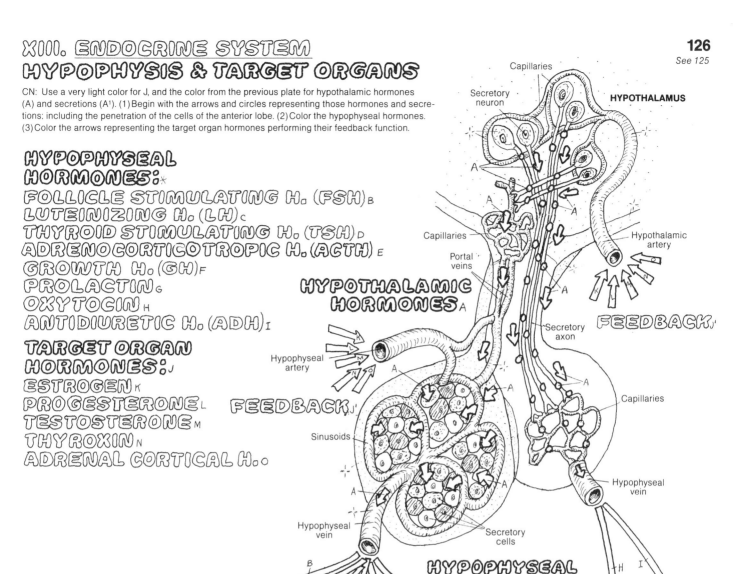

HYPOTHALAMIC HORMONES A

HYPOPHYSEAL HORMONES:*

FEEDBACK J¹

TARGET ORGAN J

TARGET ORGAN HORMONES: J

OVARY   TESTIS   THYROID   ADRENAL   BONE   BREAST   KIDNEY

STRUCTURAL/FUNCTIONAL EFFECT/FEEDBACK J¹

The hypothalamus has been seen in the previous plate to be intimately connected to the hypophysis by blood vessels and secretory axons. Here we look at *hypothalamic (releasing) hormones*, hypophyseal (pituitary) hormones, their *target organs*, and the hormones of the target organs. The secretory (glandular) cells of the pars distalis (anterior lobe) are stimulated by hypothalamic hormones which are released by secretory axons into the hypophyseal portal system in the median eminence (infundibulum). The secretions of these cells are adenohypophyseal hormones which enter the circulation and mediate the activities of a number of target organs. *FSH* drives ovarian follicular growth and secretion of *estrogen* and testicular spermatogenesis; *LH* stimulates *testosterone* secretion, ovulation, development of the corpus luteum, and estrogen/progesterone secretion. *TSH* stimulates secretion of *thyroxin*, a hormone of the thyroid gland. *Adrenocorticotropic hormone* (*ACTH*, corticotropin) stimulates the secretion of *adrenal cor-*

tical hormones; ACTH also has melanocyte-stimulating properties (dispersal of pigment in the skin). *GH* stimulates body growth, especially *bone*. *Prolactin* mediates milk secretion. *Oxytocin* and *antidiuretic hormone* (ADH; vasopressin) are products of secretory neurons in the supraoptic and paraventricular nuclei of the hypothalamus. These hormones are transported down long axons to be released among capillaries in the pars nervosa (posterior lobe) of the neurohypophysis. Oxytocin induces ejection of milk and stimulates uterine contractions. Vasopressin causes retention of body water (antidiuresis) by the kidneys; given exogenously, vasopressin is a significant vasoconstrictor.

Control of hypothalamic and hypophyseal secretions is accomplished by feedback mechanisms. Negative feedback control is reflected by the relationship of estrogen to FSH; as estrogen concentration in the plasma increases, production of FSH diminishes. Positive feedback responses are less common.

CN: Use red for H, blue for I, light colors for E, F, G, and the same colors as on Plate 124 for A and D. (1) Color the three upper views simultaneously, taking note of the arteries and veins that penetrate the thyroid. (2) Color the microscopic sections of hypoactive and hyperactive thyroid follicles; normal tissue lies between the two extremes. (3) Color the diagram of thyroid and parathyroid function.

THYROID A
THYROID FOLLICLE:⁻
FOLLICLE CELL B
COLLOID C
THYROXIN A'
PARATHYROID (4) D
PARATHORMONE D'
RELATED STRUCTURES: *
TRACHEA E
PHARYNX F
ESOPHAGUS G
ARTERIES H
VEINS I

The *thyroid gland*, covering the anterior surfaces of the 2nd to 4th *tracheal rings*, is bound by a fibrous capsule whose posterior layer encloses the four *parathyroid glands*. The thyroid gland, composed of right and left lobes connected by an isthmus, consists of clusters of *follicles* (like grapes) supported by loose fibrous tissue rich in blood vessels. A microscopic section through a follicle reveals a single layer of cuboidal epithelial *cells* forming the follicular wall. The follicle contains *colloid*, a glycoprotein (thyroglobulin), produced by the follicle cells. These cells take up thyroglobulin and dismantle it to form a number of hormones, primarily *thyroxin* (T4, tetraiodothyronine). Thyroxin is then secreted into the adjacent capillaries. Thyroid hormones contain iodide (a reduced form of iodine) which is absorbed by the follicle cells from the blood. Thyroxin formation and secretion is encouraged by thyroid-stimulating hormone (TSH) from the hypophysis. The relationship operates on a negative feedback mechanism: increased secretions of thyroxin inhibit further secretion of TSH. Thyroxin increases oxygen consumption in practically all tissues, and thus maintains the metabolic rate. It is involved at many levels in growth and development. Excessive secretion of thyroxin generally results in weight loss, extreme nervousness, and an elevated basal metabolic rate. Congenital thyroid insufficiency is manifested by dwarfism and mental retardation; in late onset, mental activity is diminished, the voice changes, and accumulation of mucous material in the skin and fascia gives a puffy appearance.

The *parathyroids* consist of small buttons of highly vascular tissue containing two cell types, one of which secretes *parathormone*. Parathormone maintains plasma calcium levels by inducing osteoclastic activity (bone breakdown), freeing calcium ions. Normal muscle activity and blood clotting depend on normal calcium levels in the plasma. Reduced parathyroid function lowers calcium levels and below certain levels causes muscle stiffness, cramps, spasms, and convulsions (tetany).

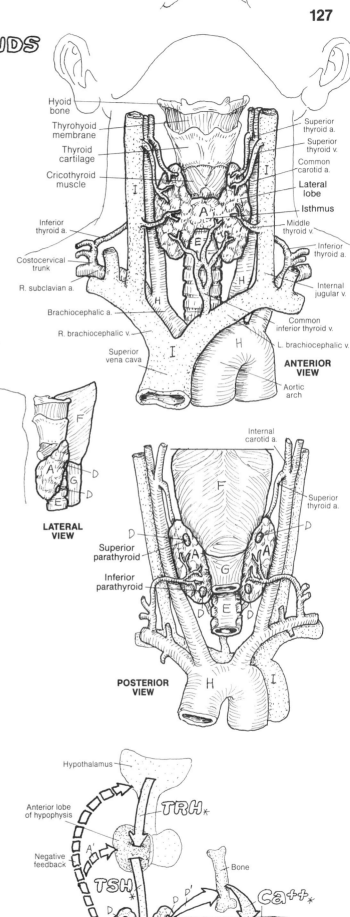

ANTERIOR VIEW

LATERAL VIEW

POSTERIOR VIEW

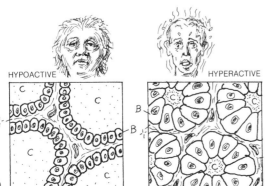

HYPOACTIVE

HYPERACTIVE

Cross sections through follicles (Approximately x 200)

THYROID FOLLICULAR ACTIVITY

THYROID & PARATHYROID FUNCTION

XIII. ENDOCRINE SYSTEM
ADRENAL (SUPRARENAL) GLANDS

CN: Use red for F, blue for G, yellow for H, and a very light color for E. (1) In the upper view, only those vessels with subscripts are to be colored. (2) Color the cross section through the adrenal, and related arrows and hormones. (3) Color the various organs associated with the "fight or flight" reaction, noting the listed effects.

ADRENAL GLAND A
  CAPSULE A'
  CORTEX: ∴
    ZONA GLOMERULOSA B
    ZONA FASCICULATA C
    ZONA RETICULARIS D
  MEDULLA E

ARTERIES: F
SUPERIOR SUPRARENAL A. F'
MIDDLE SUPRARENAL A. F²
INFERIOR SUPRARENAL A. F³

VEINS: G
R. & L. SUPRARENAL V. G²

SUPRARENAL PLEXUS: H
GREATER SPLANCHNIC N. H'
CELIAC GANGLION H²
  PLEXUS H³

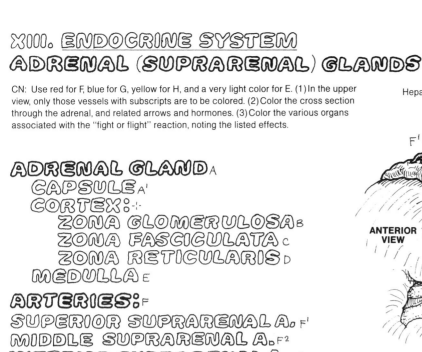

HORMONES OF THE CORTEX

ALDOSTERONE B

CORTISOL C

SEX STEROIDS D

HORMONES OF THE MEDULLA

EPINEPHRINE E'
NOREPINEPHRINE E²

The *adrenal* (suprarenal) *glands* lie in the retroperitoneum within the renal fascia on the superior and medial aspects of each kidney (T11–T12 vertebral levels). As with other endocrine glands, the adrenals are abundantly vascularized. The adrenals are two different glands encapsulated as one: the outer *cortex* and the inner *medulla*.

The adrenal cortex is organized into three regions: the outer *zona glomerulosa* (secreting hormones dealing with fluid/electrolyte balance, such as *aldosterone* and other mineralocorticoids); and the *zona fasciculata* and *zona reticularis* (secreting hormones influencing carbohydrate metabolism, such as *cortisol* and other gluco-corticoids; also low levels of *sex hormones*). Hypophyseal ACTH stimulates secretion of the glucocorticoids. Aldosterone is secreted in response to certain enzymes in the blood (renin-angiotensin system). All these hormones play roles involving all aspects of protein, carbohydrate, electrolyte, and water metabolism; thus the adrenal cortex is necessary for life.

The medulla consists of cords of secretory cells supported by reticular fibers, and an abundant collection of capillaries. Fibers of the greater splanchnic nerve (from spinal cord segments T5–T10; a major preganglionic nerve of the sympathetic division of the autonomic nervous system) pass through the celiac ganglia without synapsing to enter the adrenal gland. These fibers terminate on and stimulate the medullary secretory cells, 80% of which produce and release *epinephrine*; the rest secrete *norepinephrine*. These secretory cells are, in fact, modified post-ganglionic neurons. Their secretions elicit the "fight or flight" reaction in response to life-threatening situations as diagrammatically represented at right.

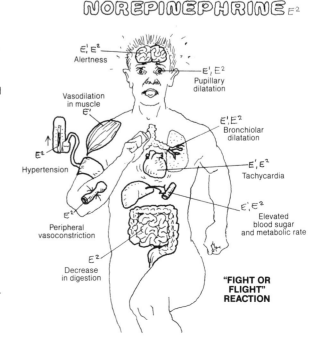

XIII. ENDOCRINE SYSTEM
PANCREATIC ISLETS

CN: Use purple for M, and light colors for K and L. (1) In coloring the upper drawing, include the broken lines representing arteries within or on the posterior surface of the pancreas. (2) Color the microscopic section of the pancreas and the enlarged view of an islet. Color the arrows and the diagram reflecting the role of glycogen and insulin receptors in liver cells with respect to glucose and glycogen.

ARTERIES TO THE PANCREAS:*
GASTRODUODENAL & BRS. A
ANT. PANCREATICO-DUOD. B
POST. PANCREATICO-DUOD. C
SPLENIC & BRS. D
DORSAL PANCREATIC E
INFERIOR PANCREATIC F
GREAT PANCREATIC G
SUPERIOR MESENTERIC H
INF. PANCREATICO-DUOD. I

The pancreas is supplied by numerous arteries from sources springing from the celiac and superior mesenteric arteries. The extensive capillary networks of the pancreas are drained by tributaries of the hepatic portal vein which conducts the secreted hormones of the pancreatic islets to the liver and beyond for general circulation.

BLOOD SUPPLY TO PANCREAS

PANCREATIC ISLET: J
ALPHA CELL K
GLUCAGON K¹/RECEPTOR K²
BETA CELL L
INSULIN L¹/RECEPTOR L²
BLOOD CAPILLARY M
GLYCOGEN N GLUCOSE N¹

MICROSCOPIC SECTION

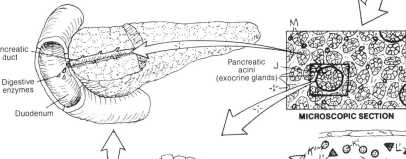

EXOCRINE GLAND (Acinus)

ENDOCRINE GLAND (Pancreatic Islet)

LIVER CELL (Low blood sugar conditions)

LIVER CELL (High blood sugar conditions)

The islands (*islets*) of endocrine tissue (and their *capillaries*) in the pancreas are surrounded by masses of grape-like clusters/follicles of exocrine gland cells. The secretions of these cells enter ducts that are tributaries of the pancreatic duct(s) opening into the duodenum.

The islets are characterized by three or four different cell types. *Alpha (A) cells*, generally located in the periphery of the islet, secrete *glucagon*, a polypeptide hormone that binds to glycogen *receptors* on liver cell membranes. Glucagon induces the enzymatic breakdown of glycogen to glucose, a process called glycolysis. Glucagon also facilitates the formation of glucose from amino acids in the liver, a process called gluconeogenesis. As a result of these processes, blood glucose levels increase.

Beta (B) cells, constituting 70% of the islet cell population, occupy the central part of the islet and secrete *insulin*, a polypeptide, primarily in response to increased plasma levels of *glucose*. Most insulin is taken up by the liver and kidney,

but almost all cells can metabolize insulin. Insulin expedites the removal of glucose from the circulation by increasing the number of proteins that transport glucose across cell membranes (glucose carriers; not shown) in muscle cells, fat cells, leukocytes, and certain other cells (not including liver cells). Insulin increases the synthesis of *glycogen* from glucose in liver cells. Uptake of insulin is facilitated by *insulin receptors* (proteins) on the external and internal surfaces of many—but not all—cell membranes. Decreased insulin secretion or decreased numbers or activity of insulin receptors leads to glucose intolerance and/or diabetes mellitus. The effects of insulin activity are far-reaching: mediating electrolyte transport and the storage of nutrients (carbohydrates, proteins, fats), facilitating cellular growth, and enhancing liver, muscle, and adipose tissue metabolism. The islets also reveal two other secretory cell types: D and F cells (not shown). These cells are in the periphery of the islet, and constitute about 5-10% of the islet cell population.

CN: Use very light colors for A and C. (1) The spinal cord has been placed behind the vertebral column in the main illustration to show the length of the cord and corresponding spinal cord regions in relation to the length and regions of the vertebral column. Note the descending spinal nerve roots (arrows coming off the cord) in the lumbar regions and below. (2) In coloring the spinal nerves and their peripheral branches at lower right, color over the lines representing them. (3) Color the motor ganglia of the autonomic nervous system (L and M) in the lower right drawing.

CENTRAL (CNS) NERVOUS SYSTEM:*
BRAIN:·:·
CEREBRUM A
BRAINSTEM B
CEREBELLUM C
SPINAL CORD D / REGIONS: D
CERV. G THOR. H LUM. I SAC. J CO. K

The nervous system consists of neurons arranged into a highly integrated central part (central nervous system, or CNS) and bundles of neuronal processes (nerves) and islands of neurons (ganglia) largely outside the CNS making up the peripheral part (peripheral nervous system, or PNS). These neurons are supported by neuroglial cells, and a rich blood supply. Neurons of the CNS are interconnected to form centers (nuclei; gray matter) and axon bundles (tracts; white matter). The brain is the center of sensory awareness and movement, emotions, rational thought and behavior, foresight and planning, memory, speech, language and interpretation of language.

The *spinal cord*, an extension of the brain beginning at the foramen magnum of the skull, traffics in ascending/descending impulses, and is a center for spinal reflexes, source of motor commands for muscles below the head, and receiver of sensory input below the head.

PERIPHERAL (PNS) NERVOUS SYSTEM:*
CRANIAL NERVES (12 PAIR) E
SPINAL NERVES/BRANCHES F
CERVICAL (8) G'
THORACIC (12) H'
LUMBAR (5) I'
SACRAL (5) J'
COCCYGEAL (1) K'
AUTONOMIC NERV. SYS.:·:*
SYMPATHETIC DIV. L
PARASYMPATHETIC DIV. M

The PNS consists largely of bundles of sensory and motor axons (nerves) radiating from the brain (*cranial nerves*) and spinal cord (*spinal nerves*) segmentally and bilaterally and reaching to all parts of the body (visceral and somatic) through a classic pattern of distribution. *Branches* of spinal nerves are often called peripheral nerves. Nerves conduct all sensations from the body to the brain and spinal cord, and conduct motor commands to all the skeletal muscles of the body. The *autonomic nervous system* (ANS) is a subset of ganglia and nerves in the PNS dedicated to visceral movement and glandular secretion, and the conduction of visceral sensations to the spinal cord and brain.

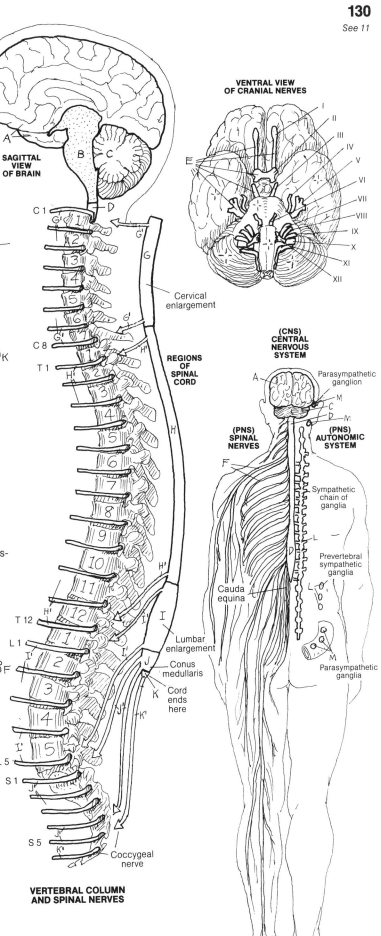

SAGITTAL VIEW OF BRAIN

VENTRAL VIEW OF CRANIAL NERVES

Cervical enlargement

REGIONS OF SPINAL CORD

(CNS) CENTRAL NERVOUS SYSTEM

Parasympathetic ganglion

(PNS) SPINAL NERVES

(PNS) AUTONOMIC SYSTEM

Sympathetic chain of ganglia

Prevertebral sympathetic ganglia

Parasympathetic ganglia

Cauda equina

Lumbar enlargement

Conus medullaris

Cord ends here

Coccygeal nerve

VERTEBRAL COLUMN AND SPINAL NERVES

CN: Use light colors throughout the plate.
Do not color the summary diagram at the top
of the page until completing the rest of the plate.

Neurons generally function in one of three modes: they conduct impulses from receptors in the body to the central nervous system or CNS (sensory or afferent neurons); or they conduct motor command impulses from the CNS to muscles of the body (motor or efferent neurons); or they form a network of interconnecting neurons in the CNS between motor and sensory neurons (interneurons). If the sensory or motor neurons relate to musculoskeletal structures or the skin and fascia, the prefix "somatic" may be applied (somatic afferent/somatic efferent). If these neurons are related to organs with hollow cavities (viscera), the prefix "visceral" may be applied (visceral afferent/visceral efferent).

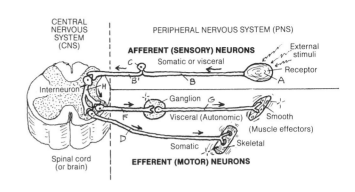

PNS:
SENSORY NEURON *
RECEPTOR A
AXON
(PERIPHERAL PROCESS) B
CELL BODY C
AXON
(CENTRAL PROCESS) B'

Sensory neurons conduct impulses from sensory *receptors* to synapses in the CNS. The receptors may be sensitive to touch, pressure, pain, joint position, muscle tension, chemical concentration, light, or other mechanical stimulus, basically providing information on the external or internal environment and related changes. Sensory neurons are unipolar neurons with certain exceptions, and are characterized by *peripheral processes* ("axons"), *cell bodies*, and *central processes* ("axons").

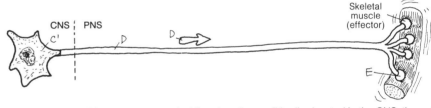

PNS:
SOMATIC MOTOR N. *
CELL BODY C'
AXON D
MOTOR END PLATE E

Motor neurons conduct impulses from *cell bodies* located in the CNS, through *axons* that leave the CNS and subsequently divide into branches, each of which becomes incorporated into the cell membrane of a muscle cell (*motor end plate*). Here the neuron releases its neurotransmitter that induces the muscle cell to shorten.

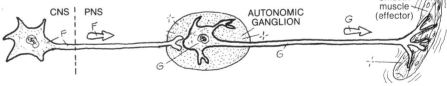

PNS:
AUTONOMIC MOTOR N. *
PREGANGLIONIC NEURON F
POSTGANGLIONIC NEURON G

Autonomic motor neurons function as paired units connected at a ganglion by synapse. The first or *preganglionic neuron* arises in the CNS, and its axon embarks for a ganglion located some distance from the CNS. There it *synapses* with the cell body or dendrite of a *postganglionic neuron* whose axon proceeds to the effector organ: smooth muscle, cardiac muscle, or glands.

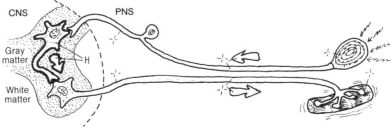

CNS:
INTERNEURON
(ASSOCIATION N.) H

Interneurons are found mostly in the CNS. They make up the bulk of the neurons of the brain and spinal cord. They come in a variety of shapes and sizes. Many of them are directly related to incoming (sensory) impulses and others to outgoing motor commands. Others serve to integrate sensory or ascending input with higher centers to effect an appropriate motor output.

SYNAPSES & NEUROTRANSMITTERS

CN: Use light colors for A, B, and C. (1) In the upper drawing, each of the synapses shown has two parts. Color only the ones labeled with subscripts (A, B, C). Color the nerve impulse title (D) and the related directional arrows. (2) Color the numbered steps in the lower drawing. Note the change of color in the presynaptic membrane between exocytosis (H) and endocytosis (K).

BASIC TYPES OF SYNAPSES:∗
AXO∆AXONIC∆
AXO∆SOMATIC∆
AXO∆DENDRITIC∆

Connections between and among neurons are called *synapses*. They are, for the most part, non-contact, chemical transmissions between one part of a neuron and another. There are other less common types of synapses (some of which are not shown) as well. For example, we show a complex of synapses between three axons and a dendrite, all surrounded by a neuroglial sheath (glomerulus). Synapses permit the conduction of electrochemical impulses among a myriad of neurons almost instantly. In the simplest example, a monosynaptic reflex arc between a sensory neuron and a motor neuron requires one synapse. At the other extreme, there are polysynaptic pathways in the brain and spinal cord involving millions of synapses. Certain neurons are known to receive more than 5000 synapses . . . each! Multiple synapses greatly increase the available options of nervous activity. The ability to integrate, coordinate, associate, and modify sensory input and memory to achieve a desired motor command is directly related to the number of synapses involved in the pathway.

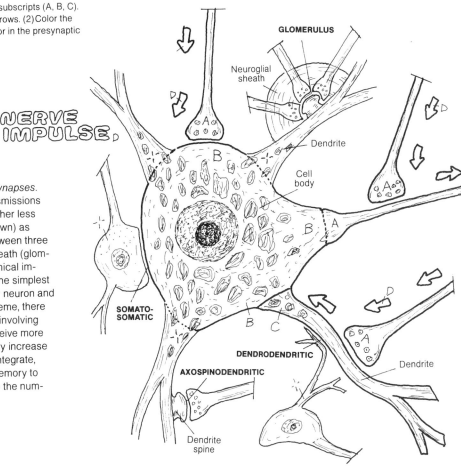

NERVE IMPULSE

GLOMERULUS

Neuroglial sheath

Dendrite

Cell body

SOMATO-SOMATIC

DENDRODENDRITIC

AXOSPINODENDRITIC

Dendrite spine

Dendrite

TYPES OF SYNAPSES

TYPICAL SYNAPSE:∗
PRESYNAPTIC AXON∆
PRESYNAPTIC MEMBRANE∊
SYNAPTIC VESICLE∊
NEUROTRANSMITTER∊
FRAGMENT∊'
EXOCYTOSIS∆
SYNAPTIC CLEFT∆
POSTSYNAPTIC MEMBRANE∆
RECEPTOR∊'
ENDOCYTOSIS∆

Here we present a typical axo-dendritic synapse. The axon (1) is *presynaptic* (in front of the synapse). Within the cytoplasm of the axon terminal are *synaptic vesicles* (2) transporting molecules of *neurotransmitter*. These vesicles migrate toward and fuse with the *presynaptic membrane* (3). Neurotransmitter is spilled into the tiny *synaptic cleft* by a process of *exocytosis*. The neurotransmitter molecules interact with the *receptors* on the *postsynaptic membrane* of the postsynaptic dendrite (4). Inactivated neurotransmitter *fragments* are taken up by the presynaptic membrane (5; *endocytosis*), enclosed in a synaptic vesicle, and re-synthesized (6).

The neurotransmitter may enhance (facilitate) or depress (inhibit) the electrical activity of the postsynaptic membrane. Sufficiently excited by multiple facilitory synapses, the postsynaptic neuron will depolarize and transmit an impulse to the next neuron or effector; sufficiently inhibited by multiple inhibitory synapses, the postsynaptic neuron will not be depolarized and will not transmit an impulse. All of this occurs very rapidly and globally throughout the nervous system.

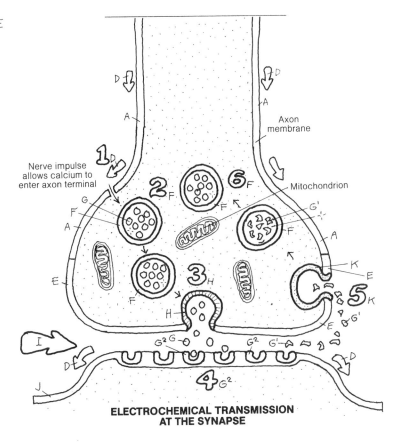

Axon membrane

Nerve impulse allows calcium to enter axon terminal

Mitochondrion

ELECTROCHEMICAL TRANSMISSION AT THE SYNAPSE

DEVELOPMENT OF CENTRAL NERVOUS SYSTEM

CN: Use light colors for A and C. (1) Begin with the two dorsal views of the 20 day-old embryo. Color as well the large arrows pointing to the surface locations. Simultaneously color the diagrammatic, cross section to its right. Follow the same procedure for the later views of the growing embryo. (2) Color the stages of brain development in the head end of the neural tube.

NEURAL PLATE_A
FOLD_A'
TUBE_A²
NEURAL GROOVE_B
NEURAL CREST_C

The nervous system develops from the dorsal surface of the ectodermal germ layer (future skin) of the embryo. In the 20-21 day embryo, a longitudinal groove (*neural groove*) begins to form on this thickened layer (*neural plate*). In the central part of the plate, the groove deepens, forming *neural folds* on either side. Deepening of the neural groove proceeds toward the head and tail ends of the embryo. By 22 days, the dorsal part of the folds fuse in the central part of the groove, forming a *neural tube*. During this process, the neural tube separates from the ectoderm. By 24 days, formation of the neural tube has progressed to the extreme ends of the embryo. Much of the neural tube will form the spinal cord; the head end of the tube will form the brain. The *neural crest* cells, formed from the neural folds, will develop into certain nerve cells of the peripheral nervous system and Schwann cells. The surrounding mesoderm will form the cranium and the vertebral column and related muscles. The notochord (a primitive supporting rod for the embryo) will be absorbed by the developing vertebral column, and remnants of it will remain as the core of the intervertebral discs (nucleus pulposus). The endoderm will contribute to the development of the digestive tract.

FOREBRAIN_D
TELENCEPHALON_E
DIENCEPHALON_F
MIDBRAIN
(MESENCEPHALON)_G
HINDBRAIN_H
METENCEPHALON_I
MYELENCEPHALON_J
SPINAL CORD_K

By the end of three weeks of embryonic development, three regions of the developing brain are apparent: *forebrain, midbrain,* and *hindbrain*. With further growth, the forebrain expands to form the massive *telencephalon* (endbrain; future cerebral hemispheres) and the more central *diencephalon* ("between" brain; future top of the brain stem). The midbrain retains its largely tubular shape as the *mesencephalon* (midbrain; future upper brain stem). The hindbrain differentiates into the upper *metencephalon* ("change" brain; future middle brain stem) and a large dorsal outpocketing (future cerebellum), and the lower *myelencephalon* (spinal brain; lowest part of the future brain stem). The brain stem narrows to become the *spinal cord* at the level of the foramen magnum of the skull.

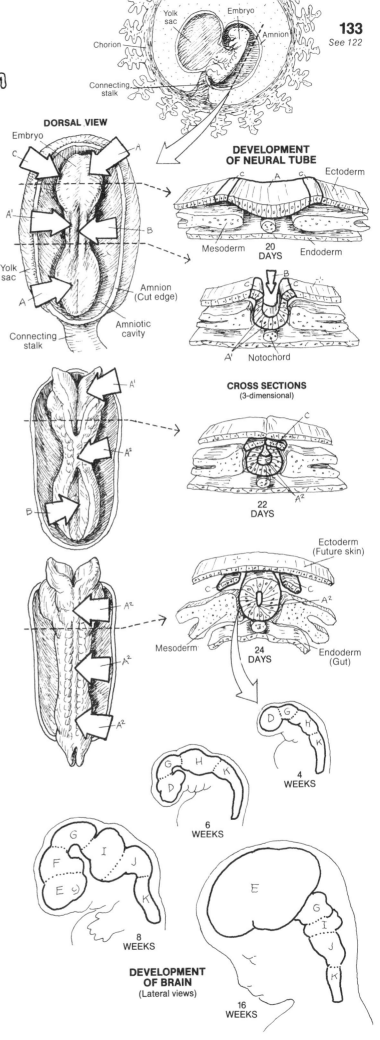

133
See 122

DORSAL VIEW

DEVELOPMENT OF NEURAL TUBE

CROSS SECTIONS
(3-dimensional)

DEVELOPMENT OF BRAIN
(Lateral views)

CN: Use light colors for B, E, I, and J. (1) Color the two large hemispheres first. Note that the stippled areas of specialized function are parts of lobes, but receive their own colors. Color the arrows identifying the major fissures and sulcus. (2) Color the coronal section and posterior portion of the brain. The coronal section of the cerebral cortex is colored gray. (3) Color gray the stretched-out cortex and the convoluted cortex illustrating how the latter provides increased surface area in a smaller space.

CEREBRAL CORTEX: A*
FRONTAL LOBE B
PRINCIPAL SPEECH AREA C
PRIMARY MOTOR AREA
(PRECENTRAL GYRUS) D
PARIETAL LOBE E
PRIMARY SENSORY AREA
(POSTCENTRAL GYRUS) F
TEMPORAL LOBE G
AUDITORY AREA H
OCCIPITAL LOBE I
VISUAL AREA J

MAJOR FISSURES/SULCUS: *
LONGITUDINAL FISSURE K
CENTRAL SULCUS L
LATERAL FISSURE M

CORONAL SECTION (Modified)

Longitudinal fissure
Central sulcus
Corpus callosum
White matter
Lateral ventricles
RIGHT HEMISPHERE
Gray matter
Basal nuclei
LEFT HEMISPHERE

RIGHT HEMISPHERE
Thalamus
Parietooccipital sulcus
POSTERIOR
Language area
Cingulate gyrus
Part of limbic lobe
Corpus callosum
ANTERIOR
Taste area
LEFT HEMISPHERE
Broca's area of speech (Left hemisphere only)

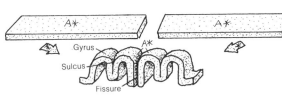

Gyrus
Sulcus
Fissure

CORTICAL CONVOLUTIONS: INCREASED SURFACE AREA

The paired cerebral hemispheres (cerebrum), derivatives of the embryonic telencephalon, consist of four major elements: (1) an outer cerebral cortex of gray matter, the topography of which reveals fissures (deep grooves), gyri (hills), and sulci (furrows); (2) underlying white matter consisting of numerous tracts destined for or leaving the cortex and oriented along three general directions; (3) discrete masses of gray matter at the base of the cerebrum (basal nuclei) that subserve motor areas of the cortex; (4) paired cavities called lateral ventricles. The cerebral cortex is the most highly evolved area of the brain. About 2-4 mm (roughly 1/6 inch) thick, the cortex is divided into lobes distinctly bordered by sulci; the lobes are generally related to the cranial bones that cover them: frontal, parietal, temporal, occipital. The exception is the limbic lobe (part of which is shown); it incorporates parts of other (frontal, temporal, parietal) lobes.

Cortical mapping experiments (based on electrical stimulation) and clinical/pathologic data have been the principal methods by which functions of the cortex have been discovered. All parts of the cortex are concerned with storage of experience (memory), exchange of impulses with other cortical areas (association), and the two-way transmission of impulses with subcortical areas (afferent/efferent projections).

The *frontal lobe* is concerned with intellectual functions such as reasoning and abstract thinking, aggression, sexual behavior, olfaction or smell, articulation of meaningful sound (*speech*), and voluntary movement (*precentral gyrus*). The *central sulcus* separates the frontal lobe from the parietal lobe. The *parietal lobe* is concerned with body sensory awareness, including taste (*postcentral gyrus*), the use of symbols for communication (language), abstract reasoning (e.g., mathematics), and body imaging. The *temporal lobe* is partly limbic and here is concerned with the formation of emotions (love, anger, aggression, compulsion, sexual behavior); the non-limbic temporal lobe is concerned with interpretation of language, awareness and discrimination of sound (hearing; *auditory area*), and constitutes a major memory processing area. The *occipital lobe* is concerned with receiving, interpreting, and discriminating visual stimuli from the optic tract, and associating those visual impulses with other cortical areas (e.g., memory).

In evolutionary terms, the limbic lobe or system is the oldest part of the cortex. It is the center of emotional behavior. The limbic neurons occupy parts of the inferior and medial cortices of each hemisphere, and some subcortical areas as well. Certain limbic areas are closely related topographically to the olfactory tracts.

The cerebral hemispheres appear structurally as mirror images of one another; functionally they are not. The speech area develops fully only on one side, usually the left. In general, the left hemisphere tends to deal with certain higher functions (mathematical, analytical, verbal) while the right concentrates on visual, spatial, and musical orientations. The matter of cerebral "dominance" (left hemisphere, left speech center, right handed) or vice versa is quite controversial.

CN: Use very light colors for F and G. (1) Color gray the various sections of cerebral cortex without coloring the cortical surfaces. (2) Among the many views shown, color each structure wherever it appears before going on to the next title/structure.

CEREBRAL CORTEX A*
SUBCORTICAL AREAS:·
BASAL NUCLEI:*
CAUDATE NUCLEUS B
LENTICULAR NUCLEUS C
LATERAL VENTRICLE D
WHITE MATTER TRACTS:·
COMMISSURES: E
CORPUS CALLOSUM E¹
PROJECTION TRACTS: F
CORONA RADIATA F¹
INTERNAL CAPSULE F²
ASSOCIATION TRACTS G

The subcortical areas of the cerebral hemispheres include the white matter below the cortex, the basal ganglia (nuclei), and the lateral ventricles.

The *basal nuclei* are discrete islands of gray matter amidst a sea of white matter at the base of the hemispheres on either side of and above the diencephalon (see next plate). The major basal nuclei are the *caudate nucleus* and the *lenticular nucleus*. Both are named according to their shape: the caudate, when viewed laterally, appears to have a head and a progressively narrowing and curving tail (cauda, tail) ending at the amygdala (a nucleus of the limbic system). The lenticular nucleus is arguably lens-shaped. The basal ganglia have connections with the cerebral cortex, the thalamus, and nuclei near the thalamus. The basal ganglia are involved in the planning, initiation, maintenance, and termination of movement (motor activity). They monitor and mediate descending commands from the motor cortex. They are instrumental centers in maintaining muscle tone and programming sequential postural movements and adjustments.

Their influence is manifested as appropriately gated impulses influencing the lower motor neurons of the cranial and spinal nerves in their innervation of skeletal muscle. Diseases of the basal ganglia include dystonias and dyskinesias and, perhaps most well known of all, Parkinson's disease (abnormal gait, rigidity, tremors).

The white matter of the hemispheres consists of tracts oriented in six general directions. Tracts connecting left and right hemispheres are called *commissures*, of which the largest is the *corpus callosum*. This massive tract spans the two hemispheres, roofing over the lateral ventricles. It makes possible communications between centers in the paired hemispheres. Bundles of white matter, both long and short, connecting anterior and posterior cortical areas, are called *association tracts*. An emotional response to a visual stimulus is made possible, in part, by association tracts.

Perhaps the most spectacular mass of white matter in the brain is that *projection tract* of myelinated axons (the *corona radiata*) radiating from the level of the basal ganglia to and from all parts of the cortex. It is continuous inferiorly with the compact band of fibers *(internal capsule)* passing between and partly encapsulating the two basal nuclei. All motor commands are conducted here; all sensory input reaching the cortex passes through these fibers. Very importantly, the thalamus and cortex communicate by this pathway.

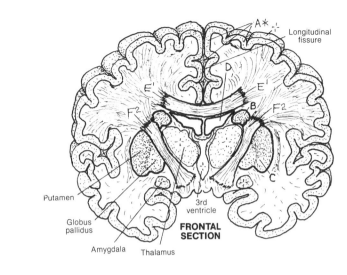

Longitudinal fissure

Putamen

Globus pallidus

Amygdala

Thalamus

3rd ventricle

FRONTAL SECTION

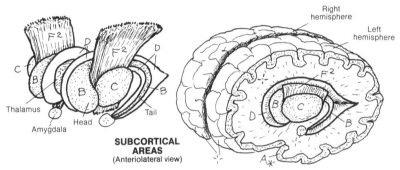

Right hemisphere

Left hemisphere

Thalamus

Amygdala

Head

Tail

SUBCORTICAL AREAS
(Anteriolateral view)

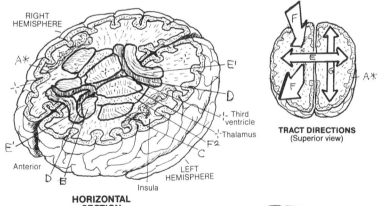

RIGHT HEMISPHERE

Third ventricle

Thalamus

Anterior

Insula

LEFT HEMISPHERE

HORIZONTAL SECTION

TRACT DIRECTIONS
(Superior view)

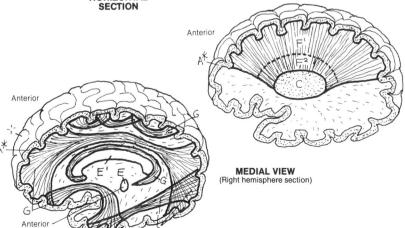

Anterior

Anterior

Anterior commissure

MEDIAL VIEW
(Right hemisphere section)

CN: Use light colors for A and B, and a very bright color for C. (1) Color each structure wherever it appears before going on to the next title. (2) Although not colored, the neighboring relations of the diencephalic structures are important and have been identified by name. These should be given special attention.

DIENCEPHALON: *
THALAMUS A
HYPOTHALAMUS B
EPITHALAMUS
(PINEAL GLAND) C
THIRD VENTRICLE D

The *diencephalon*, the smaller of the two derivatives of the early forebrain, fits between (di, between) but is not part of the massive, surrounding cerebral hemispheres.

It consists of masses of nuclei at the top of (and indeed is part of) the brain stem. The diencephalic structures are paired and are arranged around the thin *third (III) ventricle* and its infundibular recess. On each side, the diencephalon consists of the *thalamus* and the subthalamus inferior to it, the *hypothalamus* embracing the infundibular recess of the third ventricle, and the *epithalamus* or pineal gland suspended from the posterior wall of the third ventricle. The structural relationship of these structures to the white matter of the cerebrum, the third ventricle, the basal ganglia, and the midbrain should be carefully studied to ensure orientation.

The thalamus consists of several groups of cell bodies and processes that, in part, process all incoming impulses from sensory pathways (except olfactory). It has broad connections with the motor, general sensory, visual, auditory, and association cortices. Not surprisingly, the corticothalamic (cortex to thalamus) fibers contribute significantly to the corona radiata. Still other thalamic nuclei connect to the hypothalamus and other brainstem nuclei. Thalamic activity (1) integrates sensory experiences resulting in appropriate motor responses, (2) integrates specific sensory input with emotional (motor) responses (e.g., a baby crying in response to hunger), and (3) regulates and maintains the conscious state (awareness), subject to facilitating/ inhibiting influences from the cortex. Subthalamic nuclei are concerned with motor activity and have connections with the basal ganglia.

The hypothalamus consists of nuclear masses and associated tracts on either side of the lower third ventricle. The hypothalamus maintains neuronal connections with the frontal and temporal cortices, thalamus, neurohypophysis, and brainstem. Its neurosecretions (hormones) are also directed to the adenohypophysis via the hypophyseal portal system. In addition, the hypothalamus is concerned with emotional behavior, regulation of the autonomic (visceral) nervous system and related integration of visceral (autonomic) reflexes with emotional reactions, activation of the drive to eat (hunger) and the subsequent feeling of satisfaction (satiety) following fulfillment of that drive. Finally, it mediates descending impulses related to both reflexive and skilled movement. All of this in an area the size of four peas!

The epithalamus (pineal gland) consists primarily of the pineal body and related nuclei and tracts that have connections with the thalamus, hypothalamus, basal nuclei, and the medial temporal cortex. It produces melatonin (a pigment-enhancing hormone), the synthesis of which is related to diurnal cycles or rhythms (body activity in day or sunlight as opposed to dark or nocturnal periods). It may influence the onset of puberty through inhibition of testicular/ovarian function. Remarkably, the pineal is the only unpaired structure in the brain.

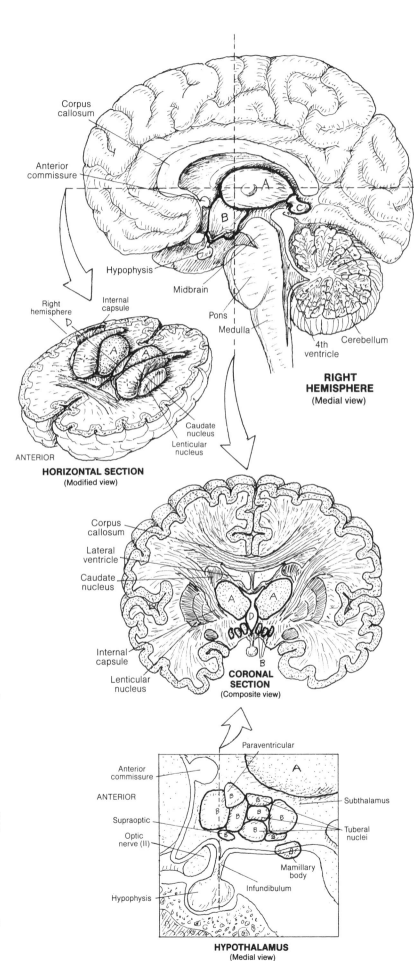

RIGHT HEMISPHERE
(Medial view)

HORIZONTAL SECTION
(Modified view)

CORONAL SECTION
(Composite view)

HYPOTHALAMUS
(Medial view)

CN: Use darker colors for C, E, M, and the lightest for K. (1) As you color each structure in as many views as it is shown, take particular note of of the orientation of the view. (2) Note that the fourth ventricle is located in both parts of the hindbrain and receives the same color in both parts. The diencephalon has been presented on the previous plate and is shown here only for orientation.

DIENCEPHALON A
MIDBRAIN: B
CEREBRAL AQUEDUCT C
SUPERIOR COLLICULUS B¹
INFERIOR COLLICULUS B²
SUP. CEREBELLAR PEDUNCLE D
HINDBRAIN: *
4TH VENTRICLE E
PONS F
MID. CEREBELL. PED. G
MEDULLA H
INF. CEREBELL. PED. I

CEREBELLUM J
ARBOR VITAE K
CEREBELLAR CORTEX L*
DEEP CEREB. NUCLEUS M

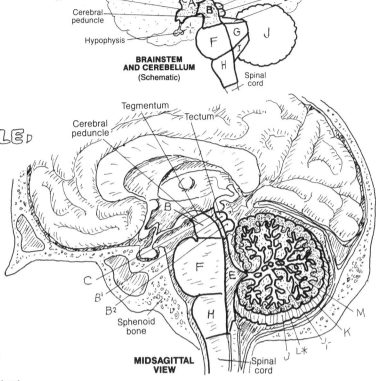

BRAINSTEM AND CEREBELLUM (Schematic)

MIDSAGITTAL VIEW

The *brainstem* consists of all portions of the brain less the cerebrum and cerebellum.

The *midbrain* includes two columns of tracts (cerebral peduncles) and mixed nuclei/tracts posterior to these columns (tegmentum), the *cerebral aqueduct*, *superior cerebellar peduncles* (conducting tracts from thalamus, midbrain, and medulla), and the more posterior *superior and inferior colliculi* (tectum of the midbrain). The peduncles consist of descending axons continuous with the internal capsule above and destined for lower motor neurons in the brain stem (cranial nerve) and spinal cord (spinal nerves) below. The deeper tegmentum contains nuclei of the reticular formation, nuclei/tracts concerned with cranial nerves III and IV, relay of impulses between lower and higher centers, and other centers/ tracts concerned with somatic and visceral motor-related impulses. The superior colliculi are centers for visual reflexes; the inferior colliculi make possible auditory reflexes (e.g., involuntary movements in response to visual and auditory stimuli, respectively).

The upper hindbrain is the *pons*. Massive stalks of white matter, the *middle cerebellar peduncles*, cross the *fourth ventricle* (pons, bridge) to reach the cerebellum. The pons consists of (1) tracts descending from the midbrain to lower centers; (2) masses of cell bodies that synapse with certain tracts of cortical origin and whose axons constitute the middle cerebellar peduncle; (3) nuclei that relate to cranial nerves V, VI, VII, and VIII; (4) several ascending tracts arising from the medulla and spinal cord; and (5) a network of polysynaptic neurons (part of the reticular formation) that facilitate/inhibit (mediate) somatic and visceral reflexes and form a mechanism for arousal, wakefulness, and alertness.

The *medulla*, continuous with the deep pons above and the spinal cord below, consists of much the same organization as the pons. Life-sustaining control centers for respiration, heart rate, and vasomotor function exist here. It contains nuclei concerned with cranial nerves VIII, IX, X, XI, and XII. The *inferior cerebellar peduncle* carries tracts from the spinal cord and vestibular centers (head balance) in the medulla. Two particularly evident bundles of fibers are seen on the anterior surface of the medulla. These pyramids consist of corticospinal fibers conducting voluntary movement-related impulses to lower motor neurons of the spinal cord. 80% of these fibers cross (decussate) to the contralateral side.

The *cerebellum* consists of two hemispheres, with a cortex of gray matter on its surface (*cerebellar cortex*), central masses of motor-related (*deep cerebellar*) *nuclei*, and bands of white matter forming a treelike appearance (*arbor vitae*, tree of life) when the cerebellum is cut in section. The cerebellum is attached to the brain stem by the three cerebellar peduncles. The cerebellum is concerned with equilibrium and position sense, fine movement, control of muscle tone, and overall coordination of muscular activity in response to proprioceptive input and descending traffic from higher centers.

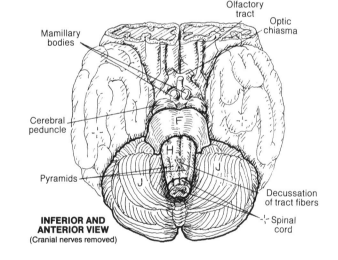

INFERIOR AND ANTERIOR VIEW (Cranial nerves removed)

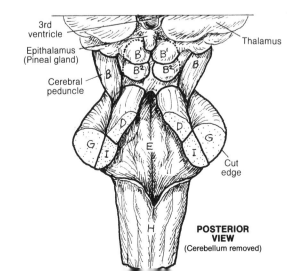

POSTERIOR VIEW (Cerebellum removed)

CN: Use bright colors for A–C (except where indicated by asterisk (∗) or no-color symbol (-⁄-). Use medium dark colors for D–I, and light colors for K–M. (1) In the upper drawing, color B∗ gray over the nerve roots within the dura mater and outside the spinal cord. (2) Color the cord sections taken at various levels. (3) Color the meninges of the spinal cord. What is not shown (because of size limitations) is the presence of the pia mater, subarachnoid space, and arachnoid around the nerve roots. (4) Do not color the structures within the subarachnoid space or the central canal in the drawing at the bottom of the plate.

SPINAL CORD A
MENINGES: -⁄-
PIA MATER A¹
FILUM TERMINALE A²
SUBARACHNOID SPACE B∗
ARACHNOID B¹
DURA MATER C
EPIDURAL SPACE C¹-⁄-

The *spinal cord* begins at the foramen magnum of the skull and ends as the conus medullaris at a vertebral level of L1 or L2. It bulges slightly in the lower cervical and lumbar regions (enlargements) in relation to the presence of large numbers of axons associated with the upper and lower limbs. Ensheathed in coverings called meninges, the spinal cord is awash in cerebrospinal fluid (CSF) within the *subarachnoid space*. The spinal meninges and related spaces include, from inner to outer: a thin, vascular *pia mater* that ensheathes the cord and nerve rootlets/roots, and continues inferiorly from the conus as the *filum terminale* to join the *dura mater* at the 2nd sacral vertebra (S2); a significant CSF-containing subarachnoid space surrounding cord and nerve roots, whose largest portion is the lumbar cistern from L2 to the end of the dural sac at S2, filled with nerve roots (cauda equina); a filmy *arachnoid*, the consistency of which is similar to a dense spider's web; a potential subdural space; and the tough, protective, fibrous *dura mater* (thecal sac). Between the dura and the ligament-covered, periosteum-lined vertebral bones/intervertebral discs is the fat-filled *epidural space* containing a plexus of veins.

GRAY MATTER D∗
POSTERIOR HORN E
ANTERIOR HORN F
LATERAL HORN (T1-L2) G
INTERMEDIATE ZONE H
GRAY COMMISSURE I
WHITE MATTER J-⁄-
POSTERIOR FUNICULUS K
LATERAL FUNICULUS L
ANTERIOR FUNICULUS M

The spinal cord consists of a central mass of gray matter arranged into the form of an H and a peripheral array of white matter (*funiculi*) consisting of descending and ascending tracts. The amount of white matter decreases as the cord progresses distally, seen especially well in the sacro-coccygeal region. The gray *posterior horns* (actually columns when seen in three dimensions) receive the central processes of sensory neurons (recall Plate 131) and directs incoming impulses to the adjacent white matter for conduction to other cord levels or higher centers. *The anterior horns* include lower motor neurons that represent the "final common pathway" for motor commands to muscle. *Lateral horns* exist only in the thoracic and upper lumbar cord and include autonomic motor neurons supplying smooth muscle (in vessels and viscera) and glands. It is in the gray matter that spinal reflexes occur in conjunction with facilitory and inhibitory influences from higher centers.

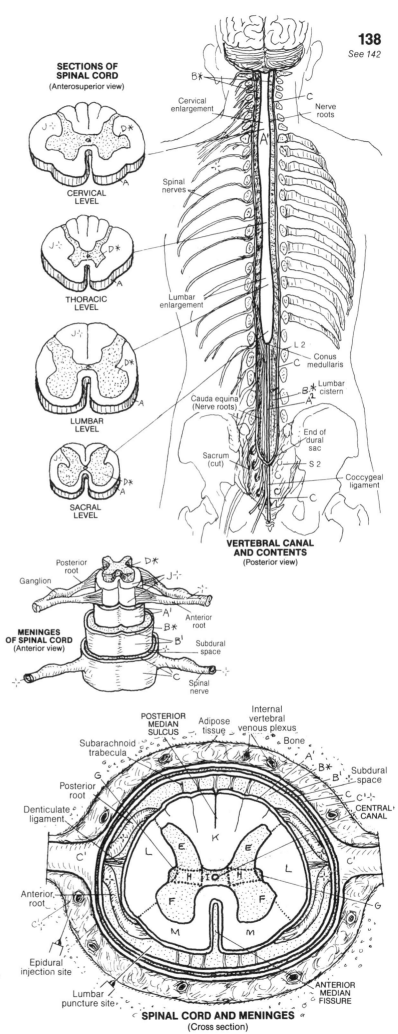

SECTIONS OF SPINAL CORD (Anterosuperior view)

CERVICAL LEVEL

THORACIC LEVEL

LUMBAR LEVEL

SACRAL LEVEL

Cervical enlargement
Nerve roots
Spinal nerves
Lumbar enlargement
L 2
Conus medullaris
Cauda equina (Nerve roots)
Lumbar cistern
Sacrum (cut)
End of dural sac
S 2
Coccygeal ligament

VERTEBRAL CANAL AND CONTENTS (Posterior view)

Posterior root
Ganglion
Anterior root
Subdural space
Spinal nerve

MENINGES OF SPINAL CORD (Anterior view)

POSTERIOR MEDIAN SULCUS
Adipose tissue
Internal vertebral venous plexus
Bone
Subarachnoid trabecula
Subdural space
Posterior root
Denticulate ligament
CENTRAL CANAL
Anterior root
Epidural injection site
Lumbar puncture site
ANTERIOR MEDIAN FISSURE

SPINAL CORD AND MENINGES (Cross section)

XIV. NERVOUS SYSTEM
CNS: ASCENDING TRACTS

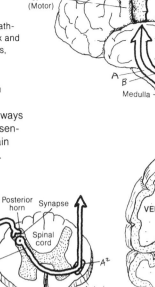

CN: Use bright colors for A-C and a light color for F. (1) Color the pain/temperature pathway which is shown on one side only for visual simplicity. Note that the sensory cortex and the thalamus are to be colored gray. (2) In the muscle stretch/position sense pathways, note there are two different cerebellar peduncles, each receiving a different color.

Ascending pathways consist of linearly arranged neurons, the axons of which travel in a common bundle (tract) conducting impulses toward the thalamus, cerebral cortex, or cerebellum. In the examples shown here, each of the pathways begins with a sensory neuron. These sensory pathways permit body surface sensations and muscle/tendon stretch information (below the head) to reach brain stem and cerebellar centers for response and cortical centers for awareness.

PAIN/TEMPERATURE: A
SENSORY NEURON A¹
LAT. SPINOTHALAMIC TRACT A²
THALAMUS *¹
THALAMOCORTICAL TRACT A³
SENSORY CORTEX *²

Pain and temperature receptors on the body surface and elsewhere below the head generate impulses that travel to the spinal cord by axons of *sensory neurons* (1st order neuron). The central process ("axon") of each sensory neuron enters the posterior horn and synapses with the 2nd order neuron whose axon crosses to the contralateral side, enters the lateral funiculus, and ascends as part of the *lateral spinothalamic tract*. This neuron ascends to the thalamus where it synapses with relay (3rd order) neurons, the axons of which traverse the internal capsule and corona radiata (*thalamocortical tract*) to reach the postcentral gyrus of the cerebral cortex ("sensory cortex").

TOUCH/PRESSURE: B
SENSORY NEURON B¹
N. CUNEATUS & GRACILIS B²
INT. ARCUATE FIBERS B³
MED. LEMNISCUS B⁴
THALAMUS *¹
THALAMOCORTICAL TRACT B⁵
SENSORY CORTEX *²

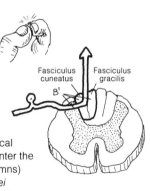

Touch and pressure receptors below the head generate electrochemical impulses that travel to the spinal cord through *sensory neurons* that enter the posterior horn and join/ascend the posterior funiculus (posterior columns) to the medulla. Here they synapse with 2nd order neurons in the *nuclei cuneatus and gracilis*. The axons of these neurons sweep to the opposite side (as *internal arcuate fibers*) to form an ascending bundle (*medial lemniscus*) in the brain stem that terminates in the thalamus. There these axons synapse with 3rd order relay neurons whose axons reach the postcentral gyrus of the cerebral cortex via the *thalamocortical tract*.

MUSCLE STRETCH/POSITION SENSE: C
SENSORY NEURON C¹
POST. SPINOCEREBELLAR TRACT C²
INF. CEREBELLAR PED. D
ANT. SPINOCEREBELLAR TR. C³
SUP. CEREBELLAR PED. E
CEREBELLAR CORTEX F

Impulses from muscle spindles and other proprioceptors (receptors responsive to muscle stretch/loads) are conducted by *sensory neurons* to the spinal cord. Single receptor input is conducted by 2nd order neurons that ascend the ipsilateral lateral funiculus (*posterior spinocerebellar tract*) and enter the cerebellum via the *inferior cerebellar peduncle*. More global proprioceptive input ascends the contralateral anterior *spinocerebellar tract* and enters the cerebellum via the *superior cerebellar peduncle*. By these and similar pathways that function in the absence of awareness, the cerebellum maintains an ongoing assessment of body position, muscle tension, muscle overuse, and movement. In turn, it mediates descending impulses from cortical and subcortical centers destined for motor neurons.

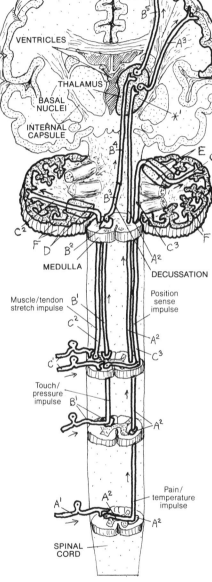

CEREBRAL CORTEX, CEREBELLUM, AND SPINAL CORD

(Schematic)

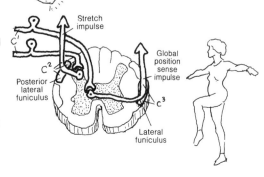

XIV. NERVOUS SYSTEM
CNS: DESCENDING TRACTS

CN: Use light colors for H, I, and K. (1) Color the pyramidal tract in the sagittal view. (2) Color the pyramidal tract in the schematic coronal section at upper right, including the percentage figures. (3) Color the extra-pyramidal system.

PYRAMIDAL TRACT_A / RELATED AREAS: *
MOTOR CORTEX *'
CORTICOSPINAL TRACT_A'
LAT._A² / ANT. CORTICOSPINAL TRACT_A³
MEDULLARY PYRAMID_A⁴
LATERAL FUNICULUS_B
ANTERIOR FUNICULUS_C
FINAL COMMON PATHWAY: *
LOWER MOTOR NEURON_D
EFFECTOR_E

Voluntary movement is initiated in the *motor cortex* (precentral gyrus) on the opposite side where the movement is desired. The axons of the cortical motor neurons (upper motor neurons) descend the corona radiata, the internal capsule, the cerebral peduncles, and the pons—without synapse—as the *corticospinal tract*. In the medulla, these axons form the anterior swellings called the pyramids (thus, pyramidal tract). Here at the decussation of the pyramids, about 80% of the fibers cross to the opposite side and enter the lateral funiculus. These fibers—and the uncrossed fibers on the same side—form the *lateral corticospinal tract*. The axons of the pyramidal tract leave the funiculus at one of many spinal levels to synapse with the anterior horn (lower) motor neurons.

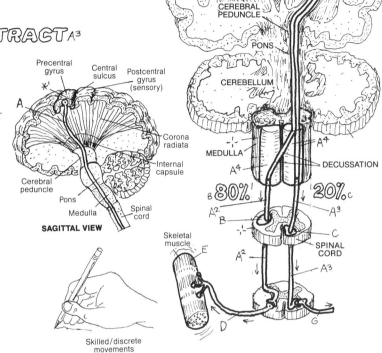

PYRAMIDAL TRACT
(Schematic)

SAGITTAL VIEW

Skilled/discrete movements

EXTRAPYRAMIDAL SYSTEM: *
PONTINE RETICULOSPINAL TRACT_F
VESTIBULOSPINAL TRACT_G
INTERNEURON_H

The desired, corticospinal-actuated command to the lower motor neuron is *not* influenced by body position, memory, and a host of other factors necessary to achieve the desired movement at the required moment. These collective inputs from the cerebral cortex, basal nuclei, cerebellum and elsewhere, arrive at the appropriate lower motor neurons by a number of descending pathways, none of which pass through the medullary pyramids (hence, extrapyramidal system or tracts). Two major extrapyramidal tracts are shown here: the *reticulospinal tract* from the brain stem reticular nuclei; the *vestibulospinal tract* from the vestibular nuclei in the brain stem. Other tracts include the rubrospinal and tectospinal tracts (not shown, but see glossary). The synaptic connections of these axons with *each* lower motor neuron (often by way of *interneurons*) are in the thousands. Depending on the neurotransmitter produced by the presynaptic neuron, the synapse may facilitate or inhibit production of an excitatory impulse from the lower motor neuron. Discharge of the lower motor neuron, or not, is dependent upon the sum of the facilitory and inhibitory impulses impinging on it at any moment. Once generated, the electrochemical impulse moving down the axon of the lower motor neuron reaches the effector without further mediation. Thus, the anterior horn motor neuron is truly the "final common pathway" for the ultimate expression of all nervous activity: muscular contraction.

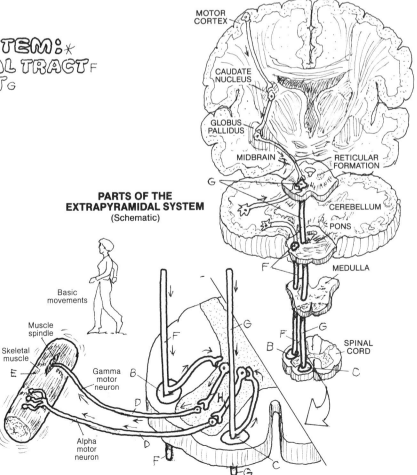

**PARTS OF THE
EXTRAPYRAMIDAL SYSTEM**
(Schematic)

Basic movements

Muscle spindle

Skeletal muscle

Gamma motor neuron

Alpha motor neuron

CN: Use a light color for A. (1) Color the three drawings of ventricular development first. (2) Color the lateral and superior views of the fully developed ventricles. (3) Color the coronal and modified sagittal sections revealing the relationship of the ventricles to surrounding structure.

VENTRICLE DEVELOPMENT: *
NEURAL CAVITY *
FOREBRAIN A
TELENCEPHALON B
DIENCEPHALON C
MESENCEPHALON D
HINDBRAIN E
METENCEPHALON F
MYELENCEPHALON G
SPINAL CORD H

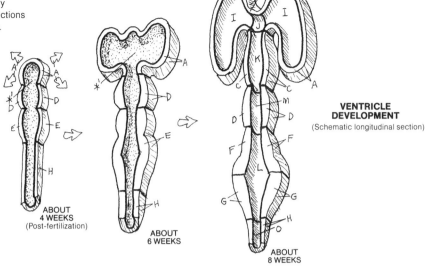

VENTRICLE DEVELOPMENT
(Schematic longitudinal section)

ABOUT 4 WEEKS (Post-fertilization)

ABOUT 6 WEEKS

ABOUT 8 WEEKS

The central nervous system develops from a hollow neural tube near the dorsal surface of the embryo. The neural cavity undergoes extraordinary revision in association with development of the brain regions. The shape of the cavity in each brain region reflects the local changes and mechanical pressures imposed by the developing brain. The ventricles may be identified by name or by roman numerals.

The cavity of the developing forebrain expands remarkably with the development of its walls into the paired cerebral hemispheres (*telencephalon*). These cavernous spaces will become the *lateral* (I, II, left to right) *ventricles*. Almost pinched off from these cavities and compressed to a thin, purse-like space by the enlarging hemispheres is the centrally located *third* (III) *ventricle*, connected to the lateral ventricles by small, paired *interventricular foramina*. The derivatives of the *diencephalon* develop in the walls of the third ventricle. The inferior recess of this ventricle dips into the infundibulum of the hypophysis (pituitary gland); the hypothalamus develops in the walls of this recess. The pineal develops in the wall of the pineal recess. The *mesencephalon* undergoes the least physical change of the early brain regions, and this is reflected in the shape of the narrow, tubular *cerebral aqueduct*. The *fourth* (IV) *ventricle* forms in the developing hindbrain and is particularly affected by the growth of the cerebellar hemispheres into which it projects lateral recesses. The tips of these recesses open into the subarachnoid spaces (lateral apertures). Another opening (median aperture) is on the floor of the IV ventricle at the cerebellar level. The ventricle narrows progressively throughout the myelencephalon. At the medullary/spinal cord junction, the cavity thins to a narrow, often occluded canal (central canal) in the spinal cord.

In each of the developing ventricles (cerebral aqueduct excepted) a highly vascular tissue forms from the pia mater covering the brain and the lining (ependymal) cells of the ventricles. Called *choroid plexus*, this tissue secretes cerebrospinal fluid or CSF (see Plate 143).

DERIVATIVES: *
LATERAL VENTRICLE (1 & 2) I
INTERVENTRICULAR FORAMEN J
3RD VENTRICLE K
4TH VENTRICLE L

CEREBRAL AQUEDUCT M
CHOROID PLEXUS N
CENTRAL CANAL O

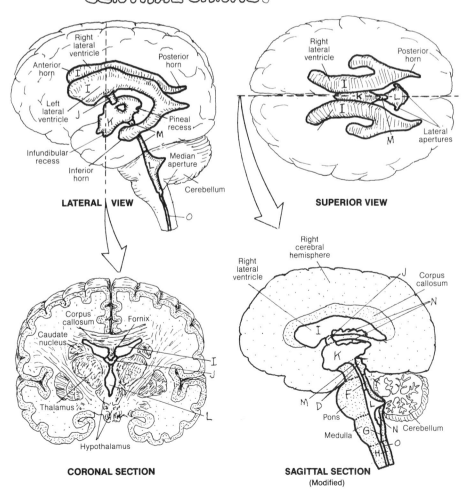

Right lateral ventricle

Anterior horn

Posterior horn

Left lateral ventricle

Infundibular recess

Inferior horn

Pineal recess

Median aperture

Cerebellum

LATERAL VIEW

Right lateral ventricle

Posterior horn

Lateral apertures

SUPERIOR VIEW

Corpus callosum

Caudate nucleus

Fornix

Thalamus

Hypothalamus

CORONAL SECTION

Right cerebral hemisphere

Right lateral ventricle

Corpus callosum

Pons

Medulla

Cerebellum

SAGITTAL SECTION
(Modified)

CN: Use very light colors. (1) Begin with the diagram of the brain and spinal cord. Note that the dura mater conceals the convolutions of the brain. (2) Color the upper two illustrations depicting the cranial meninges. The subarachnoid space (D), which is colored gray, can only be seen in the enlarged portion. Note where the inner layers of the dura mater (B) converge to form the falx cerebri (B¹). (3) Color the infoldings of the dura mater in the large cranial view below. Note the outer layer (A) adjacent to the skull bones.

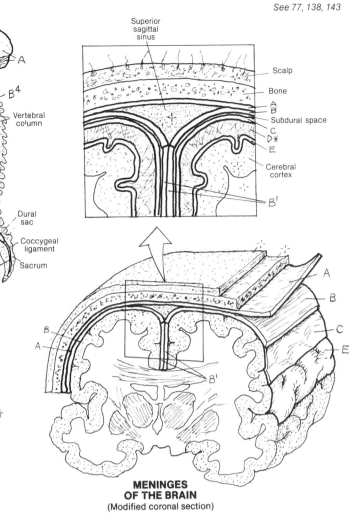

DURA MATER OF BRAIN AND SPINAL CORD

CRANIAL MENINGES:*
DURA MATER:-¹-
OUTER (PERIOSTEAL) LAYER ₐ
INNER (MENINGEAL) LAYER ᵦ
FALX CEREBRI ᵦ¹
TENTORUMI CEREBELLI ᵦ²
FALX CEREBELLI ᵦ³
ARACHNOID c
SUBARACHNOID SPACE (CSF) ᴅ*
PIA MATER ₑ

SPINAL DURA MATER ᵦ⁴

The brain and spinal cord are enveloped in fibrous coverings called meninges. The meninges of the spinal cord have been presented in Plate 138, and are the inferior extent of the cranial membranes presented here.

The outer meningeal covering (meninx) is the dense, fibrous *dura mater*. The *outer layer* is closely applied to the skull bone as periosteum. The *inner layer* splits off the cranial roof in the midline bilaterally, encloses the large venous superior sagittal sinus, and descends between the cerebral hemispheres in the longitudinal cerebral fissure as the *falx cerebri*. This sheet of dura is attached anteriorly at the crista galli and posteriorly at the internal occipital protuberance. It arches over the corpus callosum. It encloses the inferior sagittal sinus in its free border. The posterior part of the falx splits over the cerebellum to form the tent-like *tentorium cerebelli*, separating the cerebellum and other contents of the posterior fossa from the occipital lobes. The free edge of the tentorium forms a notch accommodating the midbrain: the tentorial notch or incisura. Given sufficient force, the incisura can incise the midbrain when the brain is rapidly loaded (accelerated) against the skull in serious head trauma. Extending vertically below the tentorium in the midline is the *falx cerebelli* which separates the paired cerebellar hemispheres. Note also the dural roof of the sella turcica (*diaphragma sellae*), enclosing the hypophysis. It is perforated to permit passage of the infundibulum. The inner layer of dura is continuous with the spinal dura mater.

The filmy, vulnerable *arachnoid* lies deep to and flush with the inner dura. The arachnoid is separated from the deeper *pia mater* by the *subarachnoid space* filled with cerebrospinal fluid (CSF). This space becomes voluminous at various locations (cisterns; Plate 143). The pia is a vascular layer of loose fibrous connective tissue, supporting the vessels reaching the brain (and spinal cord). It is inseparable from the surface of the brain and cord.

MENINGES OF THE BRAIN
(Modified coronal section)

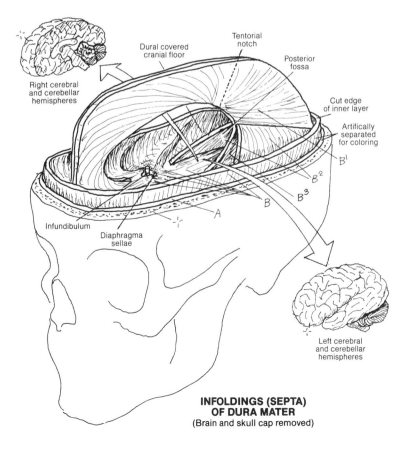

INFOLDINGS (SEPTA) OF DURA MATER
(Brain and skull cap removed)

CIRCULATION OF CEREBROSPINAL FLUID (CSF)

CN: Use the same colors as were used on the previous plate for the three meninges. Use blue for L and light colors for E through H, J, and K. (1) Color the large illustration and the coronal section simultaneously, paying close attention to the arrows of directional flow. Note that both layers of dura (A) are given one color. (2) The four cisterns, part of the subarachnoid space, all receive one color (J¹), including the lumbar cistern at lower right. (3) Color the median and lateral apertures of the IV ventricle. (4) Complete the illustration at lower right.

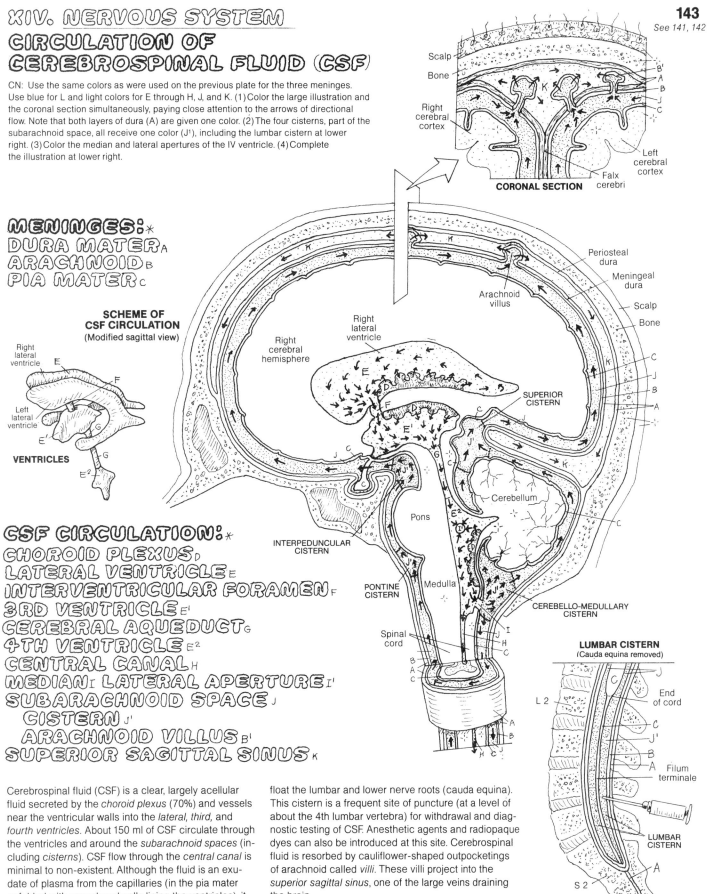

MENINGES: *
DURA MATER A
ARACHNOID B
PIA MATER C

SCHEME OF CSF CIRCULATION
(Modified sagittal view)

VENTRICLES

CSF CIRCULATION: *
CHOROID PLEXUS D
LATERAL VENTRICLE E
INTERVENTRICULAR FORAMEN F
3RD VENTRICLE E¹
CEREBRAL AQUEDUCT G
4TH VENTRICLE E²
CENTRAL CANAL H
MEDIAN I **LATERAL APERTURE** I¹
SUBARACHNOID SPACE J
   CISTERN J¹
   ARACHNOID VILLUS B¹
SUPERIOR SAGITTAL SINUS K

CORONAL SECTION

LUMBAR CISTERN
(Cauda equina removed)

Cerebrospinal fluid (CSF) is a clear, largely acellular fluid secreted by the *choroid plexus* (70%) and vessels near the ventricular walls into the *lateral, third,* and *fourth ventricles*. About 150 ml of CSF circulate through the ventricles and around the *subarachnoid spaces* (including *cisterns*). CSF flow through the *central canal* is minimal to non-existent. Although the fluid is an exudate of plasma from the capillaries (in the pia mater enfolded with ependymal cells lining the ventricles), it has significantly less density and protein than plasma. CSF drains into the subarachnoid space via *median* and *lateral apertures* (foramina of Magendie and Luschka, respectively) located in the fourth ventricle. Cisterns are dilated subarachnoid spaces formed at flexures of the brain. The most notable of the cisterns is the lumbar cistern (not the central canal!) in which

float the lumbar and lower nerve roots (cauda equina). This cistern is a frequent site of puncture (at a level of about the 4th lumbar vertebra) for withdrawal and diagnostic testing of CSF. Anesthetic agents and radiopaque dyes can also be introduced at this site. Cerebrospinal fluid is resorbed by cauliflower-shaped outpocketings of arachnoid called *villi*. These villi project into the *superior sagittal sinus*, one of the large veins draining the brain.

CSF has a shock-absorbing function of great significance; the CNS literally floats within it. Cerebral injury from blows to the head are mitigated to a high degree by this fluid cushion. On the other hand, high intracerebral pressure induced by ventricular enlargement secondary to decreased CSF absorption or ventricular blockage (hydrocephalus) can cause significant brain damage.

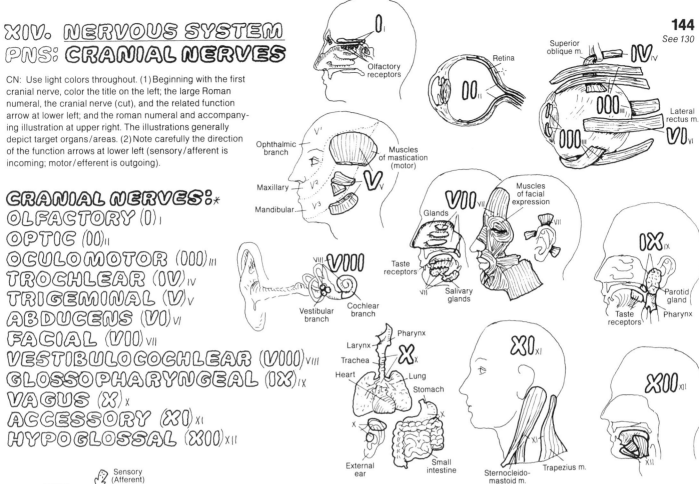

XIV. NERVOUS SYSTEM
PNS: CRANIAL NERVES

CN: Use light colors throughout. (1) Beginning with the first cranial nerve, color the title on the left; the large Roman numeral, the cranial nerve (cut), and the related function arrow at lower left; and the roman numeral and accompanying illustration at upper right. The illustrations generally depict target organs/areas. (2) Note carefully the direction of the function arrows at lower left (sensory/afferent is incoming; motor/efferent is outgoing).

CRANIAL NERVES:*
OLFACTORY (I)
OPTIC (II)
OCULOMOTOR (III)
TROCHLEAR (IV)
TRIGEMINAL (V)
ABDUCENS (VI)
FACIAL (VII)
VESTIBULOCOCHLEAR (VIII)
GLOSSOPHARYNGEAL (IX)
VAGUS (X)
ACCESSORY (XI)
HYPOGLOSSAL (XII)

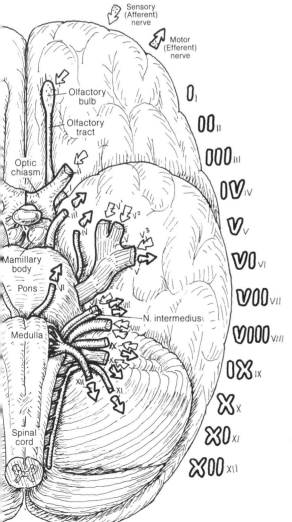

ANTERIOR-INFERIOR SURFACE
(Left brain, brainstem, and cerebellum)

Here cranial nerves and their general target organs/areas are shown. All motor nerves include proprioceptive (sensory) fibers as well. Cranial nerves I and II are derived from the forebrain; all others are brain stem-derived. Cranial nerve nuclei (neuronal cell bodies) are arranged in seven longitudinal columns in the brain stem. Functionally, these columns are general somatic afferent *(GSA)* or efferent *(GSE)*, general visceral afferent *(GVA)* or efferent *(GVE)*, special visceral afferent *(SVA)* or efferent *(SVE)*, and special somatic afferent *(SSA)*. General columns also exist in the spinal cord for spinal nerves; special columns do not. Somatic includes skin, eye, fascial, and musculoskeletal structures; visceral includes smooth muscle and glands of organs with hollow cavities.

I **SVA:** smell-sensitive (olfactory) receptors in roof/walls of nasal cavity.

II **SSA:** light-sensitive (visual) receptors in retina of eye.

III **GSE:** to extrinsic eye muscles (exc. lat. rectus and sup. oblique); **GVE:** parasympathetic to ciliary and pupillary sphincter (eye) muscles via ciliary ganglion in orbit.

IV **GSE:** to superior oblique muscle of the eye.

V **GSA:** from face via three divisions indicated; **SVE:** to muscles of mastication, tensor tympani, tensor veli palatini, mylohyoid, and digastric muscles.

VI **GSE:** to lateral rectus muscle of the eye.

VII **SVA:** from taste receptors ant. tongue; **GSA:** from ext. ear; **GVE:** parasympathetic to glands of nasal/oral cavity, lacrimal gland (via pterygopalatine ganglion in fossa of same name), submandibular/sublingual salivary glands (via submandibular ganglion in region of same name); **SVE:** to facial muscles, stapedius (mid. ear), stylohyoid, post. digastric muscles.

VIII **SSA:** cochlear part is sound-sensitive; vestibular part is sensitive to head balance and movement (equilibrium).

IX **SVA:** from taste receptors post. one-third tongue; **GSA:** from ext. ear and ext. auditory canal; **GVA:** from mucous membranes of posterior mouth, pharynx, auditory tube, and middle ear; from pressure and chemical receptors in carotid body and common carotid artery; **SVE:** to sup. constrictor m. of the pharynx, stylopharyngeus; **GVE:** parasymp. to parotid gland (via otic ganglion in infratemporal fossa).

X **SVA:** from taste receptors at base of tongue and epiglottis; **GSA:** from ext. ear and ext. aud. canal; **GVA:** from pharynx, larynx, thoracic and abdominal viscera; **SVE:** to muscles of palate, pharynx, and larynx; **GVE:** parasymp. to muscles of thoracic and abdominal viscera (via intramural ganglia).

XI Cranial root: joins vagus (**GVA** to laryngeal muscles); spinal root (C1–C5): innervates trapezius and sternocleidomastoid muscles.

XII **GSE:** to extrinsic and intrinsic muscles of tongue.

CN: Use very light colors D through G. (1) Begin with the upper illustration. Color all three pairs of spinal nerves as they emerge from the intervertebral foramens (M). (2) Color the cross sectional view in the center. (3) Color the spinal nerve axons and the arrows representing direction of impulse flow.

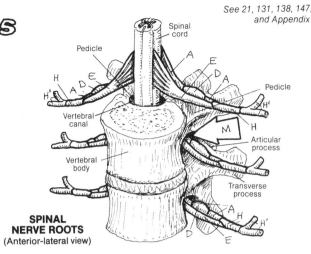

SPINAL NERVE ROOTS
(Anterior-lateral view)

SPINAL NERVE ROOTS:*
POSTERIOR ROOT A
SENSORY AXON B
CELL BODY C
POSTERIOR ROOT GANGLION D
ANTERIOR ROOT E
MOTOR AXON F
CELL BODY G
SPINAL NERVE H RAMUS H'

Spinal nerves are collections of axons of sensory and motor neurons (part of the peripheral nervous system or PNS). They are formed from nerve roots that arise directly from the spinal cord. Axons of sensory neurons make up the *posterior root*, and axons of motor neurons make up most of the *anterior root* (it has been reported that some 30% of anterior root axons are sensory). The spinal nerves and their roots are arranged segmentally and bilaterally along the length of the spinal cord. Spinal nerves are very short; formed within the *intervertebral foramina*, they branch just beyond into *anterior* and *posterior rami*. The branches of these rami are distributed bodywide below the head and provide a vehicle for acquisition by the central nervous system (CNS) of sensory information from external and internal receptors, and a means of disseminating motor commands to skeletal, smooth, and cardiac muscle, and to glands.

Spinal nerves and their roots have fairly tight quarters. The relations of these nerves and roots can best be appreciated in the cross sectional view. Nerve roots are vulnerable to irritation (radiculitis) from encroaching, hypertrophic bone in the lateral recesses (degenerative joint disease), from bulging intervertebral discs (degenerative disc disease), or from cysts, meningeal tumors, and so on. With compression of axons or blood vessels supplying the axons, functional deficits can result (radiculopathy: sensory loss, motor loss, and/or tendon reflex change).

NERVE ROOT RELATIONS:*
VERTEBRA: ÷
BODY I
LAMINA J
ARTICULAR PROCESS K
VERTEBRAL CANAL L
LATERAL RECESS L'
INTERVERTEBRAL FORAMEN M

The posterior roots of spinal nerves consist of peripheral processes (*axons*) of sensory neurons, unipolar or pseudo-unipolar neuron *cell bodies* (aggregations of which are called spinal or posterior root *ganglia*), and central processes (axons) of sensory neurons. The ganglia form obvious swellings in the area of the intervertebral foramina. The axonal endings synapse with neurons in the posterior horn of the spinal cord or enter the posterior columns (recall Plate 139). The anterior roots consist of *axons* of motor (multipolar) neurons whose *cell bodies* reside in the anterior horns of the spinal cord. These neurons are known as lower motor neurons or anterior horn cells and represent the final common pathway for motor commands to skeletal muscle. In the T1-L2 regions of the cord, motor neurons of the sympathetic division of the autonomic (visceral) nervous system reside in the lateral horns (not shown); their axons join the anterior roots.

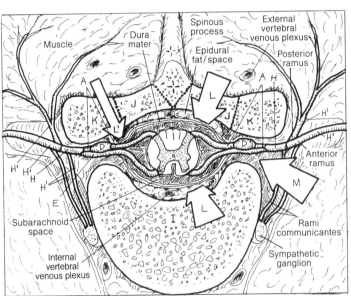

CROSS SECTION THROUGH T9
(Seen from above)

SPINAL NERVE AXONS
(Schematic lateral view)

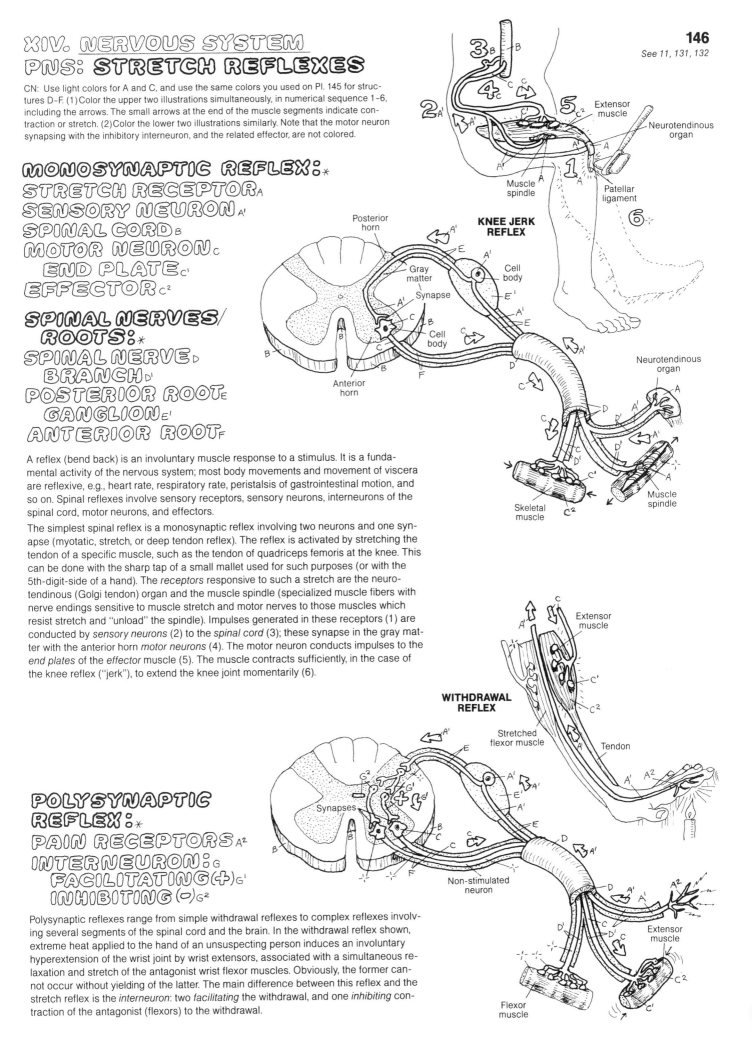

CN: Use light colors for A and C, and use the same colors you used on Pl. 145 for structures D–F. (1) Color the upper two illustrations simultaneously, in numerical sequence 1–6, including the arrows. The small arrows at the end of the muscle segments indicate contraction or stretch. (2) Color the lower two illustrations similarly. Note that the motor neuron synapsing with the inhibitory interneuron, and the related effector, are not colored.

MONOSYNAPTIC REFLEX:∗
STRETCH RECEPTOR A
SENSORY NEURON A'
SPINAL CORD B
MOTOR NEURON C
END PLATE C'
EFFECTOR C²

SPINAL NERVES/ ROOTS:∗
SPINAL NERVE D
BRANCH D'
POSTERIOR ROOT E
GANGLION E'
ANTERIOR ROOT F

KNEE JERK REFLEX

A reflex (bend back) is an involuntary muscle response to a stimulus. It is a fundamental activity of the nervous system; most body movements and movement of viscera are reflexive, e.g., heart rate, respiratory rate, peristalsis of gastrointestinal motion, and so on. Spinal reflexes involve sensory receptors, sensory neurons, interneurons of the spinal cord, motor neurons, and effectors.

The simplest spinal reflex is a monosynaptic reflex involving two neurons and one synapse (myotatic, stretch, or deep tendon reflex). The reflex is activated by stretching the tendon of a specific muscle, such as the tendon of quadriceps femoris at the knee. This can be done with the sharp tap of a small mallet used for such purposes (or with the 5th-digit-side of a hand). The *receptors* responsive to such a stretch are the neurotendinous (Golgi tendon) organ and the muscle spindle (specialized muscle fibers with nerve endings sensitive to muscle stretch and motor nerves to those muscles which resist stretch and "unload" the spindle). Impulses generated in these receptors (1) are conducted by *sensory neurons* (2) to the *spinal cord* (3); these synapse in the gray matter with the anterior horn *motor neurons* (4). The motor neuron conducts impulses to the *end plates* of the *effector* muscle (5). The muscle contracts sufficiently, in the case of the knee reflex ("jerk"), to extend the knee joint momentarily (6).

POLYSYNAPTIC REFLEX:∗
PAIN RECEPTORS A²
INTERNEURON: G
FACILITATING (+) G'
INHIBITING (−) G²

WITHDRAWAL REFLEX

Polysynaptic reflexes range from simple withdrawal reflexes to complex reflexes involving several segments of the spinal cord and the brain. In the withdrawal reflex shown, extreme heat applied to the hand of an unsuspecting person induces an involuntary hyperextension of the wrist joint by wrist extensors, associated with a simultaneous relaxation and stretch of the antagonist wrist flexor muscles. Obviously, the former cannot occur without yielding of the latter. The main difference between this reflex and the stretch reflex is the *interneuron*: two *facilitating* the withdrawal, and one *inhibiting* contraction of the antagonist (flexors) to the withdrawal.

CN: Use the same colors for E-G that were used for those structures on the preceding plate. Use light colors for A-D. (1) Begin with the nerve coverings. Note that the endoneurium (C) is shown only on the projected axons. (2) Color the typical thoracic spinal nerve and its branches in the cross section of the thorax. (3) Color the spinal nerves and their branches in the lowest drawing. Note especially the difference between intercostal nerves (one spinal nerve each) and nerves of the plexuses and their branches (combined spinal nerves).

NERVE COVERINGS:∗
EPINEURIUM A
PERINEURIUM B
ENDONEURIUM C
AXON D

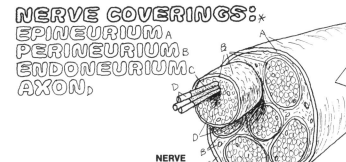

NERVE SECTION

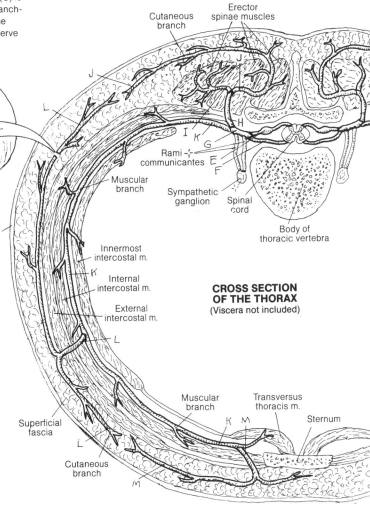

Cutaneous branch
Erector spinae muscles
Muscular branch
Rami communicantes
Sympathetic ganglion
Spinal cord
Body of thoracic vertebra
Innermost intercostal m.
Internal intercostal m.
External intercostal m.
Skin
Superficial fascia
Cutaneous branch
Muscular branch
Transversus thoracis m.
Sternum

CROSS SECTION OF THE THORAX
(Viscera not included)

Spinal nerves and their branches consist of axons of sensory and/or motor neurons ensheathed in fibrous connective tissue. Individual axons are surrounded by thin envelopes of fibrous tissue (*endoneurium*) containing nerves and capillaries that supply the living axon. Bundles (fascicles) of axons are bound by thicker, more dense coats of fibrous tissue (*perineurium*). Between (and within) the fascicles are relatively large vessels and nerve bundles. Surrounding the fascicles are circumferentially arranged loose arrays of fibrous tissue contiguous with deep or superficial fascia (*epineurium*). These supporting tissues stabilize the neurovascular elements and provide a framework for the nerve in its environment.

THORACIC SPINAL NERVE:∗
POSTERIOR ROOT E
ANTERIOR ROOT F
SPINAL NERVE G
POSTERIOR RAMUS H
LATERAL BRANCH I
MEDIAL BRANCH J
ANTERIOR RAMUS K
(INTERCOSTAL NERVE) K
LAT. CUTANEOUS BR. L
ANT. CUTANEOUS BR. M

Each *spinal nerve* leaves an intervertebral foramen and divides into *anterior* and *posterior* (primary) *rami*. The anterior rami supply all parts of the body except the deep (intrinsic) muscles and skin of the back; thus, the anterior ramus is generally larger than its posterior fellow. The anterior rami contribute to networks of interconnecting nerves (plexuses or plexi) supplying the neck, upper limb, pelvis/perineum, and lower limb. In the torso, anterior rami form the *intercostal nerves*. The distribution pattern of a typical thoracic spinal nerve is shown in the cross section of the thorax. Note the rami communicantes; these will be presented in Plate 151. Note the muscular branches of the anterior ramus passing between innermost and internal intercostal muscles, as well as the *lateral* and *anterior cutaneous branches* and their distribution in the superficial fascia. Note the areas of overlap between the cutaneous branches of the anterior rami and those of the posterior rami. This pattern occurs segmentally and bilaterally throughout the thorax; the lower thoracic spinal nerves also supply most of the abdominal wall.

The anterior rami of the cervical spinal nerves (and T1 spinal nerve) form interconnecting networks from which the nerves to the neck and the upper limb are derived (next plate). The anterior rami of the lumbar and sacral spinal nerves form interconnecting plexuses from which the nerves to the pelvis, perineum and lower limb are derived (Plate 149). Thus, the source of an intercostal nerve can be traced to the *single* spinal nerve forming it, e.g., T6 spinal nerve, whereas the source of a nerve to the limbs is traced to the *collection* of spinal nerves that form it, e.g., C5-C8 spinal nerves.

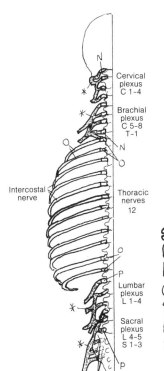

Cervical plexus C 1-4
Brachial plexus C 5-8 T-1
Intercostal nerve
Thoracic nerves 12
Lumbar plexus L 1-4
Sacral plexus L 4-5 S 1-3

SPINAL NERVES & BRANCHES:∗
CERVICAL N
THORACIC O
LUMBAR P
SACRAL Q

CN: Use light colors for A-D. (1) In the upper illustration, color the letters and numbers identifying the five roots of the brachial plexus. Note but do not color the small branches of the plexus as you color the plexus itself. Note in the lower illustration the entire plexus is colored gray. (2) As you color each of the major nerves arising from the plexus, color it in the lower illustration as well. As you color each nerve, try to visualize it on your own limb.

BRACHIAL PLEXUS & MAJOR BRANCHES:*
ROOTS C5, C6 A
UPPER TRUNK B
ROOT C7 A'
MIDDLE TRUNK B'
ROOTS C8, T1 A²
LOWER TRUNK B²
ANTERIOR DIVISION C
LATERAL CORD D
MUSCULOCUTANEOUS N. E
BR. TO MEDIAN N. F
MEDIAL CORD D'
BR. TO MEDIAN N. F
MEDIAN N. F'
ULNAR N. G
POSTERIOR DIVISION C'
POSTERIOR CORD D²
AXILLARY N. H
RADIAL N. I

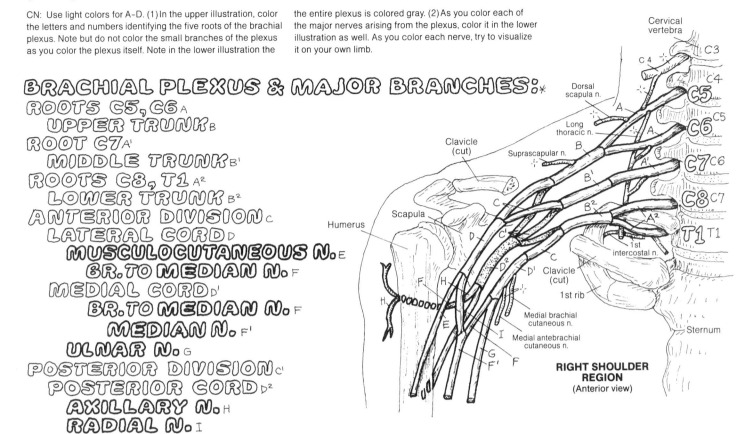

RIGHT SHOULDER REGION
(Anterior view)

The major nerves to the structures of the upper limb arise from the brachial plexus, formed from the anterior rami of spinal nerves C5-T1 (plus or minus one level). These rami form the *roots of the plexus*. In the pattern illustrated, further branching and joining of fibers in the neck, supraclavicular area, and axilla result in the formation of the five major nerves of the upper limb.

The brachial plexus is subject to injury (plexopathy) from excessive stretching or traction (e.g., rapid, forceful pulling of the upper limb) and compression (e.g., long-term placement of body weight on axillary or arm pit cushions of crutches). In such injuries, there is great variation in degree of deficit, signs, and symptoms.

The *musculocutaneous nerve* (C5-7) supplies the anterior arm muscles and is cutaneous in the forearm. Packaged in muscle, it is rarely traumatized. C5 and/or C6 nerve root compression can weaken these muscles. The *median nerve* (C5-C8, T1; "carpenter's nerve") supplies the anterior forearm muscles and the thenar muscles. It can be compressed at the carpal tunnel (recall PL. 27), resulting in some degree of sensory deficit to fingers 1-3 and

weakness in thumb movement (carpal tunnel syndrome). Similar complaints can be associated with a C6 nerve root compression.

The *ulnar nerve* (C8-T1; "musician's nerve") supplies certain muscles of the forearm and most intrinsic muscles of the hand. It is subject to trauma as it rounds the elbow in the cubital tunnel, possibly resulting in ulnar-side finger pain, hand weakness or abnormal little finger position. Similar complaints can be associated with a C8 nerve root compression. The *axillary nerve* (C5-6) wraps around the neck of the humerus to supply deltoid and teres minor. It is vulnerable in fractures of the humeral neck, possibly resulting in a weak or paralyzed deltoid muscle. The *radial nerve* (C5-8, T1) supplies the triceps, brachioradialis, and posterior forearm (extensor) muscles moving the wrist and hand. It is subject to damage as it rounds the mid-shaft of the humerus; significant nerve loss here results in "wrist drop" and loss of ability to work the hand (try moving your fingers with your wrist flexed hard). A C7 radiculopathy is characterized by a weak triceps and loss of the triceps jerk (deep tendon reflex). See the appendix for listing of upper limb muscles and their nerve supply.

MAJOR NERVES OF THE UPPER LIMB
(Right limb, anterior view)

PALMAR VIEW

Radius

Carpal tunnel

Ulna

Cubital tunnel

Medial epicondyle

Humerus

Sternocleidomastoid m.

Clavicle

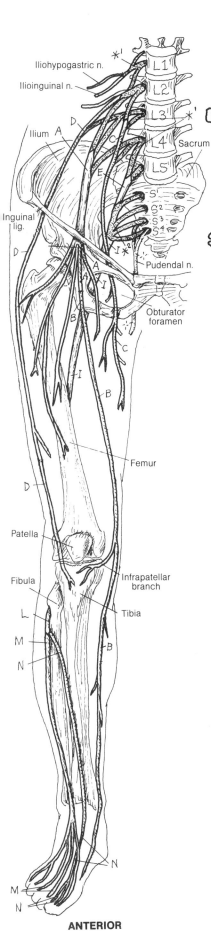

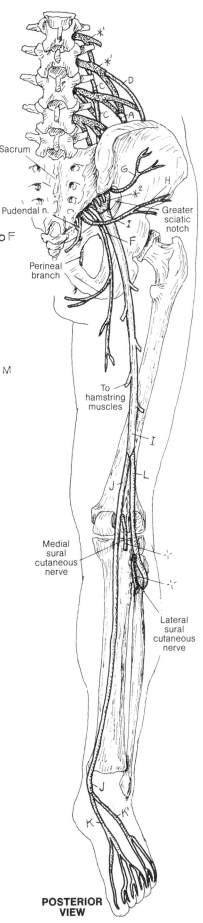

CN: Use a bright color for J. (1) Begin with the anterior view. Color the lumbar and sacral plexuses gray; note that they have been dotted for easy identification. Note the longest branch of the femoral nerve: saphenous nerve. (2) Color the posterior view which includes almost entirely the sciatic nerve and its branches.

LUMBAR PLEXUS *¹
FEMORAL N. A
SAPHENOUS N. B
OBTURATOR N. C
LAT. FEMORAL CUTAN. N. D
LUMBOSACRAL TRUNK E

SACRAL PLEXUS *²
POST. FEMORAL CUTAN. N. F
SUPERIOR GLUTEAL N. G
INFERIOR GLUTEAL N. H
SCIATIC N. I
TIBIAL N. J
MED. K LAT. PLANTAR N. K'
COMMON PERONEAL N. L
SUPERFICIAL PERON. N. M
DEEP PERON. N. N

The nerves to the lower limb arise from the lumbar (L1-L4) and sacral (S1-S3) plexuses. These plexuses are formed from anterior rami of the spinal nerves noted. The lumbar plexus is located in the retroperitoneum against the posterior abdominal wall; it is the source of two major nerves to the lower limb. The *femoral nerve* (L2-L4), giving forth an effusion of nerves just below the inguinal ligament, and in company with the (superficial) femoral artery and vein, innervates quadriceps femoris and sartorius and is sensory to the anterior thigh. Trauma to this nerve is most likely in the pelvis as it passes through or near the psoas muscle (hemorrhage, surgical misadventure, and so on). The *obturator nerve* (L2-L4) passes along the lateral pelvic wall and through the obturator foramen to break up into branches supplying the adductor muscle group. Like the femoral, it too is subject to trauma in the pelvis. Compressions of the L2-L4 nerve roots can be manifested by complaints in the anterior and medial femoral region. The sacral plexus gives rise to a number of important nerves, the most significant being the *sciatic nerve* (L4-S3). Roughly the size of your thumb, this nerve passes deep to gluteus maximus into the posterior thigh, innervating the "hamstring" muscles. Just above and behind the knee, it splits into peroneal and tibial components. The *common peroneal nerve* supplies the lateral leg muscles (*superficial peroneal*) and the antero-lateral leg muscles (*deep peroneal*). The *tibial nerve* supplies the posterior leg muscles and the plantar muscles (sole of the foot). Compression of the L4-S1 nerve roots commonly affects the sciatic distribution (e.g., sciatica or pain in the lower limb along the sciatic distribution). More significant compression results in specific leg or foot muscle weakness and sensory loss. S1 radiculopathy is characterized by a loss of the achilles (tendocalcaneus) reflex or "ankle jerk." The sciatic can be injured as it exits the greater sciatic notch or in the buttock. The common peroneal nerve is vulnerable as it rounds the subcutaneous fibular neck; trauma to this nerve may be expressed as "foot drop" (loss of ankle/toe extensors). See the appendix for listing of lower limb muscles and their nerve supply.

ANTERIOR VIEW

POSTERIOR VIEW

XIV. NERVOUS SYSTEM
DERMATOMES

CN: (1) Begin with the diagram at left, depicting sensory innervation of an area of skin (dermatome) and the degree of overlap among contiguous spinal nerve cutaneous branches and the dermatomes they supply. Color gray the three spinal nerves and the rectangular borders of the related dermatomes. Note the overlap. (2) Use very light colors for the four groups of dermatomes. Use one color for all dermatomes (represented by bordered spaces) with the letter C; another color for the dermatomes marked with a T, and so on with L and S. Suggestion: carefully outline the collection of C dermatomes with the color used for C, then color in the enclosed area, focusing on the skin areas serviced by the related spinal nerve; repeat with T, L, and S dermatomes.

SPINAL NERVE＊
DERMATOME＊'

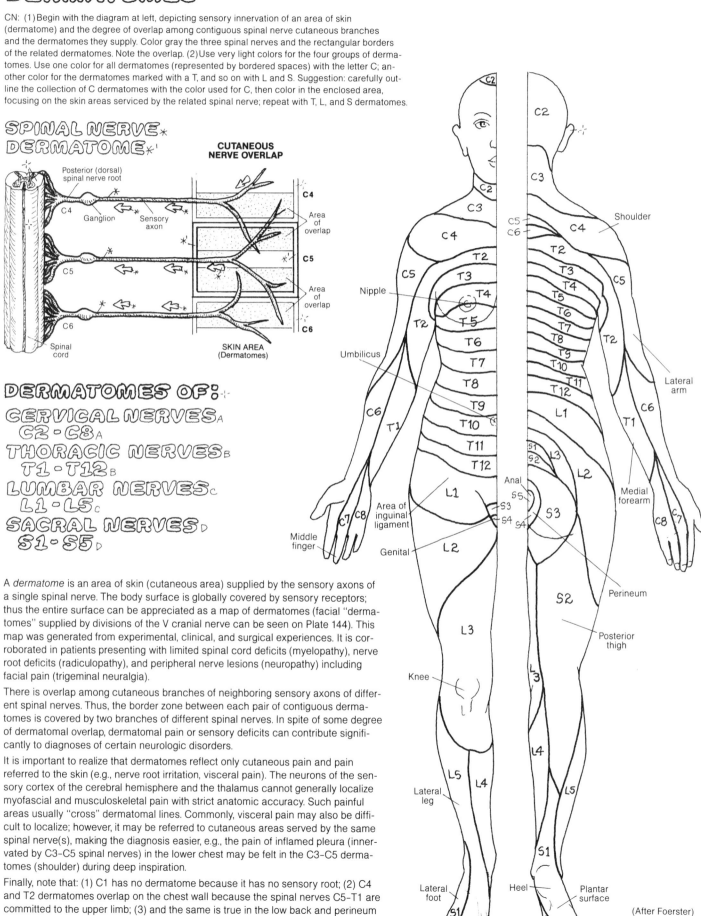

DERMATOMES OF:
CERVICAL NERVES A
C2-C8 A
THORACIC NERVES B
T1-T12 B
LUMBAR NERVES C
L1-L5 C
SACRAL NERVES D
S1-S5 D

A *dermatome* is an area of skin (cutaneous area) supplied by the sensory axons of a single spinal nerve. The body surface is globally covered by sensory receptors; thus the entire surface can be appreciated as a map of dermatomes (facial "dermatomes" supplied by divisions of the V cranial nerve can be seen on Plate 144). This map was generated from experimental, clinical, and surgical experiences. It is corroborated in patients presenting with limited spinal cord deficits (myelopathy), nerve root deficits (radiculopathy), and peripheral nerve lesions (neuropathy) including facial pain (trigeminal neuralgia).

There is overlap among cutaneous branches of neighboring sensory axons of different spinal nerves. Thus, the border zone between each pair of contiguous dermatomes is covered by two branches of different spinal nerves. In spite of some degree of dermatomal overlap, dermatomal pain or sensory deficits can contribute significantly to diagnoses of certain neurologic disorders.

It is important to realize that dermatomes reflect only cutaneous pain and pain referred to the skin (e.g., nerve root irritation, visceral pain). The neurons of the sensory cortex of the cerebral hemisphere and the thalamus cannot generally localize myofascial and musculoskeletal pain with strict anatomic accuracy. Such painful areas usually "cross" dermatomal lines. Commonly, visceral pain may also be difficult to localize; however, it may be referred to cutaneous areas served by the same spinal nerve(s), making the diagnosis easier, e.g., the pain of inflamed pleura (innervated by C3-C5 spinal nerves) in the lower chest may be felt in the C3-C5 dermatomes (shoulder) during deep inspiration.

Finally, note that: (1) C1 has no dermatome because it has no sensory root; (2) C4 and T2 dermatomes overlap on the chest wall because the spinal nerves C5-T1 are committed to the upper limb; (3) and the same is true in the low back and perineum with respect to spinal nerves L4-S2 which are committed to the lower limb.

ANTERIOR VIEW

(After Foerster)

POSTERIOR VIEW

CN: This plate is part one of a two-part presentation of the sympathetic division, and many structures on this and the next plate with the same titles and subscripts should receive the same color. (1) Begin with the schematic of the spinal cord segments containing the cell bodies of preganglionic neurons. These neurons (not shown) leave the spinal cord to enter or pass through the sympathetic chain. (2) Color the sympathetic chain and relations at upper right. (3) Color the pathways of the preganglionic and postganglionic neurons below. (4) Color the inset illustration.

SPINAL CORD SEGMENTS T1-L2 A
PREGANGLIONIC CELL BODY B
PREGANGLIONIC AXON B'
WHITE COMM. RAMUS C+
SPLANCHNIC NERVE D
PREVERTEBRAL GANGLION E
SYMPATHETIC CHAIN F
POSTGANGLIONIC CELL BODY G
POSTGANGLIONIC AXON G'
GRAY COMM. RAMUS H*

SPINAL NERVE I

The autonomic nervous system (ANS; also visceral nervous system or VNS) is a part of the peripheral nervous system (PNS) responsible for the innervation of smooth muscle and glands in viscera and skin, and specialized cardiac muscle. It is a motor system uniquely characterized by two-neuron linkages and motor ganglia (pre- and post-ganglionic neurons). Sensory impulses from viscera are conducted by typical sensory neurons not generally described with the ANS but considered part of the VNS. The sympathetic (thoracolumbar) division of the ANS is concerned with degrees of "fight or flight" responses to stimuli: pupillary dilatation, increased heart and respiratory rates, increased blood flow to brain and skeletal muscles, and other related reactions.

The *cell bodies of preganglionic neurons* are restricted to the lateral horns of the *spinal cord segments T1 through L2*. The *axons* of these neurons leave the cord via the anterior roots, join with *spinal nerves* for a very short distance, and turn medially to enter the *sympathetic chain of ganglia* via the *white communicating rami* (white because the axons are myelinated and "white"). The chain is located bilaterally alongside the vertebral column (see inset illustration). Once in the chain, the preganglionic axons can take one or more of four courses: (1) synapse with the *postganglionic neuron* at the same level it entered the chain; (2) ascend and synapse at a higher level of the chain; (3) descend and synapse at a lower level of the chain; (4) pass straight through the chain, forming a nerve that runs from the chain to the front of the vertebral column (splanchnic nerve), and synapse with a *postganglionic neuron* there (*prevertebral ganglia*).

The postganglionic neuron within the chain leaves via the *gray communicating ramus* to join the *spinal nerve*. There are gray rami bilaterally at every segment of the spinal cord; white rami exist only from T1-L2. Gray rami are so-called because the resident axons are unmyelinated and collectively have a duller color than those of the white rami. Postganglionic axons from prevertebral ganglia travel in a plexus-configuration to the viscus they supply. Plate 152 puts this division into a more meaningful perspective.

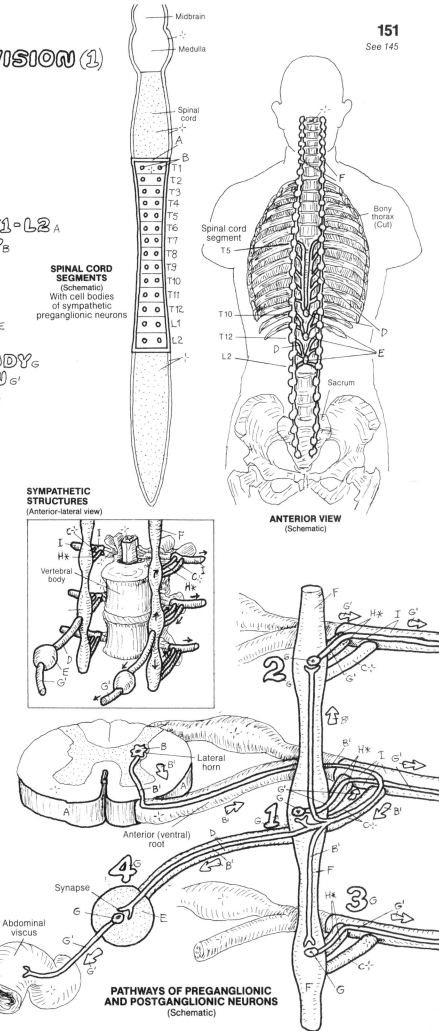

SPINAL CORD SEGMENTS (Schematic) With cell bodies of sympathetic preganglionic neurons

SYMPATHETIC STRUCTURES (Anterior-lateral view)

ANTERIOR VIEW (Schematic)

PATHWAYS OF PREGANGLIONIC AND POSTGANGLIONIC NEURONS (Schematic)

CN: Use the same colors as you used on the preceding plate for preganglionic neurons (B), splanchnic nerves (D), and postganglionic neurons (G), all of which have been given the subscripts they had on Plate 151. (1) First orient yourself to this diagram. Note the spinal cord in the center with sympathetic chains of ganglia on either side. Not all connections of both chains are shown. Here, the pathways on the left are to the skin. Pathways on the right are to viscera in the head and body cavities. (2) Start with the preganglionic neurons on the left, and color the chain and related parts on the left. Then read the related text. (3) Color the preganglionic neurons on the right and the splanchnic nerves to the abdominal viscera. (4) Color the postganglionics to the head and thorax, and then the postganglionics from the prevertebral ganglia to the abdominal and pelvic/perineal organs.

PREGANGLIONIC NEURONS B
SPLANCHNIC N. D
POSTGANGLIONIC NEURONS: G
TO HEAD & NECK G¹
TO THORACIC VISCERA G²
TO SKIN G³
SWEAT GLANDS G³
ARRECTOR PILI G³
BLOOD VESSELS G³
TO ABDOMINAL VISCERA G⁴
TO PELVIC/PERINEAL VISC. G⁵

Sympathetic innervation of *skin* (and viscera as well) begins with the *preganglionic neurons* in the thoracolumbar part of the spinal cord. The axons leave the cord via the anterior rami of spinal nerves, enter and leave the spinal nerves to join the white communicating rami. These rami bring the axons into the sympathetic chain. Axons from the upper thoracic cord ascend the chain up to the highest ganglion (superior cervical ganglion at the level of the first cervical vertebra). Axons from the lower thoracic and upper lumbar cord enter the chain and descend as far as the lowest ganglion (ganglion impar at the level of the coccyx). At every level of the chain (roughly coincident with spinal cord segments), the preganglionic axons synapse with *postganglionic neurons*. The postganglionic axons leave the chain via the gray communicating rami, enter the spinal nerves from C1 through Co1, and reach the skin via cutaneous branches of these nerves. These axons induce secretory activity in sweat glands, contraction of arrector pili muscles, and vasoconstriction in skin arterial vessels.

Postganglionics to the head (vessels and glands) leave the superior cervical ganglia and entwine about arteries enroute to the head (in the absence of spinal nerves) to reach their target organs. *Postganglionics to the heart and lungs* leave the upper ganglia of the chain, reaching these organs via cardiac nerves and the pulmonary plexus. These neurons act on heart muscle and the cardiac conduction system to increase heart rate; they induce relaxation of bronchial musculature, facilitating easier breathing.

Preganglionics to abdominal and pelvic viscera leave the cord at levels T5–L2, enter the white communicating rami, and pass through the sympathetic chain without synapsing. They form three pairs of *splanchnic nerves* between the chain and the prevertebral ganglia on the aorta. These axons synapse with the postganglionic neurons in the prevertebral ganglia. The axons of these neurons reach for smooth muscle, inducing contraction of sphincters and decreasing intestinal motility, relaxing bladder muscle and constricting the urinary sphincter. These axons stimulate the adrenal medulla to secrete mostly epinephrine and some norepinephrine, stimulate secretion of glands and muscle contraction in the male genital ducts (ejaculation), and stimulate uterine contractions.

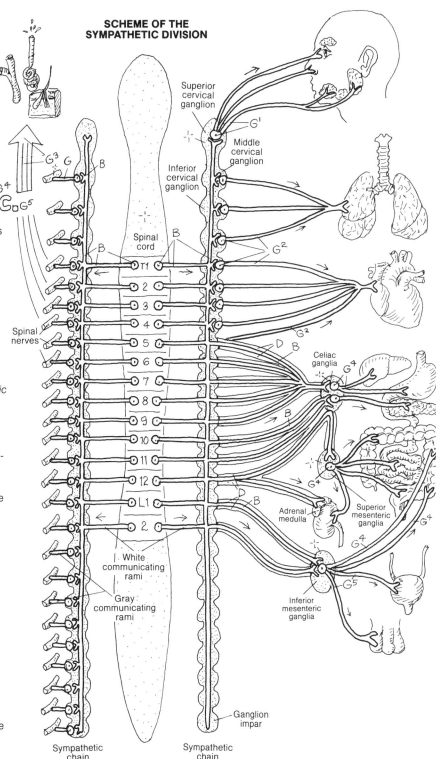

SCHEME OF THE SYMPATHETIC DIVISION

Superior cervical ganglion

G¹

Middle cervical ganglion

Inferior cervical ganglion

Spinal cord

G²

G²

B

B

Celiac ganglia

G⁴

D B

B

Adrenal medulla

Superior mesenteric ganglia

G⁴

G⁴

G⁵

Inferior mesenteric ganglia

Ganglion impar

Sympathetic chain

Sympathetic chain

Spinal nerves

G³ G

B

T1

2

3

4

5

6

7

8

9

10

11

12

L1

2

White communicating rami

Gray communicating rami

ANS: PARASYMPATHETIC DIVISION

CN: Continue using the same colors you used on Plates 151 and 152 for subscripts: B, D, and G. Use a bright color for E. This drawing shows the parasympathetic scheme on one side of the body only. (1) Start with the preganglionic neurons in the head and work through the postganglionic neurons, noting the structures innervated. Note particularly the extensive pattern associated with the vagus nerve. (2) Continue with the sacral preganglionics and postganglionics, noting the target organs. (3) Color the diagram describing ganglia location in the two ANS divisions.

PREGANGLIONIC NEURONS: B
WITH: III CRANIAL N. B¹
VII CRANIAL N. B² IX CRANIAL N. B³
X CRANIAL B⁴
PELVIC SPLANCHNIC N. D
GANGLIA: E CILIARY, E¹ PTERYGO-
PALATINE, E² SUBMANDIBULAR, E³
OTIC, E⁴ INTRAMURAL E⁵
POSTGANGLIONIC NEURONS: G
TO: EYE, G¹ NASAL/ORAL CAV-
TIES, G² SALIVARY GLANDS, G³
THORACIC/ABDOMINAL VISC., G⁴
PELVIC/PERINEAL VISCERA G⁵

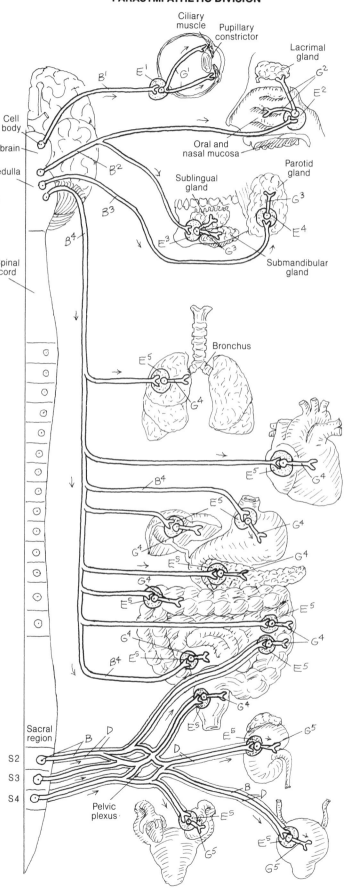

LOCATION OF GANGLIA IN THE ANS

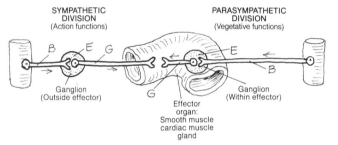

The parasympathetic division of the ANS is concerned with vegetative functions, e.g., encourages secretory activity on the body's mucous and serous membranes, promotes digestion by increased peristalsis and glandular secretion, and induces contraction of the urinary bladder.

The parasympathetic *preganglionic* neuronal cell bodies in the head are located in the brain stem associated with certain cranial nerves. The preganglionic axons leave the brain stem with their cranial nerve, and synapse at one of the cranial *ganglia*. The *postganglionic neurons* tend to be short, terminating in salivary glands and other glands of the nasal and oral cavities. The preganglionic fibers associated with the *vagus (X cranial) nerve* are unusually long, descending the neck, the esophagus, and through the esophageal hiatus to the gastrointestinal tract. The axons of these neurons extend as far as the descending colon. The ganglia are in the muscular walls of the organ they supply (*intramural ganglia*); the postganglionic axons are very short, terminating in smooth muscle and glands.

The cell bodies of the sacral preganglionic neurons are located in the lateral horns of sacral segments 2, 3, and 4 of the spinal cord. Their axons leave the cord via the anterior rami but form their own nerves called the *pelvic splanchnic nerves* (nervi erigentes). These nerves project to the pelvis, mix with sympathetic postganglionics in the pelvic plexus, and depart for their target organs. They synapse with the postganglionic neurons in intramural ganglia in the walls of the organ supplied. These fibers stimulate contraction of rectal and bladder musculature, and induce vasodilatation of vessels to the penis and clitoris (erection).

The parasympathetic and sympathetic divisions of the autonomic nervous system are not antagonistic. Their respective activities are coordinated and synchronized to achieve dynamic stability of body function during a broad range of life functions such as eating, running, fear, relaxation, and so on.

CN: Use your lightest colors for A and E. (1) Begin with the overview of a sensory pathway. (2) Color the general exteroceptors. Note that each receptor is connected to a sensory neuron (B) of a different color. (3) Color the proprioceptors in the lower illustration. Color over the entire muscle spindle, but not the surrounding muscle fibers.

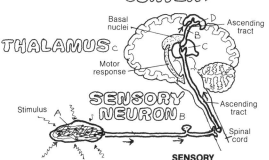

SENSORY CORTEX D

Basal nuclei

THALAMUS C

Ascending tract

Motor response

Stimulus

SENSORY NEURON B

Ascending tract

Spinal cord

RECEPTOR A

SENSORY PATHWAY (Schematic)

Sensory receptors provide information to the brain about the internal and external environment of the body. Most receptors are transducers: they convert mechanical, chemical, electrical or light stimuli to electrochemical impulses that can be conducted by the nervous system. Once generated, informational or sensory impulses travel to the CNS via sensory neurons, ultimately reaching the thalamus. Here impulses are relayed to the sensory cortex (conscious interpretation) or to motor centers for appropriate (reflexive) response.

EXTEROCEPTORS: A
SPECIAL NS
GENERAL: A
TOUCH / TACTILE R. A¹
THERMORECEPTOR A²
MEISSNER'S ENDING A³
FREE NERVE ENDING A⁴

Exteroceptors are located near the body surface. Special exteroceptors (not shown here) include the photoreceptors of the retina (light stimuli), taste receptors (chemical stimuli), and auditory receptors (sound stimuli). General exteroceptors include free nerve endings and encapsulated endings sensitive to touch, temperature change, and pain at the body surface. *Touch receptors* include those that encircle hair follicles and hair shafts, and tactile discs (of Merkel) below the epidermis. Touch-sensitive *Meissner's endings* (corpuscles) occur mainly in thick skin without hair, fitting in the cones of dermal papillae. *Thermoreceptors* (sensitive to temperature change) may be free endings or encapsulated bodies. *Free nerve endings* in the skin can be found in and deep to the epidermis. They may be sensitive to touch, heat, cold, and/ or pain (nociception). Free nerve endings sensitive to a number of different stimuli are called polymodal receptors (sensitive to more than one kind of stimulus).

PROPRIOCEPTORS: E
PACINIAN CORPUSCLE E¹
MUSCLE SPINDLE E²
NEUROTENDINOUS ORGAN E³

Proprioceptors are found in deeper tissues (superficial fascia, deep fascia, tendons, ligaments, muscles, joint capsules, and so on) of the musculoskeletal system. They are sensitive to stretch, movement, pressure, and changes in position. The *Pacinian corpuscles* are large lamellar bodies acting as mechanoreceptors: distortion of their onion skin–like lamellae induces generation of an electrochemical impulse. *Muscle spindles*, sensitive to stretch, consist of two types of special muscle fibers (nuclear bag and nuclear chain) entwined with spiral or flower-spray sensory endings. Stretch of these spindles (and the skeletal muscle in which they are located) induces discharge in the sensory fibers. These impulses reach the cerebellum. Reflexive motor commands tighten the special muscle fibers and increase resistance of the skeletal muscle to stretch. By these spindles, the CNS controls muscle tone and muscle contraction. *Neurotendinous organs* (Golgi) are nerve endings enclosed in capsules located at muscle/tendon junctions or in tendons. They are induced to generate electrochemical impulses in response to tendon deformation or stretch.

INTEROCEPTORS: NS

Interoceptors (not shown) are free or encapsulated nerve endings, often in association with special epithelial cells, located in the walls of vessels and viscera. These receptors include chemoreceptors, baroreceptors (pressure), and nociceptors. They generally are not sensitive to the same stimuli to which exteroceptors react.

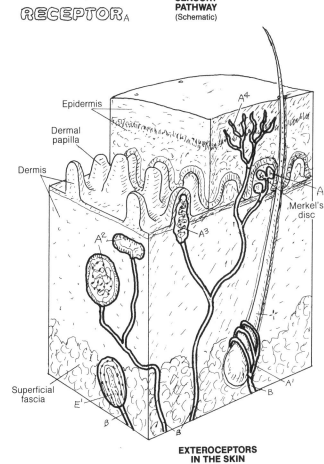

Epidermis

Dermal papilla

Dermis

A⁴

A¹

Merkel's disc

A³

A²

Superficial fascia

E¹

A¹

B

EXTEROCEPTORS IN THE SKIN

Triceps brachii muscle

Tendon

E²

E³

E²

E²

Spiral sensory ending

Nuclear chain fiber

E³

B

E²

E³

Muscle fibers

Nuclear bag fiber

PROPRIOCEPTORS IN DEEP TISSUE

SPECIAL SENSES: VISUAL SYSTEM (1)

CN: Use yellow for M, red for N, blue for O, and very light colors for C, H, I, J, and K. (1) Color the sagittal section of the eyeball and the uppermost illustration simultaneously. (2) When coloring the retinal layers, color gray the arrows (in dark outlines) representing the nerve impulse traveling opposite to the direction of the light rays.

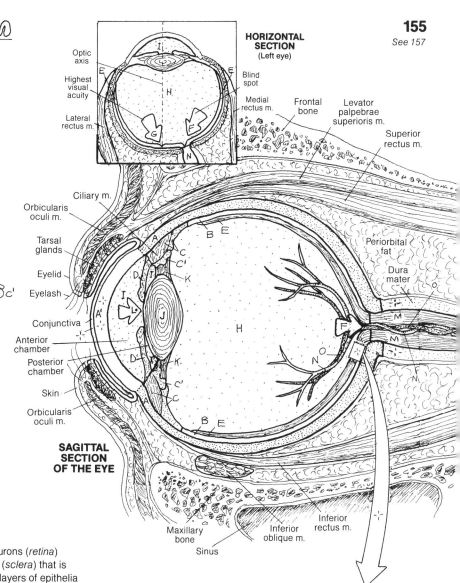

HORIZONTAL SECTION (Left eye)

SAGITTAL SECTION OF THE EYE

EYE LAYERS:*
SCLERA A /CORNEA A'
CHOROID B
  CILIARY BODY C /PROCESS C'
    IRIS D
RETINA E
  OPTIC DISC F
  FOVEA CENTRALIS G

FLUIDS:*
VITREOUS BODY H
AQUEOUS HUMOR I

OTHER STRUCTURES:*
LENS J
  SUSPENSORY LIG. K
PUPIL L
OPTIC NERVE M
RETINAL ARTERY N /VEIN O

The eye is a layer of photoreceptor cells and associated neurons (*retina*) packaged within a white, fibrous, rubberlike protective globe (*sclera*) that is transparent in front (*cornea*). The cornea, composed of five layers of epithelia and fibrous tissue, is the chief refractive medium of the eye, focusing light rays onto the retina. The *lens* (tightly packed, encapsulated non-elastic lens fibers derived from epithelial cells) also refracts light, and up to middle age, can vary its shape (and refractive index). The *aqueous humor* (extracellular fluid) filling the anterior and posterior chambers of the eye, and the more gelatinous (99% water) *vitreous humor* taking up 80% of the globe's volume, all function as refractive media. The inner surface of the posterior two-thirds of the sclera is lined with a vascular, highly pigmented layer (*choroid*) that absorbs and prevents scattering of light. The choroid thickens anteriorly as the pigmented, fibromuscular *ciliary body* that surrounds the lens. The ciliary body projects outpocketings (*processes*) to which *suspensory ligaments* from the lens attach. On the anterior aspect of the ciliary body, a thin, pigmented, epithelial and fibromuscular layer (*iris*) circumscribes the hole (*pupil*) in front of the lens.

The retina lines a bit more than the posterior half of the globe, lying superficial to the choroid and against the vitreous. It is a highly vascular membrane, complexly composed of several interlacing layers of cells. At about the center of the retina, a yellow pigmented area (macula lutea) with a clearly depressed region characterized by a dense accumulation of light-sensitive cells (cones) occurs. This site (*fovea centralis*) represents the center of greatest visual acuity (clarity of form and color) under lighted conditions. About 3mm to the nose-side of the macula lutea, the axons of the *optic tract (nerve)* and *retinal arteries / veins* stream out of the globe (*optic disc*). Absent neurons, it is an area from which no vision is possible (blind spot). The retina, derived from an evagination of the diencephalon, consists of a deep layer of photoreceptor cells (*cones* sensitive to form and color, *rods* with greatest sensitivity to light) that synapse with *bipolar cells* (neurons) that synapse with *ganglion cells* (neurons) whose axons form the optic tract (nerve) fibers. Not shown are interneurons and related multiple synapses. The visual stimulus is initiated by light rays interacting with visual purple pigment (rhodopsin) in rod cells and the pigment iodopsin in cone cells. These interactions induce an electro-chemical stimulus that can be conducted through neurons on to the CNS.

LAYERS OF RETINA: E
AXON M'
  GANGLION CELL P
    BIPOLAR CELL Q
      ROD CELL R
      CONE CELL S
PIGMENTED EPITHELIUM T

SECTION OF THE RETINA

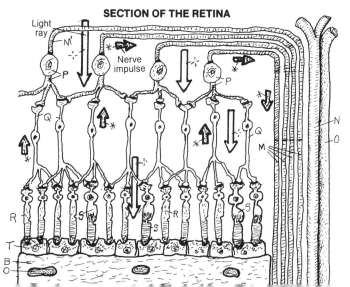

SPECIAL SENSES: VISUAL SYSTEM (2)

CN: Use the same colors as were used on the previous plate (with different subscripts) for structures J, K, L, M, N[1], and O. Use light colors for A, G, H, and I. Note that various structures in the central illustration also appear in the illustration below it.

ACCESSORY STRUCTURES: *
LACRIMAL APPARATUS: *
LACRIMAL GLAND A TEAR A'
DUCT B
LACRIMAL PUNCTA C
CANAL D
LACRIMAL SAC E
NASOLACRIMAL DUCT F
INFERIOR MEATUS OF NASAL CAVITY G
TARSAL PLATE/GLAND H
CONJUNCTIVA I

Fluid (*tears*) interfacing the *conjunctivae* of the eyelid (palpebra) and the cornea facilitate easy movement of the lids over the cornea without inducing irritation. Tears also function as a vehicle for moving epithelial debris and microorganisms from the corneal surface and undersurface of the eyelids into the nasal cavity via the lacrimal apparatus. Thus, there is an anatomic basis for blowing your nose after a good cry. The absence of tears can cause remarkable pain and even blindness. The principal gland for tears is the *lacrimal gland*, located in the anterior, superior and lateral (temporal) aspect of the orbit. Other glands and sources of tears include unicellular (goblet) glands of the conjunctiva and *tarsal glands* of the lids. Episodic blinking (rapid cycle of lid approximation and retraction) maintains a film of tears on the conjunctiva and resists "dry eye." Routine closing of the lids occurs with muscle relaxation; energetic closure requires the orbicularis oculi muscle. Retraction of the eyelids is accomplished by smooth muscle fibers (tarsal muscle of Muller; sympathetic innervation) and the levator palpebrae muscle in the upper lid.

SECRETION/DRAINAGE OF AQUEOUS HUMOR: *
FLOW OF AQUEOUS HUMOR: J
SCLERA K / CORNEA K'
CILIARY BODY L PROCESS L'
POSTERIOR CHAMBER J'
IRIS M
ANTERIOR CHAMBER J²
CANAL OF SCHLEMM N VEIN N'
VITREOUS BODY O
INTRAOCULAR PRESSURE (IOP) P

Aqueous humor is a fluid in the *anterior and posterior chambers* of the eye, secreted by cells of the *ciliary processes*. Fluid and electrolytes also enter by diffusion from the *ciliary body*. Aqueous humor is a clear, plasma-like fluid (but constituted differently). It is filtered into the *canal of Schlemm* (scleral venous sinus), a modified vein filled with fibrous trabeculae, located at the sclero-corneal junction. Fluid in the canal drains into nearby *veins*. Obstruction to drainage is one of several causes of increased *intraocular pressure*, in which the increasing pressure in the anterior/posterior chambers presses on the lens which presses on the *vitreous* (99% water). As water cannot be compressed, pressure is applied to the contiguous retina. Unrelenting pressure compresses vessels to the axons and neurons of the retina, damages neurons, and can result in blindness (glaucoma).

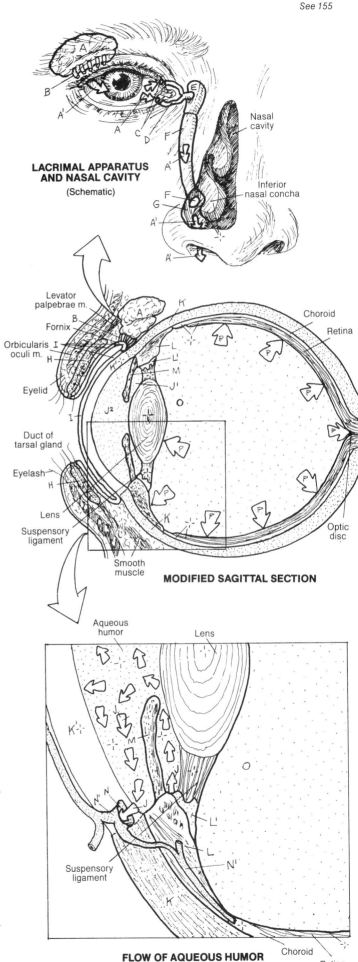

LACRIMAL APPARATUS AND NASAL CAVITY
(Schematic)

Nasal cavity

Inferior nasal concha

Levator palpebrae m.
Fornix
Orbicularis oculi m.
Eyelid
Choroid
Retina
Duct of tarsal gland
Eyelash
Lens
Suspensory ligament
Smooth muscle
Optic disc

MODIFIED SAGITTAL SECTION

Aqueous humor
Lens
Suspensory ligament
Choroid
Retina

FLOW OF AQUEOUS HUMOR

SPECIAL SENSES: VISUAL SYSTEM (3)

CN: Use light colors for A–F, H, and I. Use contrasting colors for J and K. (1) After coloring each eye muscle, color its functional arrow in the upper diagram. (2) In the drawing on ciliary action, only color the contracted ciliary muscles. (3) Carefully color the diagram below, noting that only the first titles (*visual field)*) receive J and K colors. The rest of the titles are to be colored gray, but use two colors on the structures in the diagram.

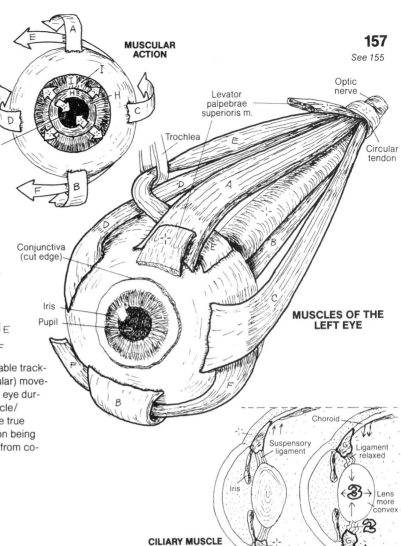

MUSCULAR ACTION

MUSCLES OF THE LEFT EYE

EXTRAOCULAR MUSCLES: *
SUPERIOR RECTUS (ELEV.)ₐ A
INFERIOR RECTUS (DEPR.)ₐ B
LATERAL RECTUS (ABD.)ₐ C
MEDIAL RECTUS (ADD.)ₐ D
SUPERIOR OBLIQUE (ROT. R.)ₐ E
INFERIOR OBLIQUE (ROT. L.)ₐ F

The extraocular (extrinsic) muscles of the eye provide for a remarkable tracking capacity of the eye. CNS mechanisms permit conjugate (binocular) movement of both eyes. Slowed, incomplete, or absent movement of one eye during tracking movements suggests cranial nerve dysfunction or muscle/tendon incarceration, as might occur in an orbital plate fracture. The true functions of these muscles is more complex than shown, one reason being eye rotation and torsion requiring multiple muscle action. Deviation from co-equal alignment of the eyes is called strabismus.

INTRINSIC MUSCLES: *
CILIARY G
SPHINCTER PUPILLAE H
DILATOR PUPILLAE I

The intrinsic muscles are located in the ciliary body (ciliary muscle) and the iris (pupillary dilator and sphincter). Contraction of the *ciliary muscles* (1) wrinkles the ciliary body tissue and puts slack in the processes, giving laxity to the suspensory ligaments (2) and permitting the lens to round up on its own accord (tension in lens fibers) (3). These muscles function (by parasympathetic innervation) during near vision in which greater refractivity is desired. The *dilator pupillae* consists of myoepithelial cells that pull the iris toward the ciliary body, dilating the pupil (sympathetic innervation). The *sphincter pupillae* circumscribes the inner iris; its contraction constricts the iris, narrowing the pupil (parasympathetic innervation). See the uppermost drawing.

CILIARY MUSCLE ACTION

VISUAL PATHWAYS: *
VISUAL FIELD J / VISUAL FIELD K
LIGHT WAVE * (J¹, K¹)
RETINA * (J², K²)
OPTIC NERVE * (J³, K³) CHIASMA * (J⁴, K⁴) TRACT * (J⁵, K⁵)
LATERAL GENICULATE BODY * (J⁶, K⁶)
SUPERIOR COLLICULUS * (J⁷, K⁷)
OPTIC RADIATION * (J⁸, K⁸)
VISUAL CORTEX * (J⁹, K⁹)

As you color the lower diagram, note that the axons (K^2) from the *retinas* on the temporal side of the optic axis do not cross at the *chiasma*. Note further that an expanding tumor of the hypophysis is likely to impair visual acuity in the temporal visual fields only ("tunnel vision"). The *thalamus* functions as a visual relay center, informing multiple memory areas and other centers of the stimulus. The *superior colliculi* are visual reflex centers, making possible rapid head and body movements in response to a visual threat. Finally, note that the image of the stimulus impinging on the *visual cortex* (K/J) is the reverse of that which was actually seen (J/K). Integration of visual and memory centers at the visual cortex makes possible perception of the image as actually seen.

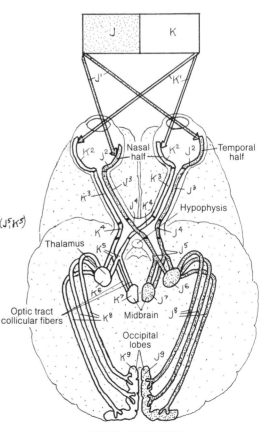

VISUAL PATHWAYS
(Horizontal brain section, schematic)

XIV. NERVOUS SYSTEM
SPECIAL SENSES: AUDITORY & VESTIBULAR SYSTEMS (1)

CN: Use yellow for Z, and light colors for A, B, G, I, M, N, W, and X. The view of the internal ear is magnified in the upper illustration for coloring purposes. Color your way down the plate, beginning with the diagram at the top.

EXTERNAL EAR: *
AURICLE A
EXT. AUDITORY MEATUS B
TYMPANIC MEMBRANE C

MIDDLE EAR: *
MALLEUS D
INCUS E
STAPES F
AUDITORY TUBE G

INTERNAL EAR: *
BONY LABYRINTH H
VESTIBULE I
OVAL WINDOW J
SEMICIRCULAR CANAL K
COCHLEA L
SCALA VESTIBULI M
SCALA TYMPANI N
ROUND WINDOW O
MEMBRANOUS LABYRINTH P
SACCULE Q /UTRICLE Q'
ENDOLYMPHATIC DUCT R
SEMICIRCULAR DUCT S
COCHLEAR DUCT T
TECTORIAL MEMBRANE U
ORGAN OF CORTI V
HAIR CELL W
SUPPORTING CELL X
BASILAR MEMBRANE Y
CRANIAL NERVE VIII Z

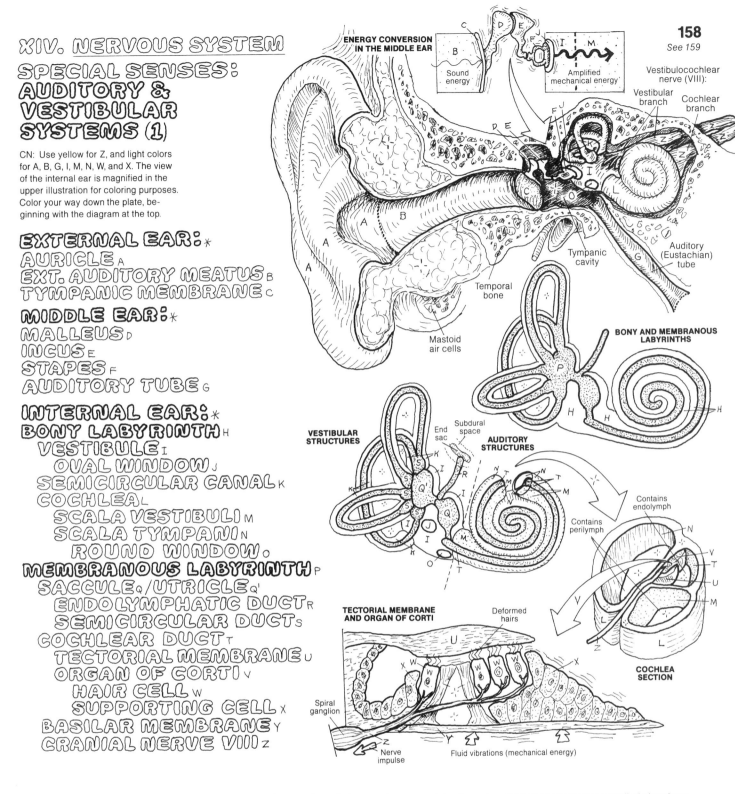

ENERGY CONVERSION IN THE MIDDLE EAR

Sound energy

Amplified mechanical energy

Vestibulocochlear nerve (VIII):
Vestibular branch
Cochlear branch

Auditory (Eustachian) tube

Tympanic cavity

Temporal bone

Mastoid air cells

BONY AND MEMBRANOUS LABYRINTHS

VESTIBULAR STRUCTURES

End sac

Subdural space

AUDITORY STRUCTURES

Contains endolymph

Contains perilymph

COCHLEA SECTION

TECTORIAL MEMBRANE AND ORGAN OF CORTI

Deformed hairs

Spiral ganglion

Nerve impulse

Fluid vibrations (mechanical energy)

The ear is the organ of hearing and equilibrium (auditory and vestibular systems). It is organized into external, middle, and internal parts. The external ear includes the *auricle* (collector of sound energy) and the *external auditory meatus* or canal (a narrow passageway conducting sound energy to the *tympanic membrane*). This membrane, lined externally by skin and internally by respiratory mucosa, converts sound energy into mechanical energy by resonating in response to incoming sound waves.

The middle ear is a small area filled with much structure, including three small bones (*malleus, incus, stapes*) joined together by synovial joints. These ossicles vibrate with movement of the tympanic membrane, amplify and conduct the mechanical energy imparted to them to the waters of the inner ear at the flexible, water-tight *oval window* (middle ear / inner ear interface). At the anterior-medial aspect of the middle ear cavity, the *auditory tube* runs to the nasopharynx, permitting equilibration of air pressure between nasal cavity (outside) and the middle ear.

The internal ear, carved out within the petrous portion of the temporal bone, consists of a series of interconnecting bony-walled chambers and passageways (*bony labyrinth: vestibule, semicircular canals*, and *cochlea*) filled with perilymph (extracellular-like) fluid. Within the bony labyrinth is a series of interconnecting membranous chambers and passageways (*membranous labyrinth: saccule, utricle, cochlear duct, and semicircular ducts*), filled with endolymph (intracellular-like) fluid. The *endolymphatic duct*, derived from the *saccule*, ends in a blind sac under the dura mater near the internal auditory meatus (see Plate 20). It drains endolymph and discharges it into veins in the subdural space. Within the coiled, membranous *cochlear duct*, supported by bone and the fibrous *basilar membrane*, a ribbon of specialized receptor (*hair*) *cells* exists integrated with supporting cells, both covered with a flexible, fibrous glycoprotein blanket (*tectorial membrane*). This device (*Organ of Corti*) converts the mechanical energy of the oscillating tectorial membrane scraping against the receptor hair cells into electrical energy. The impulses generated are conducted along bipolar sensory (auditory) neurons of the *VIII cranial nerve*. Continued on the next plate.

CN: Titles with subscripts 1, 2, and 3 require new colors; all other subscripts refer to titles and colors used on the preceding plate which should be frequently referred to when using those same colors on this plate. (1) Color the step numerals gray as you follow the sequence of events in the simplified diagram to the right. See the previous plate for the more accurate anatomical structure. (2) Color the parts of the vestibular system concerned with the maintenance of dynamic and static balance.

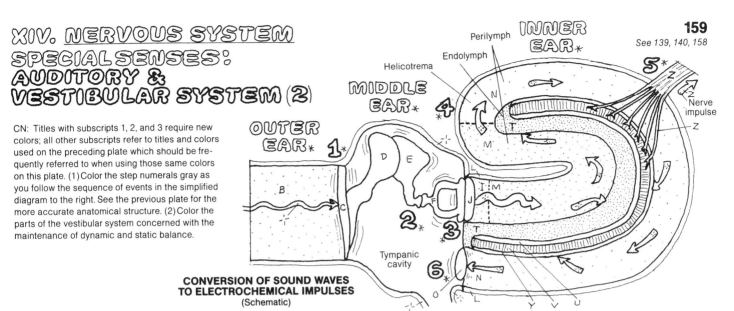

INNER EAR
MIDDLE EAR
OUTER EAR
Perilymph · Endolymph · Helicotrema · Nerve impulse · Tympanic cavity

CONVERSION OF SOUND WAVES TO ELECTROCHEMICAL IMPULSES
(Schematic)

In review: the *external ear* collects sound waves and rifles them to the *tympanic membrane* which converts the sound energy into mechanical energy. The linkage of *ossicles* increases the amplitude of the energy and transmits the force to the *oval window* of the bony labyrinth of the inner ear. Vibratory movements of the *stapes* in the window are transmitted to the perilymph of the *vestibule* of the bony labyrinth, creating wave-like motions of the fluid. These waves spread throughout the vestibule, enter and move through the *scala vestibuli* of the *cochlea* to the helicotrema at the apex of the cochlea (taking two and a half turns), and on around to the *scala tympani* which terminates at the *round win-*

dow. Here, fluid waves and vibrations are dampened. The fluid motion in the scala vestibuli vibrates the roof of the membranous *cochlear duct*, creating endolymph waves in the cochlear duct. This motion stirs the *tectorial membrane* which rubs against and bends the hair-like processes of the *receptor (hair) cells*, depolarizing them, inducing electrochemical impulses. These impulses are conducted by the sensory neurons of the cochlear division of the *VIII cranial nerve*. Stimulation of the hair cells at the base of the cochlea produces perceptions of high-pitched sounds; stimulation of the hair cells at the helicotrema produce perceptions of low-pitched sounds.

VESTIBULAR SYSTEM/EQUILIBRIUM

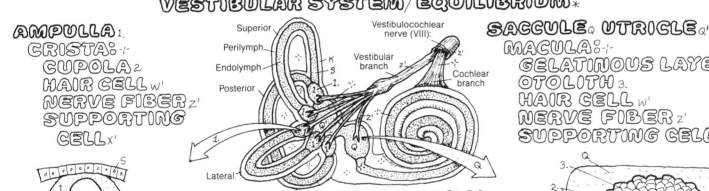

Superior · Perilymph · Endolymph · Posterior · Lateral · Vestibulocochlear nerve (VIII): · Vestibular branch · Cochlear branch

AMPULLA₁
CRISTA:
CUPOLA₂
HAIR CELL w'
NERVE FIBER z'
SUPPORTING CELL x'

SACCULE Q UTRICLE Q'
MACULA:
GELATINOUS LAYER₂
OTOLITH₃
HAIR CELL w'
NERVE FIBER z'
SUPPORTING CELL x'

SEMICIRCULAR CANAL κ
SEMICIRCULAR DUCT s

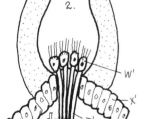
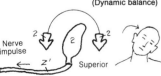
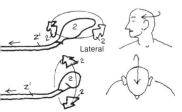

ROTATIONAL MOVEMENT
(Dynamic balance)
Nerve impulse · Superior · Lateral · Posterior

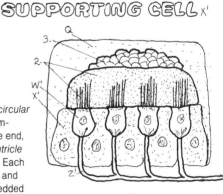
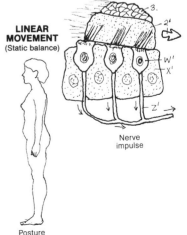

LINEAR MOVEMENT
(Static balance)
Nerve impulse · Posture

In review: the vestibular system is located in the inner ear. The bony *semicircular canals* are oriented at 90° to one another. Within these canals are the membranous *semicircular ducts*. Directly communicating with the utricle at one end, each duct terminates at the other end in an *ampulla*. Within the *saccule/utricle* and the ampullae are sensors responsive to fluid (endolymph) movement. Each ampulla has a hillock of cells (*crista* or crest) consisting of receptor (*hair*) and *supporting cells*. The hair-like processes of these receptor cells are embedded in a top-heavy, gelatinous *cupola* (like an inverted cup). Movement of endolymph in response to head turning, and especially rotation, pushes these cupolas, bending the hair cells, and causing them to depolarize, generating an electrochemical impulse. The impulses travel out the vestibular part of the VIII nerve to the vestibular nuclei in the lower brain stem. When the body is rotated rapidly, horizontal, oscillatory eye movements occur (nystagmus). These eye movements are mediated by ampullary sensory input to the brain stem. Such movements represent the brain's attempt to maintain spatial orientation (by momentary visual fixation) during head and/or body rotation. Sensations of rotational movement in the absence of body rotation is called vertigo.

Within the utricle/saccule, *hair cells* and their *supporting cells* are covered with a *gelatinous layer* in which are embedded small calcareous bodies (*otoliths*). Movement of the endolymph induces movement of the gelatinous layer against the hair cells, with responses identical to those of the ampullary receptors. Receptor activity in the utricle/saccule is influenced by linear (horizontal and vertical but non-rotational) acceleration of the body. Vestibular receptors have strong neural connections with cranial nerve nuclei concerned with eye movement and with postural motor centers.

SPECIAL SENSES: TASTE & OLFACTION

CN: Use yellow for H and light colors for A, B, C, G, and I. (1) Do not color the taste buds in the circumvallate papillae in the modified section at right. (2) In the lowest illustration, color over the neurons within the olfactory bulb.

PAPILLAE:*
CIRCUMVALLATE A
FUNGIFORM B
FILIFORM c+
TASTE BUD D+
PORE CANAL E
RECEPTOR CELL F
SUPPORTING CELL G
NERVE FIBER H

Taste receptors (*taste buds*) are located within the stratified squamous epithelial lining of the sides (moats) of *circumvallate, foliate* (not shown) and *fungiform papillae* on the tongue, and to a lesser extent, on the soft palate and lingual side of the epiglottis. They are not seen in the tiny filiform papillae. Each taste bud consists of a number of *receptor cells* and their *supporting cells*. The apex of this oval cell complex faces the moat; here it opens on to the papillary surface via a taste pore or *pore canal*. Dissolved material enters the pore, stimulating the chemoreceptor (gustatory) cells. The impulses generated are conducted along *sensory axons* which reach the brain stem via the VII, IX, and X cranial nerves (recall Plate 144). Taste interpretation occurs at the lower reaches of the sensory cortex (post-central gyrus). Basic tastes (sweet, sour, salt, and bitter) notwithstanding, interpretation of taste, as a practical matter, is a function of smell, food texture, and temperature in association with taste bud sensations.

OLFACTION (SMELL):*
OLFACTORY GLAND I
OLFACTORY MUCOSA I'
OLFACTORY NEURON J
OLFACTORY HAIR (CILIA) K
SUPPORTING CELL G'
OLFACTORY BULB H'
OLFACTORY TRACT H²

Olfactory receptors are *olfactory hairs* or cilia (actually modified peripheral processes) of *olfactory bipolar (sensory) neurons*; buried in the *olfactory mucosa* at the roof of the nasal cavity. The olfactory mucosa also has tubulo-alveolar *olfactory glands* that function to keep the chemoreceptor endings clean and, along with nasal mucous secretions, dissolve the chemicals that are sensed by these receptors. The olfactory neurons ascend the roof of the nasal cavity, through the cribriform plate of the ethmoid bone, and their central processes synapse with second order neurons in the *olfactory bulb*. The axons of these neurons form three olfactory bundles (stria) as part of the *olfactory tract*, terminating in the inferior frontal lobe and medial temporal lobe. Here exists the neural basis for olfactory relationships with memory, eating, survival, sex, and other emotional behavior.

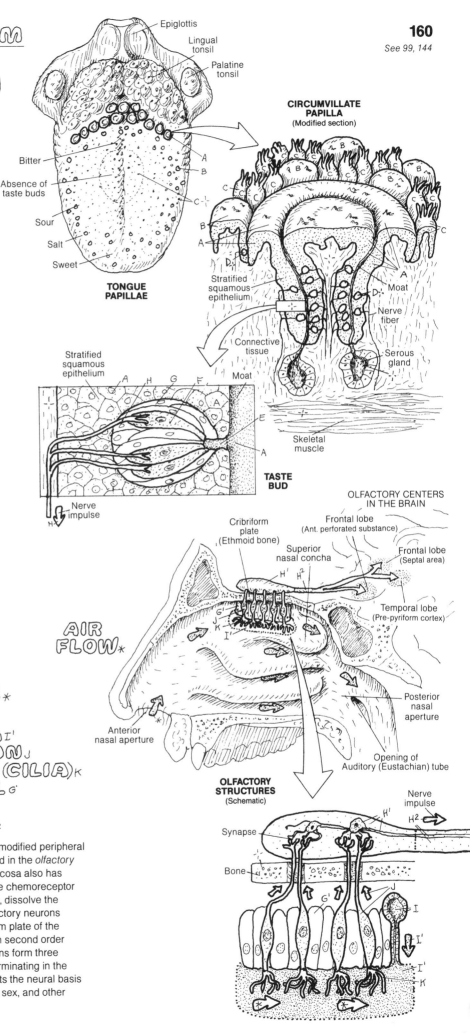

TONGUE PAPILLAE

CIRCUMVILLATE PAPILLA (Modified section)

CIRCUMVILLATE PAPILLA (Modified section)

Epiglottis
Lingual tonsil
Palatine tonsil
Bitter
Absence of taste buds
Sour
Salt
Sweet

Stratified squamous epithelium
Moat
Nerve fiber
Connective tissue
Serous gland
Skeletal muscle

Stratified squamous epithelium
Moat
TASTE BUD
Nerve impulse

OLFACTORY CENTERS IN THE BRAIN
Frontal lobe (Ant. perforated substance)
Frontal lobe (Septal area)
Temporal lobe (Pre-pyriform cortex)
Cribriform plate (Ethmoid bone)
Superior nasal concha
AIR FLOW*
Anterior nasal aperture
Posterior nasal aperture
Opening of Auditory (Eustachian) tube

OLFACTORY STRUCTURES (Schematic)
Synapse
Bone
Nerve impulse

THE INTEGUMENT

CN: Use yellow for G, red for H, blue for I, and green for J. Use light colors for A, D F, P, and Q. (1) Note that for every structure shown, there are many more within each section. (2) The stratum lucidum (B) is found only in sections of hairless skin.

EPIDERMIS:*
STRATUM CORNEUM A
STRATUM LUCIDUM B
STRATUM GRANULOSUM C
STRATUM SPINOSUM D
STRATUM BASALE E
(GERMINATING LAYER) E

DERMIS:*
CONNECTIVE TISSUE F
PAPILLAE F'
NERVE G / RECEPTOR G'
ARTERY H VEIN I
LYMPHATIC VESSEL J
HAIR:*
SHAFT K
FOLLICLE L
BULB M MATRIX M'
DERMAL PAPILLA N
ARRECTOR PILI MUS. O
SEBACEOUS GLAND P
SWEAT GLAND Q

SUPERFICIAL FASCIA R

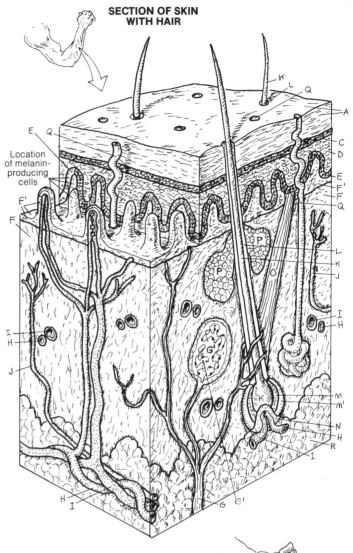

SECTION OF SKIN WITH HAIR

Location of melanin-producing cells

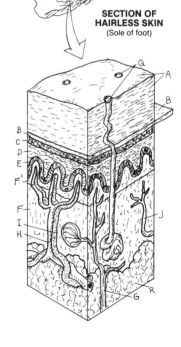

SECTION OF HAIRLESS SKIN
(Sole of foot)

"There is no magician's mantle to compare with the skin in its diverse roles of waterproof, overcoat, sunshade, suit of armour and refrigerator, sensitive to the touch of a feather, to temperature, and to pain, withstanding the wear and tear of three score years and ten, and executing its own running repairs. This vital organ of the body, 16–20 square feet in extent (the child at birth has three times the area relative to the body weight), holds the mirror to age and health even revealing general conditions such as fever, jaundice, syphilis, deficiency diseases and poisons."[1]

The integument is variably thick, from the sole of the foot with tens of layers of keratinized stratified squamous epithelia to the eyelid skin with about four layers of epidermis. Some skin is hairy, some is not. Some skin is exquisitely sensitive (face, finger tips) and some not (back, soles). It comes in a variety of colors. No matter these variations, all skin has common structural characteristics. It has an epidermal layer of stratified squamous epithelium with a number of different layers. The *stratum basale* (germanitivum) is the germinating layer from which all epidermal cells arise. This is convenient, for a significant vascular network (nutritional source) lies just deep to the avascular epidermis in the dermal cones or *papillae*. As epidermal cells get further away from this nutritional source, they dehydrate (*stratum corneum*) and die to be cast off with sweat or bath. Pigment-containing cells (melanin, carotene) are found in the stratum germanitivum layer; skin color is a function of the concentration of these pigments primarily, but is influenced by the number of blood vessels, and the degree of vasoconstriction.

The dermis is replete with thick bundles of fibrous *connective tissue, blood and lymphatic vessels, sensory receptors and related nerves*, and glands. *Sweat glands* help stabilize body temperature by excreting in response to excessive body heat. *Sebaceous glands*, associated with *hair follicles*, excrete an oily substance (sebum) that resists dehydration. Hair arises from an ingrowth of epidermal (follicle) cells that pushed down into the dermis during development. The base or bottom of the follicle is the *hair bulb* which is invaginated (*dermal papilla*) and encloses capillaries. The follicle consists of a *hair shaft* (central medulla, outer cortex, outermost cuticle) surrounded by layers of cells (root sheaths, membranes).

Non-striated *arrector pili muscles* attach to the hair follicles and to the upper parts of the dermis. They straighten the hairs, perhaps enhancing preservation of body temperature. The skin is contiguous with the *superficial fascia*, a more fatty, loose connective tissue layer. Sense receptors of the skin are presented in Plate 154.

[1] Quote taken, with permission, from Lockhart, R.D., Hamilton, G.F., and Fyfe, F.W., ANATOMY OF THE HUMAN BODY, 2nd ed., Faber and Faber, Publishers, Ltd., London, 1965.

BIBLIOGRAPHY & REFERENCES

Abbas, A., A. Lichtman, J. Pober, *Cellular and Molecular Immunology*, W.B. Saunders, Philadelphia, 1991.

Agur, A. (ed.), *Grant's Atlas of Anatomy*, 9th ed., Williams and Wilkins, Baltimore, 1991.

Basmajian, J., *Muscles Alive*, 5th ed., Williams and Wilkins, Baltimore, 1985.

Bassett, L., R. Gold, L. Seeger, *MRI Atlas of the Musculoskeletal System*, Martin Dunitz, London, 1989.

Bergman, R., S. Thompson, A. Afifi, F. Saadeh, *Compendium of Human Anatomic Variation*, Urban and Schwarzenberg, Baltimore, 1988.

Brookes, M., R. Warwick, *Nomina Anatomica*, 6th ed., Churchill Livingston, London, 1989.

Carpenter, M., J. Sutin, *Human Neuroanatomy*, 9th ed., Williams and Wilkins, Baltimore, 1992.

Cotran, R., V. Kumar, S. Robbins, *Robbins Pathologic Basis of Disease*, 4th ed., W.B. Saunders Company, Philadelphia, 1989.

Dawson, D., M. Mallett, L. Millender, *Entrapment Neuropathies*, 2nd ed., Little Brown and Co., Boston, 1990.

Diamond, M., A. Scheibel, L. Elson, *The Human Brain Coloring Book*, HarperCollins, New York, 1985.

Dickinson, R., *Human Sex Anatomy*, 2nd ed., Williams and Wilkins, Baltimore, 1949.

Dorland's Illustrated Medical Dictionary, 27th ed., W.B. Saunders, Philadelphia, 1988.

DuBrul, L., *Sicher's Oral Anatomy*, 7th ed., C.V. Mosby, St. Louis, 1980.

Fawcett, D., Bloom and Fawcett, *A Textbook of Histology*, 11th ed., W.B. Saunders, Philadelphia, 1986.

Ganong, W., *Review of Medical Physiology*, 16th ed., Appleton & Lange, San Mateo, CA, 1993.

Goodgold, J., A. Eberstein, *Electrodiagnosis of Neuromuscular Diseases*, 3rd ed., Williams and Wilkins, Baltimore, 1983.

Gosling, J., P. Harris, J. Humpherson, I. Whitmore, P. Willan, *Atlas of Human Anatomy*, 2nd ed., J.B. Lippincott, Philadelphia, 1991.

Gowitzke, B., M. Milner, *Scientific Bases of Human Movement*, 3rd ed., Williams and Wilkins, Baltimore, 1988.

Guyton, A., *Textbook of Medical Physiology*, 8th ed., W.B. Saunders and Co., Philadelphia, 1991.

Haymaker, W., B. Woodhall, *Peripheral Nerve Injuries: Principles of Diagnosis*, 2nd ed., W.B. Saunders, Philadelphia, 1953.

Higgins, C., H. Hricak, C. Helms, *Magnetic Resonance Imaging of the Body*, 2nd ed., Raven Press, New York, 1992.

Hollinshead, H., *Anatomy for Surgeons: Vol. 1: The Head and Neck*, 3rd ed., Harper and Row, New York, 1982.

___, *Anatomy for Surgeons, Vol. 3: The Back and Limbs*, 3rd ed., Harper and Row, New York, 1982.

Hoppenfeld, S., *Physical Examination of the Spine and Extremities*, Appleton-Century-Crofts, New York, 1976.

Juhan, D., *Job's Body: A Handbook for Bodywork*, Station Hill Press, Tarrytown, NY, 1987.

Junqueira, L., J. Carneiro, J. Long, *Basic Histology*, 7th ed., Appleton & Lange, Norwalk, CT, 1992.

Kapit, W., R. Macey, E. Meisami, *The Physiology Coloring Book*, HarperCollins, New York, 1987.

BIBLIOGRAPHY & REFERENCES

Kendall, F., E. McCreary, *Muscles: Testing and Function*, 4th ed., Williams and Wilkins, Baltimore, 1991.

Kessel, R., R. Kardon, *Tissues and Organs: A Text-Atlas of Scanning Electron Microscopy*, W.H. Freeman, San Francisco, 1979.

Lockhart, R., G. Hamilton, F. Fyfe, *Anatomy of the Human Body*, 2nd ed., Faber & Faber, London, 1965.

McMinn, R., R. Hutchings, *Color Atlas of Human Anatomy*, 2nd ed., Year Book Medical Publishers, Inc., Chicago, 1988.

McMinn, R., R. Hutchings, B. Logan, *Color Atlas of Head and Neck Anatomy*, Year Book Medical Publishers, Inc., Chicago, 1981.

Moore, K., *The Developing Human: Clinically Oriented Embryology*, 5th ed., W.B. Saunders, Philadelphia, 1993.

Moss, A., G. Gamsu, H. Genant, *Computed Tomography of the Body*, W.B. Saunders, Philadelphia, 1983.

Nagy, D., *Radiological Anatomy*, Pergamon Press, Oxford, 1965.

Netter, F., S. Colacino (eds.), *Atlas of Human Anatomy*, Ciba-Geigy Corp., Summit, NJ, 1989.

Nieuwenhuys, R., J. Voogd, C. van Huijzen, *The Human Central Nervous System: A Synopsis and Atlas*, 3rd ed., Springer-Verlag, New York, 1988.

Roberts, M., J. Hanaway, *Atlas of the Human Brain in Section*, 2nd ed., Lea and Febiger, Philadelphia, 1986.

Rohen, J., C. Yokochi, *Color Atlas of Anatomy: A Photographic Study of the Human Body*, Igaku-Shoin, New York, 1983.

Rothman, S., W. Glenn, *Multiplanar CT of the Spine*, University Park Press, Baltimore, 1985.

Skinner, H., *The Origin of Medical Terms*, 2nd ed., Williams and Wilkins, Baltimore, 1961.

Soderberg, G., *Kinesiology: Application to Pathologic Motion*, Williams and Wilkins, Baltimore, 1986.

Travell, J., D. Simons, *Myofascial Pain and Dysfunction: The Trigger Point Manual*, Williams and Wilkins, Baltimore, 1983.

___, *Myofascial Pain and Dysfunction: The Trigger Point Manual, Vol. II: The Lower Extremities*, Williams and Wilkins, Baltimore, 1992.

Warfel, J., *The Head, Neck, and Trunk: Muscles and Motor Points*, 6th ed., Lea and Febiger, Philadelphia, 1993.

Warfel, J., *The Extremities*, 6th ed., Lea and Febiger, Philadelphia, 1993.

White, A., M. Panjabi, *Clinical Biomechanics of the Spine*, 2nd ed., J.B. Lippincott, Philadelphia, 1990.

Williams, P., R. Warwick, M. Dyson, and L. Bannister (eds.), *Gray's Anatomy*, 37th ed., Churchill Livingstone, New York, 1989.

APPENDIX A

ANSWER KEYS

PLATE 28
JOINTS TO BE IDENTIFIED:

1. Acromioclavicular Joint
2. Glenohumeral (Scapulohumeral) Joint
3. Sternoclavicular Joint
4. Humeroulnar Joint
5. Radiohumeral Joint
6. Proximal Radioulnar Joint
7. Distal Radioulnar Joint
8. Radiocarpal Joint
9. Intercarpal Joint
10. Carpometacarpal Joint
11. Intermetacarpal Joint
12. Metacarpophalangeal Joint
13. Interphalangeal Joints

PLATE 76
ARTERIES TO BE IDENTIFIED:

A. Aortic Arch
A^1. Thoracic Aorta
A^2. Abdominal Aorta
B. Brachiocephalic
C. Common Carotid
D. External Carotid
E. Internal Carotid
F. Pulmonary Trunk
F^1. Pulmonary Artery
G. Subclavian
H. Axillary
I. Brachial
J. Radial
K. Ulnar
L. Superficial Palmar Arch
M. Deep Palmar Arch
N. Digital
O. Internal Thoracic
P. Musculophrenic
Q. Intercostals
R. Superior Epigastric
S. Inferior Epigastric
T. Celiac
U. Superior Mesenteric
V. Inferior Mesenteric
W. Renal
X. Testicular/Ovarian
Y. Common Iliac
Z. Internal Iliac
1. External Iliac
2. Femoral
3. Popliteal
4. Anterior Tibial
5. Dorsalis Pedis
6. Arcuate
7. Dorsal Metatarsal
8. Dorsal Digital
9. Posterior Tibial
10. Peroneal
11. Medial Plantar
12. Lateral Plantar
13. Plantar Arch

PLATE 81
VEINS TO BE IDENTIFIED:

A. Dorsal/Palmar Digital
B. Dorsal Network
B^1. Palmar Network
C. Basilic
D. Cephalic
E. Radial
F. Ulnar
G. Brachial
H. Axillary
I. Subclavan
J. Brachiocephalic
K. Superior Vena Cava
K^1. Inferior Vena Cava
L. External Jugular
M. Internal Jugular
N. Azygos
0. Posterior Intercostals
P. Thoracoepigastric
Q. Inferior Mesenteric
R. Superior Mesenteric
S. Splenic
T. Gastric
U. Hepatic Portal
V. Renal
W. Ovarian/Testicular
X. Hepatic
Y. Pulmonary
Z. Digital
Z^1. Plantar Venous Arch
1. Dorsal Venous Arch
2. Great Saphenous
3. Small (Lesser) Saphenous
4. Medial Plantar
4^1. Lateral Plantar
5. Dorsal
6. Anterior Tibial
7. Posterior Tibial
8. Popliteal
9. (Superficial) Femoral
10. External Iliac
11. Internal Iliac
12. Common Iliac

APPENDIX B

INNERVATION OF SKELETAL MUSCLES

The spinal cord segments give off spinal nerves the axons of which make up the peripheral nerves that supply the skeletal muscles. Those spinal nerves in bold type are the major component of the innervating nerve. Those underlined are particularly likely to reveal deficits (diminished/ absent deep tendon reflexes, sensory impairment, and/or weak muscles) in the event of nerve root injury. Muscles supplied by cranial nerves only are not included in this list. For these, see Plate 144.

REGION/ SKELETAL MUSCLE	NERVE SUPPLY	SPINAL SEGMENTS/ NERVE ROOTS
Neck:		
Sternocleidomastoid	Spinal Accessory; Ant. Rami	C1 – C3
Geniohyoid	Part of Hypoglossal	C1
Sternohyoid	Ansa Cervicalis	C1 – C3
Sternothyroid	Ansa Cervicalis	C1 – C3
Thyrohyoid	Part of Hypoglossal	C1
Omohyoid	Ansa Cervicalis	C1 – C3
Longus Colli/Capitis	Muscular Brs.	C2 – C8
Rectus Capitis	Muscular Brs.	C1 – C2
Scalene: Anterior	Muscular Brs.	C5 – C8
Scalene: Med./Post.	Muscular Brs.	C3 – C4
Upper Limb:		
Trapezius	Spinal Accessory	C2 – C4
Rhomboids	Dorsal Scapular	**C5** – C6
Levator Scapulae	Muscular Brs.	C3 – C5
Serratus Anterior	Long Thoracic	C5 – C7
Pectoralis Minor	Med. Pectoral	C6 – C8
Subclavius	N. To Subclavius	C5 – C6
Supraspinatus	Suprascapular	C5 – C6
Infraspinatus	Suprascapular	C5 – C6
Subscapularis	Subscapular	C5 – C6
Teres minor	Axillary	**C5** – C6
Deltoid	Axillary	<u>C5</u> – C6
Pectoralis Major	Med./Lat. Pectoral	C5 – T1
Latissimus Dorsi	Thoracodorsal	C5 – C8
Teres Major	Lower Subscapular	C5 – C6
Biceps Brachii	Musculocutaneous	<u>C5</u> – C6
Brachialis	Musculocutaneous	<u>C5</u> – C6
Coracobrachialis	Musculocutaneous	C5 – C6
Brachioradialis	Radial	C5 – **C6**
Triceps Brachii	Radial	C6, <u>C7</u> – T1
Anconeus	Radial	C7 – C8
Supinator	Radial	C5 – **C6**
Pronator Teres	Median	C6 – C7

APPENDIX B

REGION/ SKELETAL MUSCLE	NERVE SUPPLY	SPINAL SEGMENTS/ NERVE ROOTS
Pronator Quadratus	Median	C7 – T1
Palmaris Longus	Median	C7 – T1
Palmaris Brevis	Ulnar	C8 – T1
Flexor Carpi Radialis	Median	C6, C7, C8
Flexor Carpi Ulnaris	Ulnar	C6 – C8
Flexor Digit. Superfic.	Median	C7 – T1
Flexor Digit. Profundus	**Median**/Ulnar	C7 – T1
Flexor Pollicis Longus	Median	C7 – T1
Thenar Muscles	**Median**/Ulnar	C8 – T1
Hypothenar Muscles	Ulnar	C8 – T1
Hand Intrinsics:	Ulnar	C8 – T1
Interossei	Ulnar	C8 – T1
Lumbricales 1,2	Median	C7 – T1
Wrist Extensors	Radial	C6 – T8
Digit Extensors	Radial	C6, C7, C8
Lower Limb:		
Psoas Major	Lumbar	L2 – L3
Psoas Minor	Lumbar	L1 – L2
Iliacus	Femoral	L2 – L4
Adductors:	Obturator	L2 – L4
A. Magnus	Obturator, Sciatic	L2 – S1
Quadriceps Fem.	Femoral	L2 – L4
Sartorius	Femoral	L2 – L4
Tensor Fasciae Latae	Sup. Gluteal	L4 – S1
Gluteus Maximus	Inf. Gluteal	L5, S1, S2
Gluteus Medius/Minimus	Sup. Gluteal	L4 – S1
Hamstrings	Sciatic	L4 – S2
Lateral Hip Rotators:	Sacral Plex	L4 – S2
Piriformis	N. to Piriformis	L5 – S2
Obturator Internus	N. to Obt. Int.	L5 – S2
Obturator Externus	Obturator	L3 – L4
Gemelli	Sacral Plexus	L4 – S2
Tibialis Anterior	Deep Peroneal	L4 – S1
Ext. Hallucis Longus	Deep Peroneal	L5 – S2
Ext. Digitorum Longus	Deep Peroneal	L5 – S2
Peroneus Tertius	Deep Peroneal	L4 – S1
Peroneus Longus/Brevis	Superfic. Peroneal	L4 – S1
Gastrocnemius/Soleus	Tibial	S1 – S2
Plantaris	Tibial	L4 – S1
Popliteus	Tibial	L4 – S1
Tibialis Posterior	Tibial	L5 – S1
Flexor Hallucis Longus	Tibial	L5 – S2
Flexor Digitorum Longus	Tibial	L5 – S2
Foot Intrinsics:	Tibial/Plantar	S1 – S2
Interossei	Tibial/Plantar	S1 – S2
Ext. Digit. Brevis	Deep Peroneal	L5 – S1

APPENDIX B

REGION/ SKELETAL MUSCLE	NERVE SUPPLY	SPINAL SEGMENTS/ NERVE ROOTS
Thoracic Wall:		
Ext./Int./In. Intercostals	Intercostal	T1 – T11
Thoracic Diaphragm	Phrenic	C3 – C5
Serratus Post. Sup.	Thor. Post. Rami	T1 – T3
Serratus Post. Inf.	Thor. Post. Rami	T9 – T12
Abdominal Wall:		
Ext./Int. Oblique	Thor. Post. Rami	T6 – T12
	Lumbar Post. Rami	L1 – L2
Transv. Abdominis	Ilio-hypogastric; Ilio-inguinal	L1 – L2
Rectus Abdominis	Thor. Post. Rami	T5 – T12
Cremaster	Genito-femoral	L1 – L2
Quadratus Lumborum	Ant. Rami	T12 – L3
Pyrarnidalis	Subcostal	T12
Deep Back:		
Suboccipital Musc.	Cerv. Post. Ramus	C1
Erector Spinae, Splenius	Cerv./Thor./Lumb./Sacral	C1 – S3
Semispinalis and	Posterior Rami	
Deep Short Musc.		
Pelvis/Perineum:		
Levator Ani	Pudendal/Rectal	S2 – S4
Coccygeus	Pudendal/Rectal	S2 – S4
Perineal Muscles	Pudendal/Perin.	S2 – S4
Urethral Sphincter	Pudendal/Perin.	S2 – S4
Anal Sphincter	Pudendal/Rectal	S2 – S4

GLOSSARY

Anatomical terms as set forth and revised by the International Anatomical Nomenclature Committee of the International Congress of Anatomists, published in the 6th Edition of the *Nomina Anatomica* (1989) are included herein. For further inquiry, consult a standard medical dictionary. The terms here are compatible with those listed in *Dorland's Illustrated Medical Dictionary*, 27th Edition. Pronunciation of terms is given phonetically (as they sound, not by standard dictionary symbols). The primary accent (emphasis) is indicated by capitalized letters, e.g., ah-NAT-oh-mee, included with the definitions. The plural form is in parentheses following the term defined, e.g., alveolus(i) or alveoli. Pl. = plural. Origin of words: Gk. = Greek; L. = Latin.

A

A–, an–, without.

Ab–, away from the midline.

AB, antibody.

Abdomen, the region between the diaphragm and the pelvis.

Abscess, (AB-sess), a cavity in disintegrating tissue, characterized by the presence of pus and infective agents.

Achilles, in Greek mythology, one of the sons of Peleus, a young king, and Thetis, one of the immortal goddesses of the sea. Not wanting Achilles to be mortal like his father, Thetis dipped him into the river Styx, holding him by the heel cord (tendocalcaneus), making him invulnerable to harm except at the heel cord. Achilles later became a great Greek warrior. In the many wars between Greece and Troy, Achilles was invulnerable to harm. At last, a Trojan, aided by the god Apollo, slew Achilles with an arrow into the vulnerable heel cord. The term "Achilles heel" refers to one's vulnerabilities; the Achilles tendon is the tendocalcaneus.

Acinus(i), (ASS-ee-nus), a saclike gland.

Actin, a protein of muscle, associated with the contraction and relaxation of muscle cells.

Ad–, toward the midline.

Adeno–, (ADD-eh-no), gland.

Afferent, leading to a center.

AG, antigen.

AIDS, acquired immunodeficiency syndrome.

–algia, pain.

Alveolus(i), (al-VEE-oh-lus), grape-shaped cavity, rounded or oblong. Refers to shape of exocrine glands, air spaces within the lung, and the bony sockets for teeth.

Amino acid, (ah-MEEN-oh), a two-carbon molecule with a side chain that either contains nitrogen (in the form of NH_2) or contains a carboxyl group (—COOH).

Amorphous, (ay-MORF-us), without apparent structure at some given level of observation. What appears amorphous at 1000X magnification may be quite structured at 500,000X.

Amphi–, double, about, around, both sides.

Amphiarthosis(es), (AM-fee-ar-THRO-sis), *see* joint classification, functional.

Ampulla(e), dilatation of a tubular structure.

Anastomosis(es), (ah-NASS-toh-moh-sis), connection between two vessels.

Anatomy, (ah-NAT-oh-mee), *ana* = up, *tome* = to cut; the study of structure.

Anemia, (ah-NEE-mee-ah), a condition of inadequate number of red blood cells.

Angina, (an-JYNE-ah), pain, especially cardiac pain.

Angio–, a vessel.

Angle, the point of junction of two intersecting lines, as in the inferior angle of the scapula between the vertebral (medial) and axillary (lateral) borders of that bone.

Angulus(i), an angle.

Ankle, the tarsus. The region between the leg and foot.

Annulus(i), (AN-new-lus), a ringlike or circular structure.

Ano–, anus.

Anomaly, (ah-NOM-ah-lee), an abnormality, especially in relation to congenital or developmental variations from the normal.

A.N.S. or ANS, autonomic nervous system.

Anserine, like a goose. **Pes anserinus**, goose foot.

Ante–, (AN-tee), forward.

Antebrachium, forearm.

Antecubital, in front of the elbow (cubitus).

Anti–, against.

Antibody, a complex protein (immunoglobulin); it is a product of activated B lymphocytes and plasma cells; it is synthesized as part of an immune response to the presence of a specific antigen.

Antigen, any substance that is capable of inciting an immune response and reacting with the products of that response. Antigens may be in solution (toxins) or they may be solid structures (microorganisms, cell fragments, and so on). Particulate matter that is phagocytosed but does not incite an immune response does not constitute antigen. Formation of specific antibodies by cloning (monoclonal antibodies) may react with certain surface molecules on a cell membrane; those surface molecules constitute antigens.

GLOSSARY

Antigenic determinant, the specific part of an antigen that reacts with the product of an immune response (antibody, complement).

Aperture, (AP-er-chur), an opening.

Apical, an apex or pointed extremity.

Aponeurosis(es), a flat tendon.

Apophyseal, (app-oh-FIZZ-ee-al), refers to apophysis.

Apophysis(es), (ah-POFF-ee-sis), an outgrowth; a process.

Arborization, (ar-bor-eye-ZAY-shun), branching of terminal dendrites.

Areolar, (ah-REE-oh-lar), filled with spaces.

Arm, that part of the upper limb between the shoulder and elbow joints.

Arrhythmia(s), (a-RITH-mee-ah), a variation from the normal rhythm of the heartbeat; the absence of rhythm.

Arterio–, artery.

Arthr–, (AR-thr), joint.

Arthritis(ides), (ar-THRI-tiss), inflammation of a joint.

Articular, joint.

Articular process, an outgrowth of bone on which there is a cartilaginous surface for articulation with another similar surface.

Articulation, a joint or connection of bones, movable or not; occlusion between teeth; enunciation of words.

Aspera, rough.

Aster, Gk., star; a ray, as in rays of light; in the cell, rays of microtubules projecting from centrioles.

Atherosclerosis, a form of arteriosclerosis or hardening of the arteries; specifically, consists of yellowish plaques of cholesterol and lipid in the tunica intima of medium and large arteries.

ATP, adenosine triphosphate, a nucleotide compound containing three high-energy phosphate bonds attached to a phosphate group; the energy is released when the ATP is hydrolyzed to adenosine diphosphate and a phosphate group.

Atrophy, (AT-troh-fee), usually associated with decrease in size, as in muscle atrophy.

Avascular, (ay-VASS-kew-lur), without blood vessels or, in some cases, blood.

Avulsion, tearing a part away from the whole, as in tearing a tendon from its attachment to bone.

B

Back, the region making up the posteriormost wall of the thorax and abdomen, supported by the thoracic and lumbar vertebrae. Strictly defined, excludes the neck and sacrum/coccyx (pelvis).

Basal lamina(e), a thin layer of interwoven collagen fibrils interfacing epithelial cells (and certain other nonepithelial cells) and connective tissue. Seen only with the electron microscope.

Basement membrane, basal lamina and a contiguous layer of collagenous tissue. Seen with the light microscope, it controls diffusion and transport into/out of the cell.

Basilar, base or bottom.

Benign, nonmalignant; often used to mean mild or of lesser significance.

Bi–, two.

Bicipital, two-headed.

Bicuspid, a structure, e.g., a tooth or valve, with two cusps.

Bifurcate, (BY-fur-cate), to branch.

–blast, formative cell; immature form.

Blephar–, eyelid.

Blood-borne, refers to some structure carried by the blood.

Blood–brain barrier, a state in the CNS in which substances toxic or harmful to the brain are physically prevented from getting to the brain; it is represented by tight endothelial junctions in capillaries of the brain, tight layers of pia mater around vessels, and the presence of neuroglial endfeet surrounding vessels.

Bolus, a mass of food; any discrete mass.

Bone, immature, *see* bone, woven.

Bone, lamellar, mature bone characterized by organized layers or lamellae of bone.

Bone, mature, *see* bone, lamellar.

Bone, primary, *see* bone, woven .

Bone, secondary, *see* bone, lamellar.

Bone, woven, immature bone characterized by random arrangements of collagen tissue and without the typical lamellar organization seen in more mature bone.

Brachi–, arm.

Bronch–, bronchi or bronchioles of the respiratory tract.

Bursa(e), synovial-lined sac between tendons and bone, muscle and muscle, or any other site in which movement of structure tends to irritate or injure adjacent structure. It contains synovial fluid and is lined externally by fibrous connective tissue.

Bursitis, inflammation of a bursa.

C

Cadaver, (ka-DA-ver), a dead body.

Canaliculus(i), a small canal.

Cancellous, (KAN-sell-us), lattice- or spongylike; structure with visible holes in it.

Cancer, a condition in which certain cells undergo uncontrolled mitoses with invasiveness and metastasis (migrating from the point of origin to other sites, usually by way of the lymphatic and/or blood vascular systems). There are two broad divisions: carcinoma, cancer of epithelial cells; sarcoma, cancer of the connective tissues.

Capillary attraction, the force that attracts fluid to a surface, such as water flowing along the undersurface of a pouring tube.

GLOSSARY

Capitulum, a rounded process of bone, usually covered with articular cartilage. Synonym: capitellum.

Cardio–, heart

Carpus, carpo–, wrist.

Cauda equina (horse's tail), the vertically oriented bundle of nerve roots within the vertebral canal below the level of the first lumbar vertebra (L1). Includes nerve roots for spinal nerves L2 through the Co2, bilaterally.

Cauda equina syndrome, irritation/compression of the cauda equina resulting in bilateral symptoms and signs that may include bladder and bowel incontinence, weakness in the lower limb musculature, sensory impairment from the perineum to the toes, and reflex changes.

Cauterization, destruction of tissue by heat, as with an electrocauterizing instrument.

Cavity, potential, a space between membranes that can potentially enlarge with fluid accumulation, as can occur in the peritoneal cavity (ascites) or pericardial cavity (cardiac tamponade).

CD4, CD, "clusters of differentiation." The abbreviation refers to a collection of cell surface molecules with specific structural characteristics (markers) reflecting a common lineage. The identification of these markers is made by purebred (monoclonal) antibodies which react only with surface markers of cells of a common lineage. Cells exhibiting cell surface markers of a common lineage belong to a cluster (of differentiation), identified by number, e.g., 4. Most helper T lymphocytes have surface markers of 3 different clusters, CD3, CD4, and CD8. Cytolytic T lymphocytes are CD3, CD4, and CD8.

Cell body, the main, largest single mass of a neuron, containing the nucleus surrounded by organelles in the cytoplasm.

–centesis, puncture.

Central, at or toward the center.

Ceph–, head.

Cerebro–, brain; specifically, cerebral hemisphere.

Cerumen, (sur-ROO-men), the wax secretion of the external ear.

Cerv–, neck.

Cheil–, (KY-el), lip.

Chest, the thorax.

Chir–, (kir), hand.

Choana(e), (KOH-ah-nah), referring to a funnel, as in the nasal passageways or apertures.

Chol–, (koll), bile.

Chondro–, (KOND-row), cartilage.

Chromosome, (KRO-moh-sohm), "colored body."

–clast, (klast), disruption, breaking up.

Clearing, the process of clearing water or solvent out of a specimen in preparation for microscopic study.

Clinical, the setting in which a person is examined for evidence of injury or disease.

Clot, coagulated blood; a reticular framework of fibrin, platelets, and other blood cells; associated fluid is serum.

Cm, centimeter.

C.N.S. or CNS, central nervous system, consisting of the brain and spinal cord.

Co–, con–, together.

Coagulation, the clotting of blood.

Coelom, the embryonic body cavity.

Collagen, (KOLL-ah-jen), the protein of connective tissue fibers. Several different types. Found in fasciae, tendons, ligaments, cartilage, bone, vessels, organs, scar tissue, and wherever support or binding is needed. Formed by fibroblasts, endothelia, muscle cells, and Schwann cells.

Collateral circulation, alternate circulatory routes; vessels between two or more points that exist in addition to the primary vessels between those points; such circulation exists by virtue of anastamoses among a number of vessels.

Colli–, neck.

Colo–, colon.

Complement, a group of proteins in the blood whose activation causes their cleavage and fragmentation; these fragments have several biologic functions of which one is combining with antibody antigen complexes, enhancing the destruction of antigen.

Concentric contraction, a type of muscle contraction in which the internal contracting force of a muscle is greater than the external load imposed on it (positive work), and the muscle shortens.

Conch, (kawnk), a large spiral shell.

Concha(e), (KAWNK-ah, or KAWN-cha; pl. KAWNK-ee or KAWN-chee), a structure shaped like a conch shell.

Concretion, an inorganic or mineralized mass, usually in a cavity or tissue.

Condylar, condyloid, referring to a rounded process, as in a joint surface.

Condyle, a rounded projection of bone; usually a joint surface, covered with articular cartilage.

Contiguous, (kon-TIG-yu-us), adjoining and being in contact. The basement membrane is contiguous with the basal surfaces of certain epithelial cells.

Contra–, against.

Contraction, shortening.

Cornu(a), (KOR-new), A horn-shaped process.

Corona, crown.

Coronoid, (KOR-oh-noid), crownlike, or beak-shaped; refers to a bony process.

Corpus(ora), body

Corpuscle, (KOR-pus-il), any small body, not necessarily a cell. Red blood corpuscles lack nuclei and are not considered cells.

Costa, rib

Costochondritis, an inflammation surrounding the cartilage of a joint of a rib, usually involving the

GLOSSARY

synovium and fibrous joint capsule and perhaps related ligaments.

Coxa(e), hip; the hip (coxal) joint. Deformities of the upper femur often include the term (such as coxa varus or coxa valgus). Preceded by the term "os," it refers to the coxal or hip bone.

Crani–, cranium.

Cranium, that part of the skull containing the brain.

Cribriform, perforated; like a sieve.

–crine, (krin), separate off, referring to glands that separate from classical epithelial surfaces.

Crus (crura), leg.

Cu, cubic.

Cubital, front (anterior aspect) of elbow.

Cusp, a triangular structure characterized by a tapering projection.

Cutan–, cutaneous (kew-TANE-ee-us), referring to the skin.

Cysto–, bladder.

–cyte, (site), cell.

Cytokine, a product of a cell that facilitates destruction of antigen by inducing or enhancing an immune response.

Cytolysis, the dissolution and destruction of a cell.

Cytotoxin, a product of a cell that acts to destroy another cell; has a toxic effect.

Dachry–, tear.

Dactyl, finger, toe.

Defecation, elimination of waste material through the anal canal/anus from the rectum.

Deglutition, swallowing.

Demi–, half.

Denervation, (dee-nerv-AY-shun), a condition in which a muscle or area of the body is isolated from its nerve supply.

Dentin, (DEN-tin), the hard portion of a tooth. More dense (harder) than bone; less dense (softer) than enamel.

Depolarization, neutralization of polarity; in biological systems, it is an electrical change in stimulated excitable tissues (nerves, specialized cardiac muscle cells) from a baseline polarity (about –90 millivolts) toward neutral (0 millivolts). Such an event induces the conduction of an electrochemical wave (impulse) to move along an excitable tissue (e.g., nerve).

Derm–, skin.

–desis, fixation.

Desiccation, (dess-ee-KAY-shun), drying out; without water.

Desmo–, fibrous.

Dexterity, skillful with the hands.

Di–, twice.

Diaphragm(ae), (DIE-ah-fram), a partition separating two cavities. There are three significant fibromuscular diaphragms in the body: thoracic (separating thorax and abdomen), pelvic (separating pelvis and perineum), and urogenital (separating the anterior recesses of the ischiorectal fossa from the superficial perineal space).

Diarthrosis(es), (die-ar-THRO-sis), *see* joint classification, functional.

Differentiation, making something different; in development of a cell, it is the structural and functional changes within that cell that make it different from other cells; an increase in heterogeneity and diversification.

Diffusion, spontaneous movement of molecules without the application of additional forces.

Digit, finger or toe.

Dis–, apart.

Disc, a wafer-shaped, rounded or oval fibrocartilaginous structure; if crescent-shaped, it is called a meniscus. It may interface articular cartilage surface in a synovial joint (articular disc) or it may interface opposing cartilage endplates of vertebral bodies (intervertebral disc).

Discharge, to set off or release, to fire, to let go.

Dissect, (dis-SECT), to cut up, to take apart. In gross anatomy laboratories, the human body is studied by an ordered dissection by regions.

Dys–, a prefix meaning abnormal, painful, or difficult.

Dorsum, back. Refers to posterior aspect of hand and the "top" of the foot.

Ec–, out.

Eccentric contraction, a type of muscle contraction wherein contracted muscle is stretched and lengthened during the contraction, such as antigravity contractions by antagonists during movement directed toward gravity. Even though there is a load on the muscle, the muscle is stretched (negative work).

–ectasis(es), dilatation.

–ectomy, removal.

Efferent, leading away from a center (organ or structure).

Elbow, the region between the arm and forearm.

Electrochemical, referring to combined properties of electrical and chemical, such as the neuronal impulse.

Ellipsoid, a closed curve more oval than a perfect circle. Ellipsoid joints are reduced forms of ball and socket joints, and, broadly speaking, include condylar-shaped joints.

Em–, in.

Embalm, (em-BAHM), to treat a dead body with preservative chemicals to prevent structural breakdown by microorganisms.

–emia, blood.

Emission, an involuntary release of semen; also the movement of sperm from the epididymis to the prostate during sexual stimulation in the male.

En–, in.

GLOSSARY

Encapsulate, to surround with a capsule.

Encephalo–, brain.

Endo–, in.

Endochondral, (en-do-KON-dral), *endo* = in, *chondral* = cartilage.

Endochondral ossification, *see* ossification.

Endocrine, (EN-do-krin), *endo* = in, *crine* = separate. Glands that secrete their products into the tissue fluids or vascular system.

Endocytosis, the ingestion of matter into a cell by surrounding the material with the cell membrane and budding it off in the cytoplasm.

Endometr–, endometrium.

Endosteum(a), the lining of the medullary canal of long bones, consisting of a thin sheet of collagen fibers and large numbers of osteoprogenitor cells.

Endothelium(a), (en-do-THEE-lee-um), the epithelial lining of blood and lymph vessels and the heart cavities. They are of mesenchymal origin, not ectodermal, and have different properties than classical epithelia.

Entero–, referring to the intestines.

Enteroendocrine, cells of the epithelial layer/glands of the gastrointestinal mucosa which secrete hormones that stimulate/inhibit (regulate) intestinal/pancreatic gland secretion and/or motility of smooth muscle.

Enzyme, a protein molecule that facilitates a reaction without becoming involved (changed or destroyed) in the reaction. Enzymes are identified by the suffix –ase.

Epi–, upon, at.

Epicondyle, an elevation of bone above a condyle.

Epidid–, epididymis.

epidural, outside the dura between the dura and the skull.

Epithelium(a), (ep-ee-THEE-lee-um), *epi* = upon, *thelia* = nipple.

Erg, a unit of work.

Ergo–, a combining form meaning "work."

Ex–, exo–, out.

Excretion, (ex-CREE-shun), the discharging of or elimination of materials, such as waste matter. If the material excreted has some useful in-body function or use outside the body (e.g., semen), it has probably been secreted, not excreted. No universal agreement on this. *See* secretion.

Exocrine, (EX-oh-krin), *exo* = out, *crine* = separate off; referring to glands that separate from classical epithelial surfaces.

Exocytosis, removal of matter from a cell.

Extracellular, outside of the cell, such as the fibrous tissue supporting the cells, and the vascular spaces.

Extrinsic, coming from the outside. With reference to a specific area (e.g., thumb, hand, foot), extrinsic muscles are those with origins outside of the specific area, but which insert in the area and have an effect on the specific area. *See* intrinsic.

Facet, (FASS-et), a small plane or slightly concave surface. The flat cartilaginous surfaces of a joint may be called facets, as on the articular processes of vertebrae.

Facet joint, a joint between articular processes of adjacent vertebrae; also called zygapophyseal joints.

Facilitation, to enhance or assist an event.

Falx inguinalis (conjoint tendon), a tendon composed of fibers from transversus abdominis and internal oblique that arcs over the spermatic cord and attaches to the pectineal line of the pubic bone. *See* Plate 43.

Fascia(e), (FASH-uh, pl. FASH-ee), a general term for a layer or layers of loose or dense irregular fibrous connective tissue. Superficial fascia, often infiltrated with adipose tissue, is just under the skin. Deep fascia envelops skeletal muscle and fills in spaces between superficial fascia and deeper structure, and between/among muscle bellies (myofascial structure). Extensions of deep fasciae include (form) intermuscular septa, support for viscera (e.g., endopelvic fascia), and fibrous bands, and support neurovascular bundles. Smaller, microscopic layers of fibrous tissue (perimysium, endomysium, vascular tunics, and so on) do not constitute deep fascia, even though they may be distant extensions of it. These fibrous connective tissue investments, integrated with tendons, ligaments, periosteum, and bone, blend into a unibody construction, resistant to all but the most traumatic of forces.

Fascia, thoracolumbar, strong fascial layers enveloping the deep back or paravertebral muscles from the iliac crest and sacrum to the ribs/sternum. Plays an important role in limiting and moving motion segments of the back.

Fascicle(s), (FASS-ih-kul), a bundle.

Feedback, a communication relationship between two structures wherein the output (e.g., secretion) of one substance induces an inhibition or facilitation of the secretion of another substance. Negative feedback reflects an inhibitory effect; positive feedback reflects a facilitating relationship.

Fiber, elongated lengths of tissue, e.g., living muscle fibers (cells or their parts), connective tissue fibers (nonliving cell products), living nerve fibers (extension of cell bodies).

Fibril, (FY-brill), an elongated structure smaller than and part of a fiber.

Fibrous, (FY-brus), referring to a fiber or fiberlike quality.

Fibrosus, (fy-BROHS-us), a fibrous structure.

Filament, a small, delicate fiber; in biology, a structure of some length, often smaller than a fibril which is smaller than a fiber.

GLOSSARY

Filtration, movement of a fluid by the application of a force, such as pressure, vacuum, or gravity.

Fissure, a narrow crack or deep groove.

Fixation, a process in preparation of tissue for microscopic study. Treatment of fresh tissue with a fixative preserves structure, preventing autolysis and bacterial degradation.

Flaccid, (FLA-sid or FLAK-sid), without tone; denervated; lax or soft.

Foot, the most distal part of the lower limb; the skeleton of the foot consists of the tarsus, metatarsal bones, and the phalanges. It joins with the leg at the ankle (talo-tibio-fibular joint).

Foramen(ina), (foh-RAY-men), opening or hole.

Forearm, that part of the upper limb between the elbow and wrist (radiocarpal) joints.

Forefoot, that part of the foot anterior to the transverse tarsal (talonavicular and calcaneocuboid) joints.

Fossa(-ae), a depressed or hollow area; a cavity.

Fusiform, spindle-shaped; shaped like a round rod tapered at the ends.

Gastro-, stomach.

Genia-, origin.

Genital(s), L., belonging to birth. Refers to reproductive structures; loosely, the term refers to the external genitals of either sex.

Glia, *see* neuroglia.

Glosso-, tongue.

Glyco-, sweet, pertaining to sugar or carbohydrate, e.g., glycogen (starch), glycoprotein (sugar-protein complex).

Glycoprotein, an organic compound consisting of carbohydrate and protein.

Glycosaminoglycan, a long chain of double sugars (disaccharides) connected with a nitrogen-containing group (amine); glyco, sugar; glycan, polysaccharide. Previously termed mucopolysaccharide. Proteins combined with glycans are termed proteoglycans.

Gomphosis(es), (gom-PHO-sis); bolting together. *See* joint classification, structural.

Gray matter, brain and spinal cord substance which consists largely of neuronal cell bodies, glia, and unmyelinated processes. Collections of gray matter are generally called nuclei or centers.

Groove, a linear depression in bone.

H

Hallux, great (first) toe. **Hallucis,** genitive form.

Hand, the most distal part of the upper limb, the skeleton of which consists of carpus, metacarpus, and phalanges. It joins with the forearm at the wrist (radiocarpal) joint.

Haustra(e), sacculations of the large intestine held in tension by longitudinal bands of smooth muscle (taeniae).

Haversian system, a cylindrical arrangement of bone cells and their lacunae, named after C. Havers, a seventeenth-century anatomist; the central tubular cavity is the Haversian canal containing vessels. Seen in compact bone.

Head, that part of the body supported by the skull and superior to the first cervical vertebra.

Hem-, blood.

Hematocrit, (he-MAT-oh-krit), the measurement of red blood cell volume in a tube of centrifuged blood; the tube itself is called a hematocrit tube.

Hematoma, (hee-mah-TOE-ma) *hemat* = blood, *oma* = tumor or swelling. A collection of blood under the skin, fascia, or other extracellular membrane.

Hematopoiesis, (hee-mah-toh-po-EE-sus), blood cell formation; occurs in the bone marrow, and in early life, in the liver and spleen; blood cells include red blood corpuscles and white blood cells.

Hemi-, half.

Hemopoiesis, (hee-mo-po-EE-sus), *see* hematopoiesis.

Hemorrhage, (HEM-or-ij), bleeding; escaping of blood from blood vessels into the adjacent tissues or onto a surface.

Hemorrhoid, a varicose dilatation of a vein that is a part of the superior/inferior rectal (hemorrhoidal) plexus of veins.

Hemosiderin, (hee-mo-SID-er-in), storage form of iron.

Heparin, a glycoprotein present in many tissues that has anticoagulation "blood thinning") properties.

Hepat-, liver.

Herniation, a protrusion through a wall or wall-like structure.

Heterogenous, a mixture of nonuniform elements.

Hg, mercury (chemical symbol).

Hiatus, an opening.

Hindfoot, that part of the foot posterior to the transverse tarsal (talonavicular and calcaneocuboid) joints.

Hip, the coxal bone; the region of the hip (coxal) joint.

Histamine, a nitrogenous molecule that has many functions, including contraction of smooth muscle and capillary dilatation.

HIV, human immunodeficiency virus.

Homogeneous, of uniform quality.

Hydroxyapatite, a mineral or inorganic compound that makes up the mineral substance of bone and teeth. $Ca_{10}(PO_4)_6(OH)_2$. A structure very similar to that found in nature outside the body.

Hyper, excessive.

Hyperplasia, increased number of normal cells.

Hypertonia, increased muscle tension; increased resistance to stretching of muscle.

Hypertrophy, increase in size of muscle.

Hypo, inadequate or reduced.

GLOSSARY

Hypoesthesia, reduced sensation.

Hyster–, uterus.

0

–iasis, condition, presence of.

Ileo–, ileum of the small intestine.

Ilio–, ilium of the coxal (hip) bone.

Immuno–, immune system, or reference to some activity or part of that system.

Immunosuppression, suppression of immune (lymphoid) system activity; also called immunodepression.

Impinge, to have an effect on something; contact, irritate, strike.

Infarction, (in-FARK-shun), an area of dead tissue caused by interruption of the blood supply to the tissue.

Infection, the invasion of body cells, tissues or fluids by microorganisms, sually resulting in cell or tissue injury, inflammatory and immune response.

Inflammation, a vascular response to irritation characterized by redness, heat, swelling, and pain; may be acute, subacute (more than two weeks, or chronic).

Infra–, under.

Inhibition, restraint or restraining influence.

Injury, anatomic disruption at some level of body organization in response to an external force (e.g., blunt, penetrating, electrical, radiation, thermal).

Innate, inborn, congenital.

Innervation, (in-nerv-AY-shun), to be supplied by a nerve.

Innominate, unnamed. First applied to the coxal (hip bone) by Galen; first applied to the artery by Vesalius.

Integument, the skin.

Inter–, between; e.g., interscapular, between the scapulae.

Intercalated, inserted between.

Interface, surfaces facing one another; to face a surface.

Interstitial, interstices, interstitium, interspaces of a tissue; between two or more definitive structures.

Intima, internal.

Intra–, within; e.g., intracellular, within a cell.

Intramembranous ossification, *see* ossification.

Intravenous, within a vein.

Intrinsic, part of a specific area and not extending beyond that area (e.g., thumb, hand, foot). Muscles that arise (originate) and insert within the hand region are known as intrinsic muscles (of the hand).

Investing, surrounding or enclosing.

Isometric contraction, a contraction that involves muscle contraction without bone movement; thus the muscle maintains the same apparent length. However, there is fibril shortening in such a contraction, and this is offset by the inherent elasticity of the myofascial tissue.

–itis, inflammation. Term does not include the cause of the inflammation; therefore, it does not mean infection, but may refer to the inflammation induced by or associated with an infection.

J

Jejuno–, jejunum of small intestine.

Joint classification, functional; joints are classified according to the degree of movement, i.e., immovable, partly movable, freely movable. Immovable joints are called synarthroses; partly movable joints are called amphiarthroses; and freely movable joints are called diarthroses. Immovable joints may be fibrous (sutures, gomphoses) or cartilaginous (synchondroses). Synovial joints are never normally immovable. Partly movable joints may be fibrous (syndesmoses) or cartilaginous (symphyses). Freely movable joints are always synovial. In fact, synovial joints are limited in their motion by joint architecture and ligaments, but within those limitations, they are normally freely movable. *See also* syn–.

Joint classification, structural; joints are classified according to the material that makes the joint, i.e., fibrous, cartilaginous, bony, synovial. Fibrous joints are further classified as sutures (thin fibrous tissue between flat bones of the skull), syndesmoses (ligamentous sheets between the bones of the forearm and leg), and gomphoses (fibrous tissue between tooth and bony socket). Cartilaginous joints are further classified as synchondroses (hyaline cartilage between the end and shaft of developing bone) and symphyses (fibrocartilaginous discs between bones, as between vertebral bones and between the pubic bones). Bony joints are fibrous or cartilaginous joints which have ossified over time (synostoses). Classification of synovial joints can be seen in Plate 33.

Jugular, (JUG-yoo-lar), referring to the neck or a necklike structure. Specifically refers to the vein(s) of the neck so named.

K

Kary–, nuclear.

Keratin, a sclero-protein that is insoluble and fibrous. It is the principal constituent of the outer layer of stratified squamous epithelia in skin (stratum corneum; *see* Plate 161), hair, and tooth enamel (Plate 100).

Kerato–, outer skin.

–kine, movement.

Kinin, (KY-nin), a polypeptide (short protein) that influences reactions, such as antigen–antibody complexes.

Knee, the region between the thigh and the leg.

GLOSSARY

Kyphosis, (ky-PHO-sis), humpback. Anatomically, a curve of the vertebral column in which the convexity is directed posteriorly; in orthopaedics, it is an excessive curvature of the thoracic vertebrae.

L

Labium(i), lip, or any fleshy border.

Labyrinthine, (laba-RINTH-een), an interconnecting, winding, interwoven series of passageways.

Lacerum, (lahss-AYR-um), torn or jagged.

Lacuna(e), a cavity or lakelike pit.

Lacrimal, referring to tears.

Lamella(e), a thin, platelike structure; may be circular, as seen in Haversian system of bone.

Lamina(e), layer.

Laryngo–, larynx.

Latency, inactivity. Usually a period between moments of activity.

Latent, *see* latency.

Leg, that part of the lower limb between the knee joint and the ankle joint.

–lemma, covering or sheath.

Lepto–, slender.

Leptomeninges, pia mater and arachnoid combined.

Levator, a lifter; an elevator.

Lieno–, spleen.

Ligament, fibrous tissue connecting bone to bone; also a peritoneal attachment between organs.

Lip–, pertaining to lipids; fat; a triglyceride composed of glycerol and three fatty acids.

–listhesis, slip.

Lith–, stone.

–lithotomy, removal of a stone.

Lordosis, a curve of the back seen in the cervical and lumbar regions in hich the convexity is directed anteriorly; anatomically, it refers to any curve of the back so described; orthopaedically, it is an excessive curve as described.

Lumen(ina), (LEWM-un), a cavity, space, or tunnel within an organ.

Lunar, referring to the moon. Semi-lunar, half-moon shaped.

Lymphatic, refers to the system of vessels concerned with drainage of body fluids (lymph).

Lymphoid, refers to the tissue or system of organs (lymphoid or immune system) the basic structure of which is lymphocytes and reticular tissue.

Lymphokine, (LIM-fo-kine), a product of activated lymphocytes that enters into solution and influences immune responses, generally by enhancing destruction of antigen.

–lysis(es), (LYE-sis), destruction or dissolution.

M

Macro, large, as in macromolecule.

Magnum, great.

–malacia, softening, as in demineralization of bone; changes in matrix of a tissue resulting in a loss of turgor or fibrous quality.

Mamm–, breast.

Manual, referring to the hand.

Manus, hand.

Mastication, (*masticate*, to chew), the act of chewing.

Mastoid, breast-shaped.

Matrix(ices), (MAY-trix), the background or ground substance. Often apparently amorphous and homogeneous. Often colorless; fluid or viscous. Within may be dispersed a variety of organic compounds and minerals.

Meatus, (mee-AYT-us), an opening or passageway.

Media, middle.

Mediastinum(a), (mee-dee-ahs-TY-num), middle partition. The partition or eptum between the lungs in the thorax.

Mediate, influence.

Mediator, an influential substance; a substance that acts indirectly but influentially in a reaction or in inducing a reaction.

Medulla, inner part.

Medusa, in Greek mythology, one of the Gorgon sisters, characterized as winged monsters with heads of snakes in place of hair. When a person looked at one of them, he was turned to stone. Medusa was the only mortal Gorgon. In offering service to his tyrant king, Perseus pursued Medusa and cut off her head (which, though detached, still had the power to turn onlookers into stone). Perseus presented the head to the vile king and his men who, upon casting their eyes on the snake-covered head, promptly turned to stone. Perseus became king. The radiating, contorted, dilated venous network bulging out on the surface of the anterior abdominal wall of chronic sufferers of portal vein hypertension/obstruction has been given the name caput Medusae (head of the Medusa).

Mega–, big, great, as in megakaryocyte.

–megaly, enlargement.

Menin–, meninges.

Meninges, dura mater, arachnoid, and pia mater coverings of the spinal cord and brain, and the first part of cranial and spinal nerves.

Mento–, referring to the chin, as in mental foramen.

Mesenchyme, mesenchymal, embryonic connective tissue, often with plenipotentiary cells.

Mesothelium(ia), (meezo-THEE-lee-um), the epithelium lining the great (closed) body cavities, e.g., pleura, peritoneum, and pericardium. Of mesenchymal origin, not ectodermal. Different properties than classical epithelia.

Meta–, change.

Metr–, uterus.

Micro, small, as in microtubule.

Microorganism, a group of organisms including bacteria, viruses, fungi, protozoans, and other microscopic life forms.

GLOSSARY

Micturition, urination; discharge of urine outside the body.

Mineralization, the process of mineral (calcium complexes) deposition, especially in bone formation and remodeling, as well as formation of teeth.

Mm, millimeter.

Mm Hg, millimeters of mercury. A pressure measuring system in which the open end of an evacuated (vacuum) graduated cylinder (tube) is placed in a container of liquid mercury. The pressure of the atmosphere or fluid pressing on the mercury will push the mercury up the cylinder. The distance the mercury moves up the tube is measured in mm Hg and reflects the pressure imposed.

Modulate, to induce a change.

Modulator, a controlling element or agency.

Mortise, a recess cut to receive a part, as the talus fits into the recesses of the tibia and fibula.

Motor, referring to movement; with respect to the nervous system, term refers to that part concerned with movement.

Mucosa(e), (mew-KOS-ah), a lining tissue of internal cavities open to the outside; epithelial/gland cells secrete a mucus onto the free surface of the lining; consists of epithelial lining cells, glands, and underlying connective tissue and nerves/vessels; may have a thin layer of muscle.

Mucous, referring to mucus.

Mucus, the secretion of certain glandular cells, composed largely of glycoproteins in water, forming a slime-gel consistency. Thicker than serous fluid.

Multi–, many.

Muscularis, (muss-kew-LAHR-is), a layer of muscle.

Musculoligamentous, muscle and ligament.

Musculoskeletal, muscles, bones, ligaments, tendons, fasciae, and joints.

Musculotendinous, muscle and tendon.

Myelin, (MY-eh-lin), compressed cell membranes of Schwann cells in the PNS and oligodendrocytes in the CNS, arranged circumferentially, in layers, around axons. Composed of cholesterol, components of fatty acids, phospholipids, glycoproteins, and water.

Myelo–, marrow; usually refers to spinal cord.

Myelopathy, neurologic deficit resulting from spinal cord injury or disease.

Myo–, a prefix referring to muscle.

Myoepithelium(a), contractile epithelial cells. Usually are located at the base of gland cells, with tentacle-like processes embracing secretory cells. Particularly prominent in sweat, mammary, lacrimal, and salivary glands.

Myofascial, skeletal muscle ensheathed by fibrous connective tissue, vascular and sensitive.

Myoglobin, the oxygen-containing, pigment-containing protein molecule of muscle.

Myosin, the principal protein of muscle associated with contraction and relaxation of muscle cells.

Myriad, a great number.

Myx–, mucus.

Naso–, nose, nasal.

Neck, that part of the body inferior to the head and superior to the first thoracic vertebra; is confluent with the shoulders, upper back, and upper chest; cervical region.

Necrosis, (neh-KRO-sis), a state of cellular or tissue death.

Nephro–, Gk., kidney.

Neuro–, (NOO-roh), nervous or referring to nervous structure or the nervous system.

Neuroglia, (noo-ROHG-lee-ah), nonconducting support cells of the nervous system, including the astrocytes, oligodendrocytes, ependyma and microglia of the CNS, and Schwann cells and satellite cells of the PNS.

Neurologic, neurology, concerned with disorders of the nervous system. Also, refers to those nerve/neuronal disorders seen in a clinical setting.

Neuron, (NOO-ron), nerve cell.

Neurovascular, nerve(s) and vessels(s); implies a collection of both, as in eurovascular bundle.

Nociceptor, (no-see-SEP-tur), a receptor for pain.

Nucha–, (NOO-kaw), posterior neck.

Oculi–, (AWK-yu-lye), eye (oculus).

–oid, similar form; like.

–oma, tumor.

Ooph–, ovary.

Ophth–, eye.

Optic, eye.

Or–, mouth.

Orb–, sphere, round.

Orbicular, rounded, circular.

Orbit, the bony cavity containing the eyeball.

Orchi–, testis.

Organelle(s), (or-gan-ELL), the small functional structures within the cell cytoplasm are called organelles.

Os–, bone.

Oscilloscope, an instrument that permits visualization of baseline and waves of changes in electrical voltage.

–osis, condition or state of. Arthrosis is a generic term for a condition of a joint.

Osseous, bone.

Ossification, endochondral, formation of bone by replacement of cartilage/calcified cartilage.

Ossification, intramembranous, formation of bone directly from osteoprogenitor cells in embryonic connective tissue (mesenchyme) or in fibrous tissue adjacent to fractured bone. There is no

GLOSSARY

intermediate stage of cartilage formation or replacement.

Ossification, primary center of, the principal center of bone formation in the diaphysis or center of developing bone.

Ossification, secondary center of, a satellite center of ossification, as in the epiphysis.

Osteo–, bone.

Osteoclastic, referring to osteoclasts or bone-destroying cells.

Osteoid, (OSS-tee-oyd), bonelike; nonmineralized bone.

Osteoprogenitor, a primitive cell that has the potential, when stimulated, to become a bone-forming cell, i.e., osteoblast.

–ostomy, operation for making an artificial opening.

Ovale, oval.

Oxy–, oxygen.

Pachy–, thick.

Pachymeninx, dura mater.

Palpable, (PAL-pah-bul), touchable; by touch.

Palpate, to touch or feel; a common clinical technique.

Palsy, weakness.

Para–, alongside of.

Parenchyma, (pah-REN-keh-ma), the functional substance of an organ.

Paresis, weakness due to incomplete paralysis.

Parietal, (pah-RY-et-all), a wall or outer part.

–pathy, disease.

Ped–, foot.

Pedal, foot.

Pedicle, footlike process; narrow stalk.

Pedo–, child.

Peduncle, a narrow stalk specifically applied to masses of white matter in the CNS.

Pelvic girdle, the two coxal (hip) bones.

Pelvis(es), the ring of bone consisting of the two coxal (hip) bones and the sacrum and coccyx.

–penia, deficiency or decrease.

Penicillar, resembling a painter's brush or pencil.

Pennate, feather-shaped.

Peri–, (PAR-ee), around.

Perichondrium, (paree-KOND-ree-um), the fibrous envelope of cartilaginous structures (except articular); contains blood vessels, fibroblasts, and chondroblasts (immature cartilage cells).

Perineal, that region inferior to the pelvis.

Periodontal, around a tooth.

Periosteum, (paree-OS-tee-um), the fibrous envelope of bone, containing osteoprogenitor cells, osteoblasts, fibroblasts, blood vessels; the life support system of bone.

Peripheral, away from the center, near or toward the periphery.

Peristalsis, (paree-STAHL-sis), waves of coordinated and rhythmic muscular contractions in the walls of a cavity or tubelike organ. Induced by hormones or other secreted factors, and by nerves of the autonomic nervous system.

Peroneal, the lateral (fibular side) of the leg.

Perpendicular, a plane at right angle (90 degrees) to an adjoining plane.

Pes, foot.

Pes anserinus, goose's foot. Refers to the tendons of sartorius, gracilis, and semitendinous that collectively insert on the medial proximal tibia.

Petrous, (PEET-russ), rocky or like a rock.

–pexy, fixation or suspension.

Phagocyte, a cell that takes up cell fragments or other particulate matter into its cytoplasm by endocytosis. Phagocytes with a segmented nucleus are called polymorphonuclear leukocytes (neutrophils); mononuclear phagocytes (of the monocyte-macrophage lineage) are known by several names depending on their location, e.g., macrophages, monocytes of the blood, histiocytes of the connective tissues, Kupffer cells of the liver, alveolar (dust) cells of the lung, microglia of the central nervous system. Many cells are phagocytic under certain circumstances but are not called or considered phagocytes.

Phagocytosis, (fago-site-OH-sus), the taking into a cell fragments or other articulate matter.

Phlebo–, vein.

–physis(es), (FIE-sis), growth, growing.

–pial, pia mater.

Pinocytosis, the activity of cellular ingestion of fluid.

Pituitary (archaic), referring to mucus.

–plasia, referring to development or growth.

Plasm–, referring to the substance of some structure, e.g., cytoplasm or cell substance.

–plasty, surgical correction.

Plenipotentiary, having the capacity to develop along a number of different cell lines. Undifferentiated mesenchymal cells, pericytes, and certain other cells have such capability.

Pneumo–, air.

P.N.S. or PNS, peripheral nervous system, consisting of cranial and spinal nerves, and the autonomic nervous system.

Pole, polar, either extremity of an axis, as in south and north poles of the earth. Also refers to processes of a neuron (unipolar, etc.).

Pollex, thumb. **Pollicis,** genitive form.

Poly–, many or multi–.

Polymodal, many modalities, as in polymodal receptors which are responsive to several different stimuli.

Portal circulation, veins that drain a capillary bed and terminate in a second capillary or sinusoid network, as in the hepatic portal vein and the portal system of the hypophysis.

Post–, back of, after, posterior to.

Pre–, in front of, anterior to.

Precursor, a forerunner; its existence precedes the existence of something that is formed from it.

GLOSSARY

Pro–, in front of.

Procerus, (pro-SEH-russ), long, slender.

Process, bony, a projection sticking out from a surface.

Process, neuronal, an extension of the neuron containing cytoplasm/organelles and limited by a cell membrane; part of a living cell. May be a dendrite or an axon.

Procto–, rectum.

Prolapse, the sinking down or displacement of a structure, such as the uterus sinking into the vagina.

Propria, (prohp-REE-ah), common.

Protein, a chain of amino acids of varying length.

Proteoglycan, chain of disaccharides (carbohydrates) connected to a core of protein; a binding material.

Proteolytic, digestion or the breakdown of protein.

Protuberance, a projection or something sticking out from a surface.

Proviral, refers to viral DNA which has been integrated into the DNA of the host cell.

Pseudo, (SOO-doh), false. In anatomy or medicine, having the appearance of one structure or phenomenon but not, in fact, being such a structure or phenomenon.

Pterygoid, (TAYR-ee-goid), winglike.

–ptosis, falling, drooping.

Pulp, a soft, spongy tissue, often vascular.

Pyel–, pelvis.

Pyo–, pus.

Quad, four

Quadrant, one-quarter of a circle.

Quadrate, four-sided; rectangular; usually square.

Radi–, ray.

Radiculitis, inflammation/irritation of a nerve root.

Radiculopathy, nerve root deficit characterized by the following: change in eep tendon (stretch) reflex, sensory loss (objective numbness), and muscle weakness.

Radix, root.

Ramus, (RAY-mus), a branch.

Ratio, a fixed relationship between two things, e.g., 1:4 means that there is 1 thing for every 4 other things; indicates a proportion.

Recto–, rectum. *See also* procto–.

Reflux, backward flow.

Renal, L., kidney. *See also* nephro (Gk).

Repolarization, an electrical change in excitable tissue away from neutral polarity, e.g., increasing polarity from 0 millivolts to –90 millivolts.

Residue, (REZ-ih-doo), the material left over after processing and extraction of other parts.

Reticular, –um, a small network.

Retro–, back, behind, posterior; opposite of antero–.

Retroperitoneum, the area posterior to the posterior layer of parietal peritoneum. It lies anterior to the muscles of the posterior abdominal wall. Includes the kidneys, ureters, abdominal aorta and immediate branches, inferior vena cava and immediate tributaries, pancreas, and ascending and descending colon.

–rhaphy, suture.

Rotundum, round.

Salpingo–, uterine tubes.

Salpinx, uterine tube.

Sarco–, flesh.

Scavenger cell, *see* phagocyte.

Schwann cell, cell of the peripheral nervous system which provides myelin for some and a membranous covering for all axons. A line of Schwann cells forms a tube for axonal regeneration after axonal injury.

Sciatica, pain in the buttock radiating to the foot via the posterior and/or lateral thigh and leg; follows the distribution of the sciatic nerve, and therefore is assumed to be irritation of that nerve or its roots (radiculitis).

Scoliosis, (sko-lee-OH-sis), any significant lateral curvature of the vertebral column. Some degree of lateral curvature is seen in most spines, and is probably related to handedness.

–scopy, inspection or examination of.

Sebum, the oil lying on the surface of skin, secreted by sebaceous glands (*see* Plate 161).

Secondary sex characteristics, anatomic and physiologic changes occurring as result of increased sex hormone secretion (testosterone in the male, estrogen in the female); these characteristics develop at puberty (generally at 11–14 years of age); in the male, they include growth of body hair, change in voice due to change in laryngeal structure, increased skeletal growth, increase in size of external genitals, functional changes in internal genitals, change in mental attitude, and so on. In females, they include enlargement of breasts, change in body shape due to skeletal growth and distribution of body fat, maturation of internal and external genital structures, and so on.

Secretion, elaboration of a product from a gland into a duct, vessel, or cavity. *See* excretion.

Sella, saddle.

Sellar, saddle-shaped.

Semi–, half or partly.

Sensitive, responsive to stimuli, eliciting an awareness of touch, pressure, temperature, and/or pain; innervated.

Sensory, referring to sensation (touch, temperature, vision, and so on).

GLOSSARY

Septum(a), a wall or an extension of a wall; can also be used to separate structure.

Serosa, (sir-OH-sa), lining tissue of cavities closed to the outside; consists of layer of squamous or cuboidal cells and underlying connective tissue.

Serotonin, a nitrogenous molecule with many functions, including acting as a neurotransmitter, inhibitor of gastric secretion, and a vasoconstrictor.

Serous, clear, watery fluid; see serum.

Serum, any clear fluid; also blood plasma less plasma (clotting) proteins.

Sesamoid, pea-shaped. Generally refers to small bones of the hand and foot. he largest sesamoid bone is the patella. These bones are formed within the tendons or ligaments at points of stress.

Sharpey's fibers, fibrous bands of ligaments, tendons, and/or periosteum inserting directly into bone.

Shoulder, the part of the body where the upper limb is joined to the trunk; specifically, it is the shoulder joint and surrounding area including the upper lateral scapula and distal clavicle (acromioclavicular area).

-sial, saliva.

Sinus(es), (SY-nuss), a cavity or channel; venous sinus is a large channel, larger than an ordinary vein; air sinus is a cavity.

Sinusoid, literally, a small sinuslike vessel; usually, a thin-walled vessel, often porous, often found in glands. Generally slightly larger than capillaries, sinusoids vary in their structure depending on their location.

Soft tissue, any tissue not containing mineral, e.g., not bone, teeth. Generally refers to myofascial tissues.

Soma, the body; the body wall.

Somatic, referring to the body or body wall, i.e., the cell body of a neuron (soma); in organizational terms, contrasted with viscus or viscera (organs containing cavities).

Spasm, rapid, violent, involuntary muscle contraction, usually resulting in some contortion of the body part experiencing the spasm.

Spheno, (SPHEE-no), shaped like a wedge; a triangular-shaped structure which comes to a thin edge on one side.

Sphincter, a concentric band of muscle surrounding a narrowed cavity or passage.

Spindle, a structure which is round and tapered.

Spinosum, spiny or spinelike.

Spleno-, spleen. See also lieno-.

Spondyl-, vertebra.

Squamous, platelike, thin. Generally refers to flat, thin epithelial cells.

Stenosis, (sten-OH-sis), narrowing.

-stomy, hole or opening.

Stratified, layered; more than one layer.

Stria, striated, stripes or parallel markings.

Styloid, (STYL-oyd), the form of a pointed spike or pillar.

Sub-, under.

Subchondral, under cartilage; specifically the bone adjacent to articular cartilage.

Subcutaneous, (sub-kew-TANE-ee-us), under the skin.

Subdural, under the dura; between the dura and the brain or spinal cord.

Supra-, above.

Suture, (soo-CHUR), a type of fibrous or bony joint characterized by interlocking, V-shaped surfaces, as in the skull.

Swallowing, see deglutition.

Sym-, see syn-.

Symphysis(es), (SYM-fih-sis); see joint classification, structural.

Syn-, (SIN), together, with, alongside.

Synarthrosis(es), (sin-ar-THRO-sis); see joint classification, functional.

Synchondrosis(es), (sin-KON-dro-sis); see joint classification, structural.

Syndesmosis(es), (syn-des-MO-sis); see joint classification, structural.

Synostosis(es), (syn-os-TOH-sis); see joint classification, structural.

Synovial, (sih-NOH-vee-ul), refers to a viscous fluid similar to the consistency of uncooked egg white. This fluid and the membrane that secretes it line freely movable joints (synovial joints), bursae, and tendon sheaths.

Synthesis(es), formation of a structure from smaller parts; integration of parts.

T

Taenia(e) coli, strips of longitudinal muscle of the muscularis externa of the large intestine (excluding rectum and anal canal).

Tarsal, tarso-, the ankle.

Tendinitis, inflammation of a tendon.

Tendinous, (TEN-dih-ness), referring to tendon.

Tendon, fibrous tissue connecting skeletal muscle to bone or other muscle. May be cordlike or sheetlike (aponeurosis).

Thigh, that part of the lower limb between the hip joint and the knee joint.

Thorax, the region between the neck and the abdomen.

Thrombosis(es), a condition of clots or thrombi within a vessel or vessels.

Thrombus(i), a clot within a blood vessel, obstructing flow.

-tomy, incision.

Tone, normal tension in muscle, resistant to stretch.

Torso, the part of the body less the limbs and head; the trunk.

Transcriptase, an enzyme (polymerase) directed by DNA to facilitate synthesis of a single strand of RNA which is structurally complementary to a strand of DNA.

GLOSSARY

Transcriptase, reverse, a polymerase (enzyme) directed by RNA to facilitate synthesis of a single or double strand of DNA which is structurally complementary to the strand of RNA. In HIV infection of cells, the viral RNA complex has the polymerase called reverse transcriptase. Once in the cell, the RNA-directed enzyme makes possible the transcribing of viral RNA sequences into double-stranded DNA; this is then integrated into the host cell's DNA. This combined DNA is called proviral DNA.

Trauma, an anatomic or psychic response to injury.

Trochanter, a large process. Specifically, the two trochanters of the upper emur.

Trochlea, (TROHK-lee-ah), a pulley-shaped structure.

–trophic, relating to nutrition.

Truss, a collection of members (beams) put together in such a way as to create a supporting framework.

Tubercle, (TOOB-er-kul), a rough, small bump on bone.

Tuberosity, (toob-eh-ROSS-eh-tee), a bump of bone, generally larger than a tubercle, smaller than a process.

Tunica, referring to a coat or sheath; a layer.

Turcica, (TUR-sih-kah), Turkish, as in sella turcica or Turkish saddle.

U

Uni–, one. A unicellular gland is a one-cell gland.

Unibody, of one body; a structure with parts integrated into one unit.

Unit, a single thing or quantity; the basic part of a complex of parts.

Urogenital, referring to structures of both the urinary and genital (reproductive) systems.

Urogenital diaphragm, a layer in the perineum consisting of the sphincter urethrae and deep transverse perineal muscles and their fasciae. Also called the deep perineal space.

V

Vacuolation, (vac-u-oh-LAY-shun), formation of small cavities or holes; part of a degenerative process in cartilage during bone development.

Vacuum, a space devoid of air, with, therefore, no pressure. In the relative sense, decreased pressure in the thoracic cavity during inspiration represents a partial vacuum, drawing air from a space with air of higher pressure.

Varix(ces), an enlarged, tortuous (twisted) vessel.

Varicosity(ies), an enlarged and irregular-shaped, highly curved (tortuous) ein(s). Most often seen in superficial veins of the lower limbs and the testes/scrotum.

Vas(a), vessel.

Vasa vasorum, vessel that supplies a larger vessel.

Vascular, referring to blood or lymph vessels; also to blood supply.

Vasorum, of the vessels.

Ventricle, a cavity.

Vessel, a tubelike channel for carrying fluid, as in blood or lymph.

Vestibule, an entrance-way, cavity, or space.

Villus(i), a finger, fingerlike projection(s) of tissue, as in the intestinal tract, placenta, and so on.

Viral, refers to a virus.

Virion, a single virus, also called virus particle, consisting of genetic material (DNA or RNA) and a protein shell (capsid). It is infectious.

Virus, one of a group of extremely small infectious agents, consisting of genetic material and a protein shell. It is not capable of metabolism, and thus requires a host to replicate. On attachment to a surface molecule on a cell membrane, it is enveloped by the cell membrane and brought into the cytoplasm; such constitutes infection.

Viscous, a fluid or semi-fluid state wherein molecules experience significant friction during movement.

Viscus(era), an organ with a cavity in it.

Vomer, a plowshare or plowshare-shaped structure.

W

White matter, a substance of the brain and spinal cord consisting of largely myelinated axons arranged in the form of bundles or tracts. Appears white in the living or preserved brain.

Wrist, the carpus. The region between the forearm and hand.

Wrist drop, a condition in which the extensors of the wrist are weak or paralyzed; the wrist cannot be extended and therefore the wrist "drops" when one attempts to hold the hand horizontally or vertically upward. This condition is usually due to radial nerve denervation.

X

Xeno–, foreign.

Xero–, dry

Z

Zygapophysis(es), (zi-gah-POFF-ee-sis), an articular process of a vertebra; also a joint between vertebrae (zygapophyseal articulation); such synovial joints may be called facet joints. *See also* facet.

Zygo–, (ZY-go), referring to a yoke or union; joined.

INDEX

Words are indexed by generic terms (process, foramen, ligament, artery and so on), although there is an incomplete listing by name (articular, intervertebral, and so on). Principal references are bold (dark type).

It is recognized that instructors and texts often employ a variety of terms for the same structure. We have used here the internationally accepted standard for anatomical terminology: the *Nomina Anatomica*, 6th edition (1989). Where that failed us, we have employed *Dorland's Illustrated Medical Dictionary*, 27th edition (1988). Synonyms of anatomical terms are included here, especially medical terms which are referred to the related anatomical term. For example, the dural sac is known to radiologists as the thecal sac. Both terms are included here.

INDEX

INDEX

INDEX

INDEX

INDEX

INDEX

INDEX

INDEX

INDEX

INDEX

INDEX

INDEX

INDEX

INDEX

INDEX

Vein(s), (continued)
gastro-epiploic, left/right, 80, 103
gluteal, inferior/superior, 79
hemiazygos (inferior), 78
    accessory (superior), 78
hemorrhoidal (rectal), 80, 105
hepatic, 78, **80**, 106, 107, 109
    portal, 80, 106, 107
hypophyseal, 125, 126
iliac, common/external/internal, 78, 79, 109
intercostal, 42
    first posterior (highest), 78
    posterior, 78
    superior, 78
interlobar, 112
interlobular, 112
intestinal, to villus, 104
jugular, anterior, 77
    external/internal, 77, 78
lingual, 77
lumbar, 78
    ascending, 78, 80
marginal, lateral/medial (foot), 79
maxillary, 77
median, of forearm, 71
mesenteric, inferior/superior, 80
metatarsal, 79
plantar, 79
obturator, 79
ophthalmic, superior, 77
ovarian, 78, 117
pancreatic, 80
of the penis, 115
phrenic, 78
plantar arch, deep, 79
    lateral/medial, 79
popliteal, 79
portal, hepatic, 80, 106, 107
    hypophyseal, 125, 126
pulmonary, 65
rectal, inferior/middle/superior, 80
renal, 78, 109–112, 128
retromandibular, 77
saphenous, great/small, 79

segmental, renal, 112
sinusoid, see Sinusoid
splenic, 80
structure of, 64
subclavian, 71, 77
subcostal, 78
suprarenal, 78, 109, 128
temporal, superficial, 77
testicular, 78, 109, 113
thoracoepigastric, 81
thoracic, internal, 78
thyroid, inferior, 78, 127
    middle/superior, 77, 78, 127
tibial, anterior, 79
    posterior, 79
types of, 64
umbilical (fetal), 82
uterine, 117
varicose, discussion of, 79
vena cava, see Vein, caval
venous arch, dorsal, 79
vertebral, 77
Vena cava; see Vein, caval
Ventral, defined, 2
Ventricle, development of, 141
    fourth (IV), 137, **141**, 143
    lateral (I, II), 135, **141**, 143
    third (III), 125, 136, **141**, 143
        infundibular recess, 125, 141
        pineal recess, 141
    of the heart, 66
Venule, structure of, 64
Vermiform appendix; see Appendix, vermiform
Vertebra(e), cervical, 21, 22
    coccygeal, 21, 23, 29
    coccyx, 21, 23, 29
    lumbar, 21, 23
    lumbarized, defined, 23
    parts of, 21
    prominens, 22, 24
    sacral, 21, 23, 29
    sacralized, defined, 23
    sacrum, 21, 23, 29
    thoracic, 21, 22, 24

transitional, defined, 23
vertebral column, 21
    regions, 21
Vertigo, 159
Vesicle, pinocytotic, 3
    synaptic, 132
Vessels, great (of the heart), 65
Vestibule, of the bony labyrinth, 158, 159
    of the perineum, 116
Villus(i), arachnoid, 143, 144
    chorionic, 122, 123
    intestinal, 104
Virus, HIV, 90
Viscus(-era), defined, 12
    general structure of, 12
Vision, tunnel, 157
Visual pathway, 157
Vitreous, see Body
Volume, in relation to pressure, 97

W

Wall, bony thoracic, 24
Walls, of cavities, 16
Wave, P, 67
    QRS, 67
    T, 67
Window, oval, 158, 159
    round, 158, 159
Wrist, 15

Y

Yolk sac, 122, 123

Z

Z line, of skeletal muscle, 10
Zona fasciculata, 128
    glomerulosa, 128
    pellucida, 121
    reticularis, 128
Zygoma (zygomatic bone), 19, 20
Zygote, 121